ANTARCTICA AS AN EXPLORATION FRONTIER--HYDROCARBON POTENTIAL, GEOLOGY, AND HAZARDS

Edited by

Bill St. John

AAPG Studies in Geology #31

Published by
The American Association of Petroleum Geologists
Tulsa, Oklahoma, U.S.A.

Published December, 1990

ISBN: 0-89181-039-0
ISSN: 0149-1377

Association Editor: Susan A. Longacre
Science Director: Gary D. Howell
Publications Manager: Cathleen P. Williams
Special Projects Editor: Anne H. Thomas
Project Editor: Mary E. Horne

Cover photos by John B. Anderson, Rice University, Houston, Texas.
Cover design by Elizabeth A. Mitchell.

About the Editor

Bill St. John was born in 1932 in an oil boom town, Wink, Texas, and grew up in oil camps around Olney, Kamay, and Wichita Falls, Texas. He served in the U.S. Marine Corps from 1951 to 1954, including Korea in 1953. Bill received a B.S. degree in 1958 and an M.A. degree in geology in 1960 from the University of Texas (Austin). He worked in Libya, Mauritania, and Senegal from 1960 to 1963, returned to the University of Texas in 1963 and received a Ph.D. degree in 1965. From 1965 through 1972 he worked for Exxon in Norway, England, and Morocco, and in Houston, Texas, where he did international basin analysis. In 1973 he joined an independent in Tulsa, Oklahoma, as international manager; that job evolved into his becoming president of Agri-Petco International. In 1981 he moved again—to Houston, as President of Primary Fuels, Inc., a wholly owned subsidiary of Houston Industries Incorporated. Primary Fuels was sold in 1989 and Bill began consulting internationally, most recently as Technical Advisor to the Ethiopian Institute of Geological Surveys, Ministry of Mines and Energy, in Addis Ababa, Ethiopia. His career has been divided between international basins studies and exploration and production management. He has worked in or visited more than 40 countries during his 28-year oil industry career.

Foreword

The Symposium on Antarctica, presented at the 1987 annual meeting of the American Association of Petroleum Geologists, resulted from a recommendation by the AAPG Marine Geology Committee in 1986. At that time, several Antarctica symposia had been held but none had directly involved the members of the nongovernment petroleum community. Research and interest in Antarctica had been the private domain of geographic explorers, academic institutions, government agencies, and national oil companies. It was time to stake a claim.

Most of the papers in the present volume were given at the symposium; however, a few were solicited later to fill in what were believed to be critical gaps in the original program. All of the papers are oriented toward the hydrocarbon potential of Antarctica. Contents include regional seismic surveys involving tectonic and stratigraphic interpretations extending from the Adelie Coast margin (Wannesson), over the Ross Sea (Cooper et al.) and Bellingshausen Sea, through the Bransfield Strait (Gambôa and Maldonado, and Jeffers and Anderson) and along the northern Antarctic Peninsula (Anderson et al.). Another paper compares in detail the Mesozoic sedimentary basins of Antarctica and includes source rock analyses (Macdonald and Butterworth). A tectonic synthesis of Antarctica and the surrounding southern seas is presented (Royer et al.). Hazards to petroleum exploration and production offshore Antarctica are described (Reid and Anderson). Specific source rock information from a single borehole is given (Collen and Barrett). A solicited post-symposium paper inventories the relevant offshore seismic surveys that could be used to make a regional offshore seismic interpretation (Behrendt).

The convener/editor and the authors trust that the papers presented here will serve to advance the geologic knowledge of this last truly frontier exploration area, and that they will stimulate further interest and research by the petroleum industry.

-Bill St. John

Table of Contents

Chapter 1

Evolution and Hydrocarbon Potential of the Northern Antarctic Peninsula Continental Shelf

John B. Anderson
Peter G. Pope
Mark A. Thomas
Department of Geology and Geophysics
Rice University
Houston, Texas, U.S.A.

ABSTRACT

During the 1987 United States Antarctic Program-*Polar Duke* Cruise, 3200 kilometers of seismic reflection profiles were collected on the northern Antarctic Peninsula shelf. These data, plus the results of land-based studies (from Polish and U.S. scientists) and ocean drilling (DSDP Leg 35) were used to reconstruct the tectonic and climatic development of the shelf and to assess possible hydrocarbon prospects of the region.

The study area consists of a foredeepened shelf, typical of the Antarctic continental shelf. The continental margin has evolved from an active margin to a tectonically passive one as the Aluk Ridge was gradually subducted at the Antarctic Plate Boundary. This transition was diachronous, as the timing of ridge subduction proceeded from south to north (oldest to youngest). Thus, the shelf is segmented both tectonically and sedimentologically as the extent of tectonic deformation and post-tectonic sedimentation varies correspondingly.

Besides the obvious tectonic controls, major changes in shelf sedimentation also took place due to climatic changes during the Cenozoic. Antarctica is unique in that as its climate cooled and ice sheets formed, the source of terrigenous organic carbon to the shelf was completely eliminated. Also, streams and rivers were gradually eliminated (by early Neogene) so that running water was no longer contributing to the transport of terrigenous sediments to the shelf. The shelf was later overdeepened by glacial erosion (middle Miocene?). By the late Miocene there was an intensification of oceanic circulation and an increase in the flux of organic-rich, siliceous biogenic sediments to the continental margin.

Seismic records show four sequences reflecting different episodes of shelf development. Sequence 4, the oldest sequence, consists of folded and faulted pretectonic and syntectonic (subduction) deposits, presumably volcaniclastic material deposited as subduction occurred. Sequence 3 is an accretionary sequence unconformably overlying S_4 and reflecting efficient sediment transport across the shelf after subduction ceased. Sequence 2 rests unconformably on Sequence 3 and is interpreted as glacial deposits that are bounded by erosional surfaces. This sequence marks the onset of glaciation sufficient to overdeepen the shelf. Sequence 1 is believed to consist of glacial-marine deposits.

The hydrocarbon potential for that portion of the continental shelf situated north of the Tula Fracture Zone is low, but is slightly higher for that portion of the shelf situated south of the Tula Fracture Zone. This is because the age and thickness of sedimentary deposits increases to the south, and the time window for formation and burial of suitable source and reservoir rocks increases in that direction. Sequence 4 is believed to include carbonaceous marine shales deposited when the climate was temperate and is considered to be the most likely hydrocarbon source, at least south of the Anvers Fracture Zone. The siliciclastic deposits of Sequence 3 are the most probable reservoir rocks.

INTRODUCTION

The Pacific-Antarctic continental margin has the most complex tectonic history of any part of the Antarctic margin (Lawver et al., 1988). Continental margin development has been influenced by a series of ridge-trench collisions between the Pacific and Antarctic plates and involving the Aluk Ridge. The age of ridge-trench collisions decreases from southwest to northeast along the margin, with major fracture zones separating segments of sea floor with different subduction histories (Herron and Tucholke, 1976; Barker, 1982) (Figure 1). Thus, each segment of the margin provides a "snapshot" in time of continental margin evolution in this type of tectonic setting. Our work is based on an analysis of 3200 km of seismic reflection data from the Antarctic Peninsula continental shelf (Figure 2). The study area spans several major fracture zones (Figure 1), so our seismic records provide a means of examining continental shelf development within several different tectonic segments.

No drilling has been conducted on the shelf, so the timing of sequence boundaries must be inferred from the geological record derived through outcrop studies conducted on nearby Seymour and King George Islands (Figure 2) and on DSDP Leg 35 drill sites from the Bellingshausen continental rise and abyssal floor (Figure 1).

METHODS

The continental shelf of the northern Antarctic Peninsula has a relatively thin postsubduction sedimentary cover, especially on the inner shelf. Therefore, the study area provides an ideal setting in which to examine the tectonic features that form the different subbasins on the shelf and the sediments which fill these subbasins by using high resolution seismic reflection techniques.

The seismic data acquired in this investigation were collected during the austral summer of 1987/1988 onboard the United States research vessel *Polar Duke*. A total of 3200 km of seismic data were acquired along 15 traverses extending from the upper slope to within 10 km of the coast, and along two strike lines (Figure 2). The data were collected with either one or two 100 in.^{3} Hamco water guns fired at an 8 second shot rate. Data were acquired with a single channel, 600-meter-long Teledyne/Litton streamer, and the unfiltered data were digitally recorded. The results presented in this chapter are based on our initial investigation of the unprocessed data.

TECTONIC SETTING

Ridge-trench collisions occurred progressively northeastward along the Antarctic Peninsula Coast (Herron and Tucholke, 1976; Barker, 1982). Initially, the fracture fabric of the subducted plate was oblique to the direction of subduction; however, during the Late Cretaceous the Antarctic Peninsula began to subduct sea floor with fracture zones trending parallel to the direction of subduction. Thus, the boundaries between segments of different age sea floor stayed at the same point on the margin, allowing tectonic segmentation of the margin (Hawkes, 1981; Barker et al., 1988; Jeffers et al., in press). As each segment of the ridge approached the trench, magmatism in the corresponding arc segment stopped, and the arc and forearc were tectonically uplifted and eroded (Barker, 1982; Larter and Barker,1989). After each segment of the ridge was subducted, the corresponding margin became tectonically passive. The youngest anomaly at the margin provides a date for the cessation of subduction within a particular segment of the sea floor (Figure 3). Southwest of the Tula Fracture Zone the ridge was subducted in the middle to late Eocene (Kimura, 1982) (Figure 3). The ridge segment between the Tula and Adelaide Fracture Zones was subducted in the late Oligocene, the ridge between the Adelaide and Anvers Fracture Zones was subducted in the early Miocene, and the segment between the Anvers and Hero Fracture Zones was subducted during the late Miocene (Figure 3). The segment north of the Hero Fracture Zone, the Bransfield Basin segment, is situated within the 4 Ma collision zone (Figure 3) and is the sole

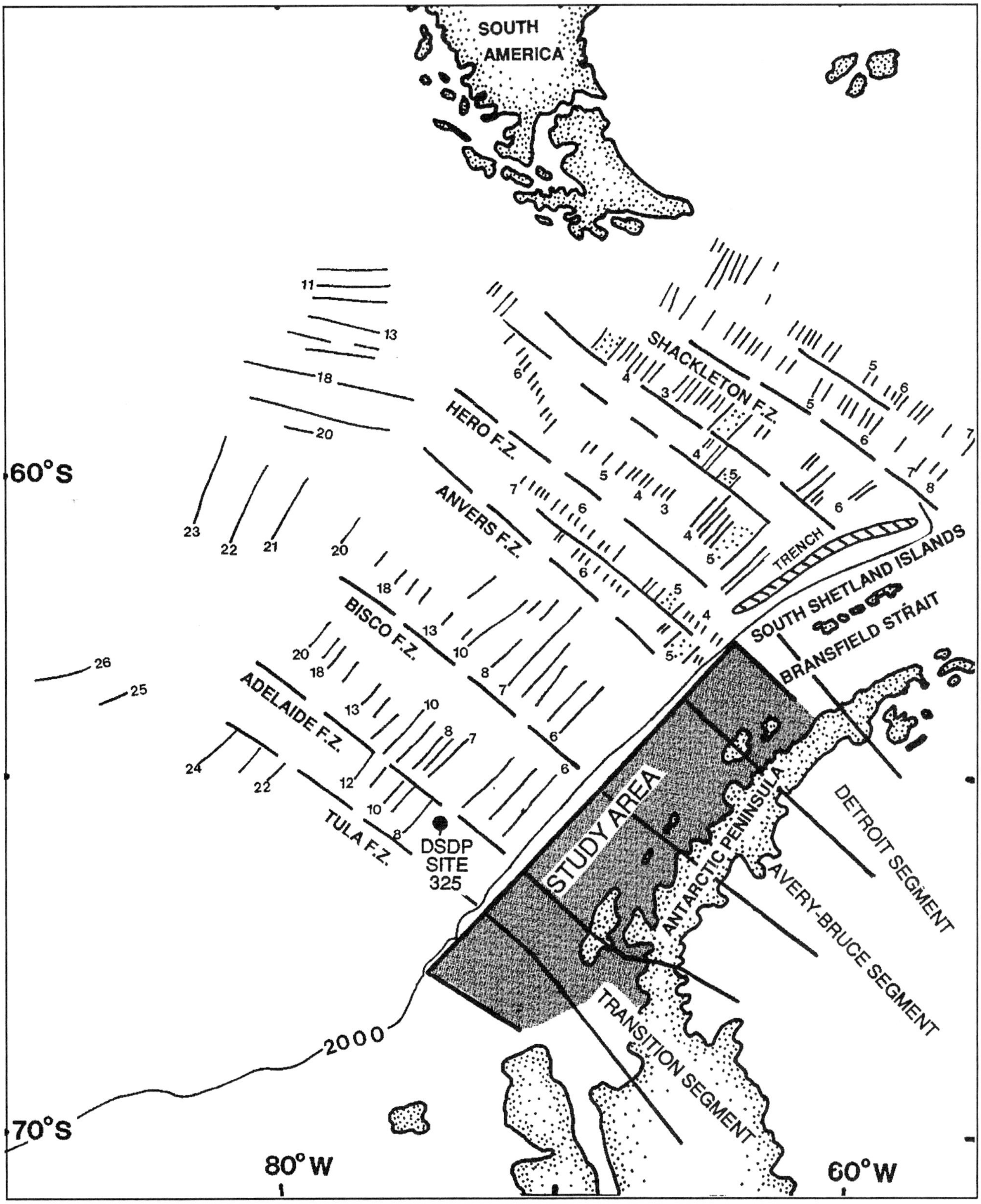

Figure 1—Tectonic setting of the southeast Pacific Basin. Sea floor magnetic anomalies and fracture zone locations are from Barker (1982) and the transverse megafractures of the Antarctic Peninsula are from Hawkes (1981). Hawkes subdivides the peninsula into four segments: the Transition Segment is situated between the Tula and Adelaide Fracture Zones, the Avery-Bruce Segment between the Adelaide and Anvers Fracture Zones, the Detroit Segment between the Anvers and Hero Fracture Zones, and the Bransfield Segment is situated north of the Hero Fracture Zone. Also shown is the location of DSDP Site 325.

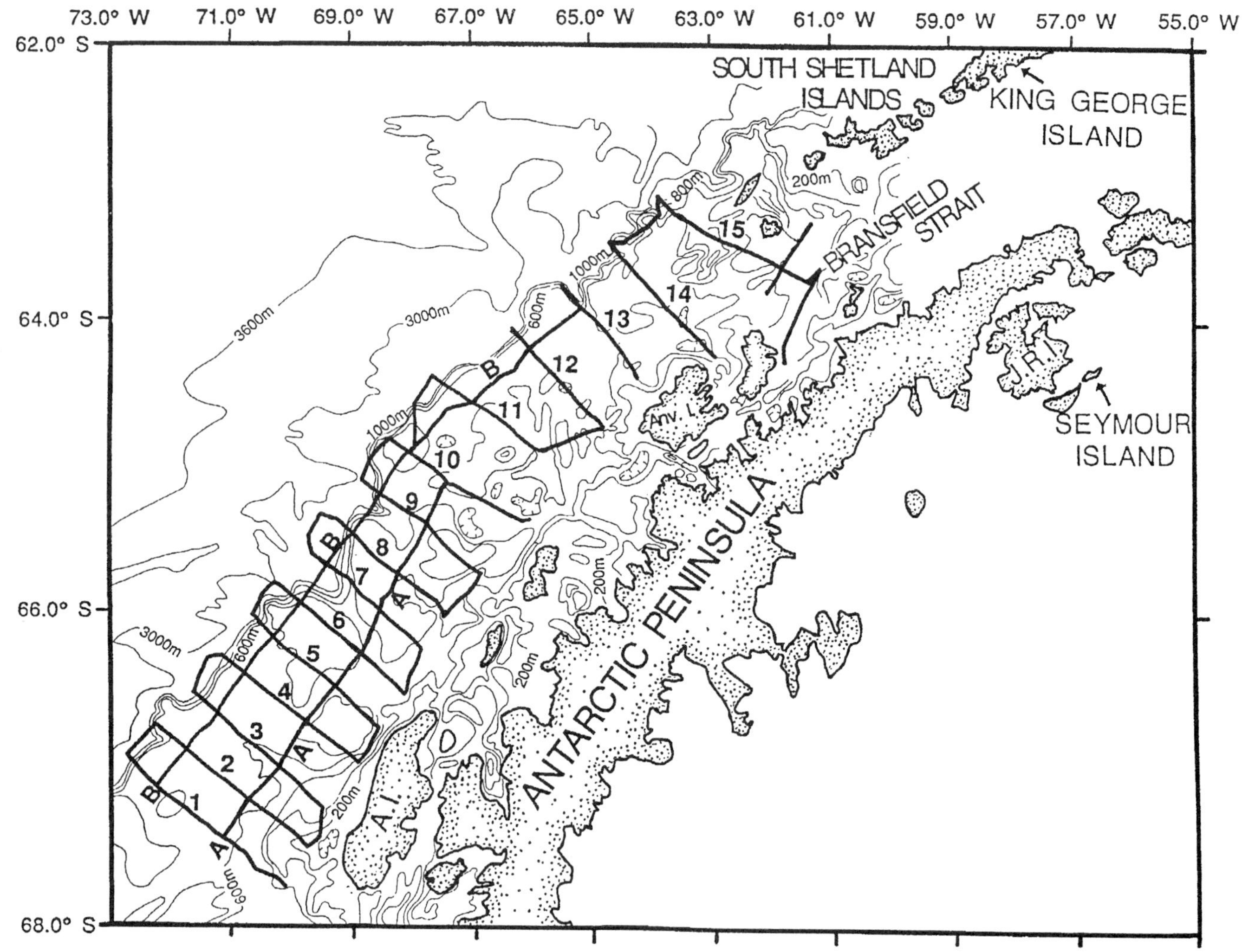

Figure 2—The locations of seismic lines collected during USAP-87. A.I. = Adelaide Island, Anv. I = Anvers Island, J.R.I. = James Ross Island. Bathymetry is in meters.

surviving segment of the spreading ridge; it is the only area with trench topography. This segment apparently stopped spreading at about 4 Ma, before the ridge reached the margin.

The Antarctic Peninsula is tectonically segmented by "transverse megafractures" (Hawkes, 1981) which appear to correspond to fracture zones (Figure 1). These transverse megafractures divide the northern Antarctic Peninsula into segments having different histories of magmatism, uplift and erosion, and postorogenic sedimentation (Hawkes, 1981). Hawkes divides the peninsula into four principal segments, from south to north: the Transition, Avery-Bruce, Detroit, and Bransfield segments (Figure 1).

Crystalline basement rocks of the Antarctic Peninsula range from late Paleozoic to Cenozoic age (Dalziel and Elliot, 1982; Birkenmajer et al., 1989). Mid-Jurassic calc-alkaline plutons were emplaced in the northern portion of the peninsula and this plutonic activity was accompanied by block faulting, tilting, and gentle folding (Hawkes, 1981). On King George Island (Figure 2), there is a 3000-meter-thick suite of basaltic through andesitic, and locally rhyolitic-dacitic rocks, which implies nearly continuous Late Cretaceous through late Paleogene subduction-related, calc-alkaline island arc volcanism (Birkenmajer et al., 1989). Cenozoic volcanics occur dominantly in the south. Recent volcanism is restricted to the northern portion of the peninsula and to Bransfield Basin.

CLIMATIC AND GLACIAL EVOLUTION OF THE REGION

No drilling has been conducted on the Antarctic Peninsula continental shelf. So, we must rely on information from outcrop studies on nearby King George Island and Seymour Island and on DSDP Leg 35 drill sites on the nearby deep sea floor for clues as to the timing of major seismic stratigraphic events. Understanding of the Antarctic Peninsula's early Cenozoic climatic record has recently been expanded by studies of stratigraphic sequences on Seymour Island (Figure 2) (Zinsmeister, 1982; Askin and Fleming, 1982). Field studies on King George Island (Figure 2) by Polish scientists have led to significant breakthroughs in our understanding of the Antarctic Peninsula's Cenozoic glacial/climatic record. The Polish group has succeeded in obtaining good isotopic ages for important climatic and glacial events. Figure 4 summarizes the significant episodes of climatic and

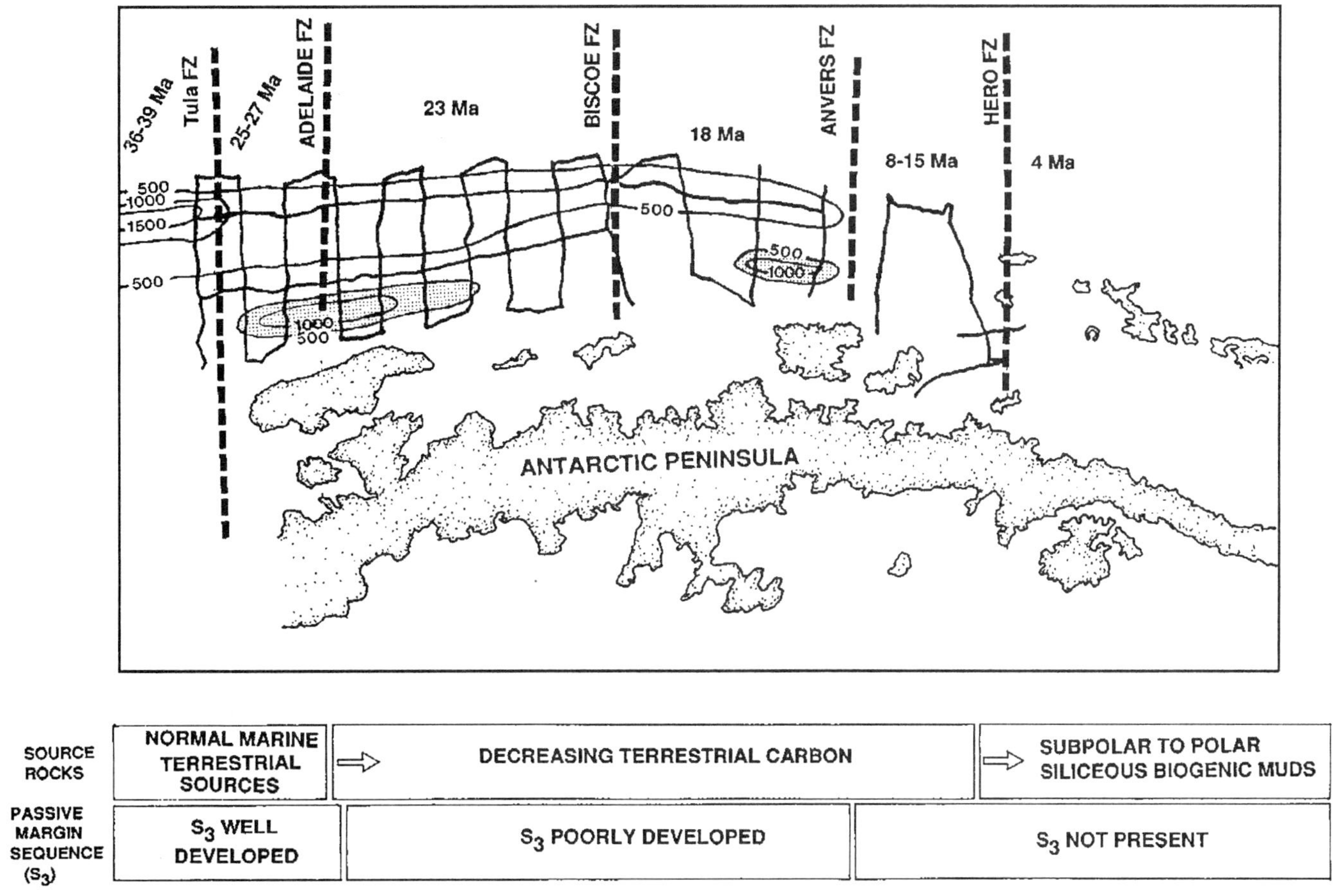

Figure 3—Seismic track lines and the major fracture zones along with the ages of fracture zone-bounded segments of the continental shelf. Also shown is an isopach map (in meters) of Sequence 3 on the outer shelf and of sediments which fill forearc basins on the inner shelf (dotted areas). Sequence 3 is the most likely sequence to contain suitable reservoir rocks, so the level of its development, along with the probable source rocks for various portions of the shelf, are presented.

glacial changes in the Antarctic Peninsula region. This table and the following discussion are based largely on the work of Polish scientists (Birkenmajer 1981a,b, 1982a,b, 1984, 1985a,b, 1987, 1989; Birkenmajer et al., 1985a,b; Gazdzicki and Pugaczewska, 1984; Biernat et al., 1985; Zastawniak et al., 1985) who have worked on King George Island, by United States scientists who have worked on Seymour Island (Zinsmeister, 1982; Askin and Fleming, 1982), and by the scientific party of DSDP Leg 35 (Hollister and Craddock, 1976).

Antarctica has occupied a high latitude position since late Paleozoic time, but for reasons that are still not altogether clear, ice sheets did not develop on the continent until much later. In fact, during the Late Cretaceous through Paleocene, the Antarctic Peninsula was covered by forest with plush undercover. There is a relatively continuous sequence of Late Cretaceous through Paleocene deposits on Seymour Island which includes the older (upper Campanian to possibly lowermost Tertiary) Marambio Group and younger (Paleocene) Cross Valley Formation (Zinsmeister, 1982). The Marambio Group has a total thickness of just over 1200 m and includes sandstones, siltstones, and mudstones which are typically carbonaceous and contain abundant plant debris and coalified logs (Zinsmeister, 1982). The Cross Valley Formation consists largely of nonmarine sandstones and volcaniclastic deposits. It too contains abundant carbonized plant material, and at least one coal seam has been described (Zinsmeister, 1982). Paleobotanical and palynological studies of these deposits have yielded a diverse fossil flora indicating relatively warm climatic conditions (Askin and Fleming, 1982). The Cross Valley Formation is overlain by predominantly marine strata of the late Eocene (and possibly earliest Oligocene) La Meseta Formation. This formation contains an abundant and diverse marine fauna and bears no evidence of glacial activity (in the form of ice-rafted debris) at that time (Zinsmeister, 1982).

On King George Island, older Tertiary strata include the Ezcurra Inlet Group (Paleocene), Fildes Peninsula Group (Eocene?), and Dufayel Island Group (Eocene). These strata contain abundant plant remains and a coal seam has been described from the Petrified Forest Member of the Ezcurra Inlet Group (Birkenmajer, 1987). The spore and pollen assemblage of the Petrified Forest Member consists of 36 individuals belonging mainly to the Nothofagus-Pteridophyta assemblage (Stuchlik, 1981). These Nothofagus forests had dense undercover consisting mainly of ferns. The total assemblage is similar to that of the rain forests that now occupy the north island of New Zealand and indicates moist and warm climatic conditions (Stuchlik, 1981).

During Eocene time the climate in the Antarctic Peninsula region began to cool. On King George Island sedimentary deposits of the Polinia Glacier Group and the Point Hennequin Group contain plant fossils which include species of

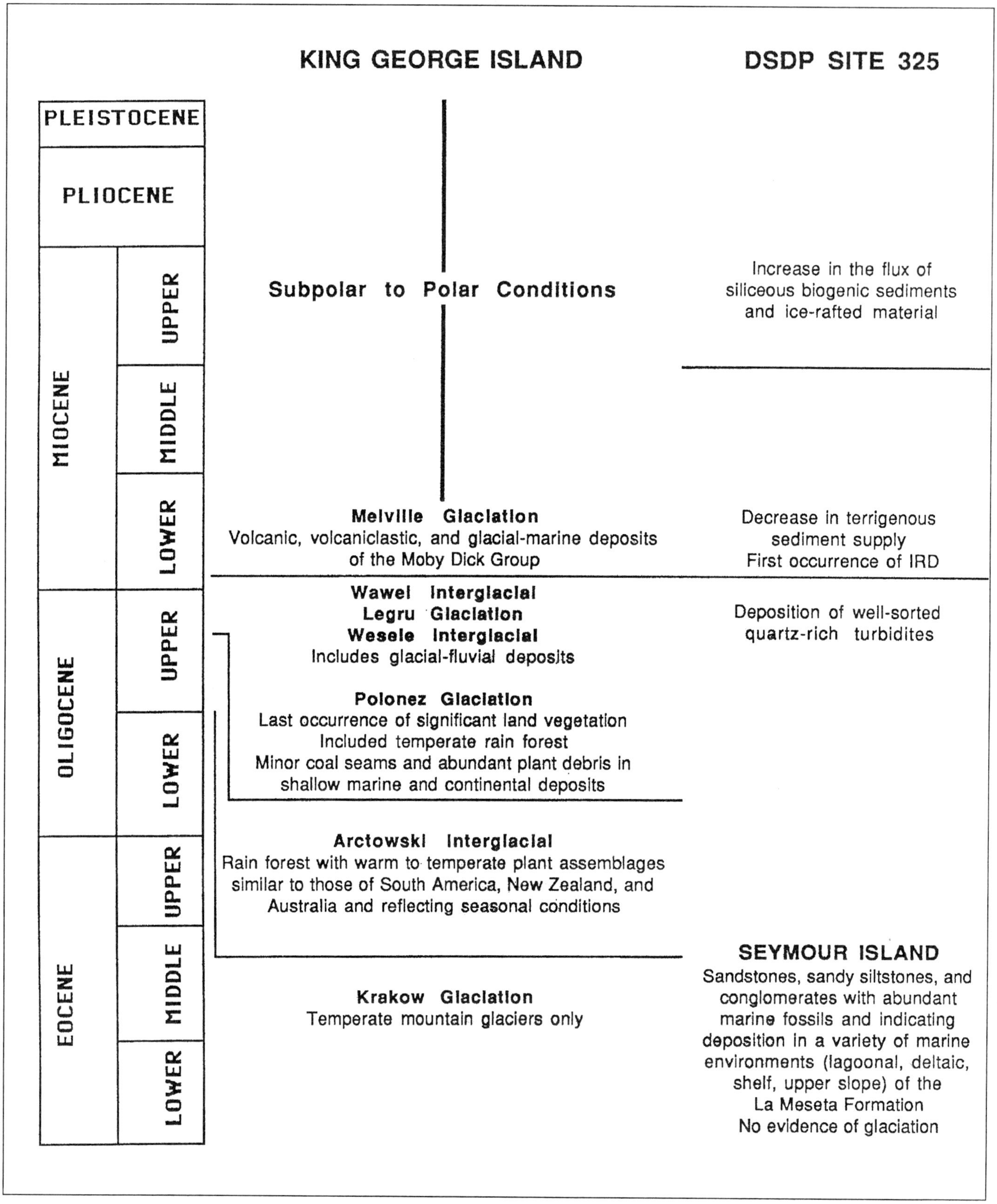

Figure 4—The glacial and climatic record of the northern Antarctic Peninsula region. See text for details and sources of information.

Nothofagus and Podocarpaceae and reflect cool, temperate (seasonal) conditions which persisted into the late Oligocene (Birkenmajer, 1987). These temperate rain forests were the last vestiges of abundant land vegetation in the region.

The oldest glacial deposits of the Antarctic Peninsula region consist of Eocene glacial sediments of the Krakow Glaciation (Birkenmajer et al., 1989). The apparent absence of glacial sediments of this age on Seymour Island (Figure 3) suggests that the Krakow glacial sediments were deposited by mountain glaciers. The Krakow Glaciation was followed by a prolonged interglacial, the Arctowski Interglacial, during which temperate rain forests existed on the island. The next episode of glaciation on King George Island occurred during the Oligocene (Polonez Glaciation). This glacial event was followed by the Wesele Interglacial, Legru Glaciation, and Wawel Interglacial, respectively, which together span the late Oligocene. The Wesele and Wawel interglacial deposits include fluvial deposits, which implies a temperate climatic setting. Another glacial sequence is situated stratigraphically above the Wawel Interglacial beds (Melville Glaciation), whose age is early Miocene (Birkenmajer et al., 1985a).

The land record of climatic events that occurred after the Melville Glaciation is sparse and poorly understood. However, drill core from the deep sea floor (DSDP Site 325, Figure 1) provides indirect evidence for climatic changes since the late Oligocene. During the late Oligocene?-early Miocene, turbidites were shed from the continental shelf and deposited on the Bellingshausen continental rise (Site 325). Sands in these turbidites are well sorted, display little vertical size grading, and consist dominantly of quartz. This implies derivation from fluvial/fluvial deltaic and/or coastal source beds as apposed to glacial deposits, and is consistent with the occurrence of coeval fluvial deposits within the Wesele and Wawel Interglacial sequences.

Between the early and middle Miocene, sedimentation rates at Site 325 decreased from 5 to 10 cm/1000 yr to less than 1 cm/1000 yr (Hollister and Craddock, 1976). This decrease in the sedimentation rate corresponds to a decrease in turbidite deposition at the site and implies a dramatic reduction in the amount of terrigenous sediment being shed from the Antarctic Peninsula. This event perhaps signifies the onset of subpolar climatic conditions on the Antarctic Peninsula, and perhaps overdeepening of the shelf by glacial erosion. Beginning in the late Miocene, there was a steady increase in the flux of siliceous biogenic sediments to the continental rise (Site 325), which may have resulted, in part, from the coincident decrease in the flux of terrigenous sediment to the sea floor. This event spread to the abyssal floor by the beginning of the Pliocene (DSDP Site 322). The increase in the rate of biogenic sedimentation is attributed to increased surface productivity related to more vigorous surface circulation at this time (Tucholke et al., 1976). The amount of ice-rafted debris deposited on the deep sea floor also increased during this time (Tucholke et al., 1976), but this too may have resulted from a decrease in the flux of fine-grained terrigenous material to the deep sea floor.

SEISMIC STRATIGRAPHY

Seismic records show four sequences which indicate different episodes of shelf development (Figures 5, 6). The oldest of these sequences consists of folded and faulted strata (S_4, Figures 5, 6). These are precollision deposits (precollision accretionary prism of Larter and Barker, 1989). Larter and Barker suggest that following ridge-trench collision the forearc was uplifted and eroded. This implies that the age of the S_4-S_3 Sequence boundary, marked by onlap of S_3 onto S_4, marks an episode of postcollision thermal subsidence. The oldest magnetic anomaly situated at the edge of the continental margin provides an approximate age for the culmination of ridge-trench collision (Figure 3). The magnitude of deformation of S_4 varies along the margin, and old forearc basins are still evident in some of the profiles (Figures 5, 6). However, differences in the magnitude of deformation of S_4 does not appear to correspond to the major fracture zones, as was previously predicted.

The sediments that comprise S_4 are presumed to consist of volcaniclastic material, similar to that which occurs on the sea floor north of the South Shetland Islands today (Jeffers et al., in press), and in Tertiary outcrops on King George Island. North of the Tula Fracture Zone, S_4 should include glacial marine sediments deposited during the Krakow, Polonez, and Legru Glaciations.

Sequence S_3 is an accretionary sequence separated from S_4 by a strong angular unconformity (Figures 5, 6). Sequence S_3 thins from southwest to northeast, consistent with a younger age of ridge subduction in that direction and the reduced interval of postsubduction margin development. The basal age of S_3 decreases from southwest to northeast across the shelf. South of the Tula Fracture Zone, S_3 may be as old as Eocene and span that interval of time when temperate glaciers existed on the peninsula in harmony with Nothofagus forests (Figure 4). Perhaps the maritime setting then was similar to that of the Gulf of Alaska today. Anderson and Molnia (1989) emphasize that temperate glacial marine settings, such as the Gulf of Alaska, are characterized by high rates of terrigenous sedimentation, whereas subpolar and polar glacial marine settings, such as exist in Antarctica today, are characterized by slow rates of terrigenous sedimentation. Thus, thinning of S_3 from southwest to northeast (Figure 3) may also reflect a change in the climate with time and reduced sedimentation on the shelf.

Sequence S_3 was deeply eroded to form the U_2 unconformity, which is present everywhere on the shelf and extends to the shelf edge (Figures 5, 6). This unconformity is the first evidence of widespread glacial erosion on the continental shelf, and therefore the first evidence for grounded ice caps which extended to the edge of the shelf.

The thickness of Sequence S_2 varies widely across the shelf. The seismic expression of S_2 is dominated by discontinuous, often hyperbolic, reflectors and considerable crosscutting of acoustic units (Figures 5, 6). These features are characteristic of glacial deposits (King and Fader, 1986). The strike line from the outer shelf (Figure 5, Line B) shows that S_2 includes many erosional surfaces, and some very large U-shaped troughs can be recognized. These large troughs presumably mark the former positions of large ice streams.

The erosion that occurred during S_2 time contributed to the deep and rugged topography of the shelf. This episode of shelf evolution may correspond to the Melville Glaciation (lower Miocene) on King George Island. This important episode of shelf development corresponds to a decrease in terrigenous sediment flux from the continent to the deep sea floor (Site 325, Figure 4). This decrease in sediment flux probably resulted from shelf overdeepening and foredeepening, and from a change from temperate to subpolar glacial conditions.

The youngest sequence is a laminar, draped sequence (S_1) that thins landward (Figures 5, 6). This sequence is generally conformable with Sequence S_2, but locally downlaps onto S_2 in a landward direction (Figures 5, 6). As the marine ice sheet eroded the shelf, it created a setting that was less suited for long-term grounding by thin (<2000 m thick) marine ice sheets. Sequence S_1 is believed to consist of glacial marine sediments that were deposited on the

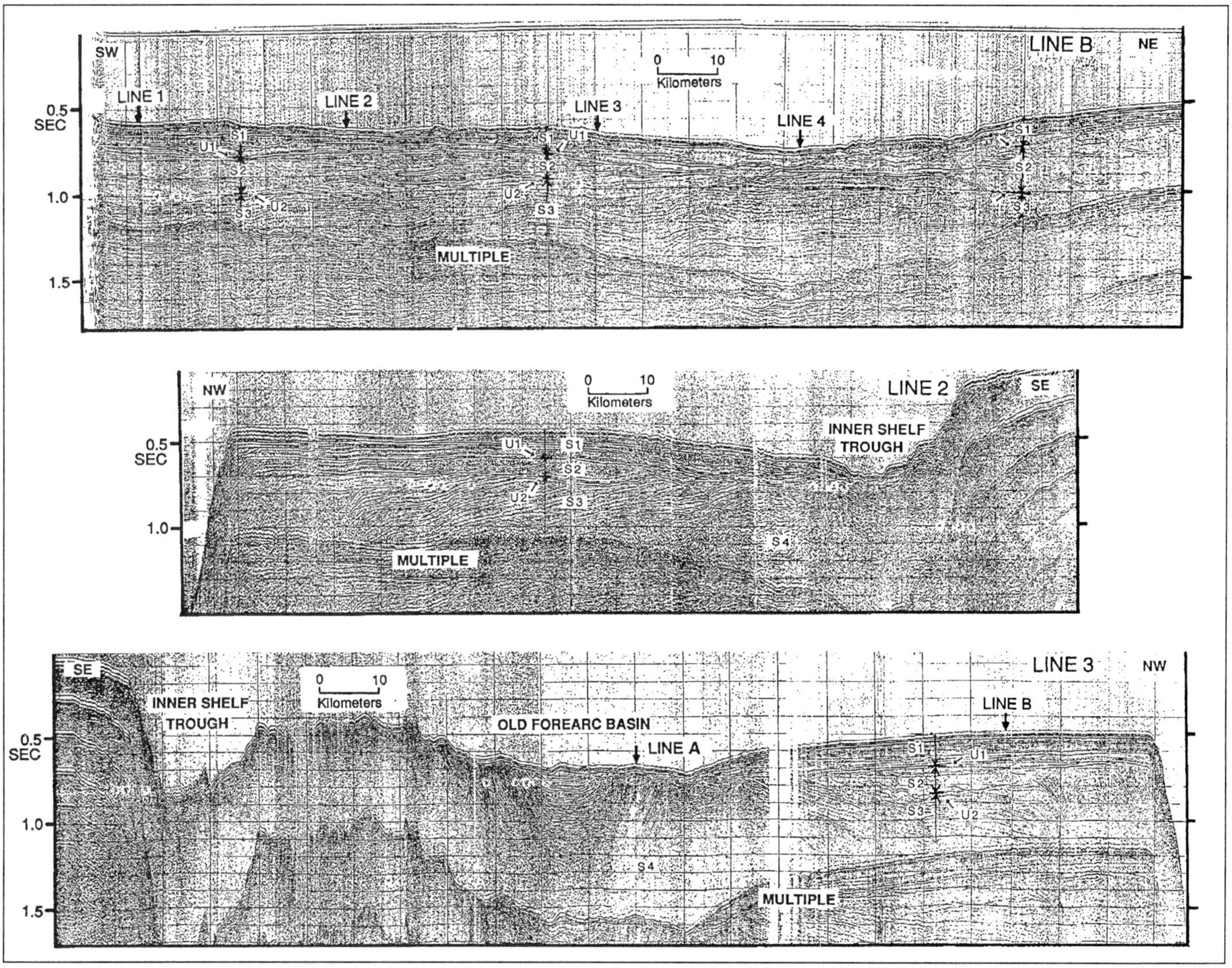

Figure 5—Representative seismic lines showing sequence stratigraphy of the study area. Lines 2 and 3 are dip lines and Line B is a strike line. See Figure 2 for the locations of these lines. The oldest sequence in the area is Sequence 4, which is a pretectonic and syntectonic sequence. Line 3 crossed an old forearc basin within Sequence 4. Sequence 3 is a prograding sequence which is best illustrated in Lines 2 and 3. Sequence 3 is cut by an erosional surface (U_2), which is the oldest glacial unconformity. Sequence 2 is a glacial sequence that includes many erosional surfaces. This sequence is characterized by discontinuous, often hyperbolic reflectors (Line 2) and large-scale cut and fill (Line B). The youngest sequence in the study area (S_1) is a glacial marine sequence that thins in an onshore direction and locally onlaps S_2 in that direction. Note that the inner shelf has been stripped of its sediment cover by glacial erosion.

overdeepened and foredeepened continental shelf. Sequence S_1 does include scattered and discontinuous intervals in which the reflection character resembles that of S_2. These intervals may reflect advances of thick (>2000 m) marine ice sheets onto the shelf; ice sheets would have to be at least this thick to ground on the present deep shelf.

As is typical of the Antarctic continental shelf, the inner shelf of the study area has been stripped of its sediment cover by glacial ice (Figures 5, 6). This implies that the ice sheet has grounded on the inner shelf during recent time.

POSSIBLE HYDROCARBON PROSPECTS

In order to assess hydrocarbon prospects on the Antarctic continental shelf, several factors have to be considered.

1. When did shelf basins first form?
2. What are the principal structural components of each basin and how have they influenced basin evolution?
3. What are the potential source rocks of each basin, what is the source of organic carbon, and when were these source rocks deposited?
4. What are the potential reservoir rocks and when were they formed?
5. What has the thermal/subsidence history of the basin been?
6. Are there suitable conduits for hydrocarbon migration into reservoirs, and are there seals to trap these hydrocarbons?

Antarctica is unique relative to other continents in that the source of terrigenous organic carbon (land vegetation) was eliminated as the climate cooled and ice sheets began to form. But this cooling was associated with intensified oceanic circulation and increased flux of marine biogenic sediments to the sea floor. Presently, shelf basins of the Antarctic are being filled with sediments which have relatively high concentrations of siliceous biogenic material (Anderson et al., 1983) and the organic carbon content of these sediments is typically in the range of 1 to 3% (Dunbar et al., 1985).

The period of time between ridge subduction and the onset of subpolar glacial conditions in the peninsula region is crucial to the possible hydrocarbon prospect within each segment of the shelf. This is the time window during which conditions were most suited for the formation of siliciclastic reservoir rocks. During the active margin phase, volcaniclastic sediments were deposited on the shelf. The development of subpolar glacial conditions on the peninsula resulted in a dramatic change in the style of sedimentation there, most importantly the deposition of poorly sorted glacial-marine sediments.

The development of ice sheets ultimately reached a stage where they had spread to the coast; and the climate became so cold that running water no longer contributed to the delivery of terrigenous sediments to the sea. Thus, those sedimentary environments which are the main production lines for reservoir rocks elsewhere in the world (fluvial, fluvial-deltaic, eolian, and coastal environments), no longer existed in Antarctica. In general, glacial and glacial-marine sedimentary environments are not thought to produce good reservoir rocks. To our knowledge, the Antarctic climate was not suited for significant carbonate and evaporite deposition on the continental shelf.

Because the changes from active to more tectonically passive margin development, and from a temperate to a subpolar climate, were diachronous (south to north) along the peninsula, determining the coexistence of suitable basin settings and favorable reservoir rocks is the key to this analysis. The major sediment accumulations are associated with syntectonic forearc deposits (S_4) and the post-subduction wedge (S_3).

There are two forearc basins preserved on the Antarctic Peninsula continental shelf (Figure 3). At this time, pinpointing the maximum age for the forearc basins is not possible. It is probable that throughout the active subduction phase forearc basins have been transient features that periodically formed and were eventually destroyed. We do not speculate as to what process accounts for the cycles of basin formation and destruction, but can only infer the youngest basins were preserved as subduction ceased. As defined by seismic data, these basins are relatively small and shallow. The maximum depth is approximately 700 m (assuming a velocity of 2000 m/sec), and their width is less than 20 km. The southern basin is 56 km long and the northern basin is 30 km long. Sediments filling these basins are assigned to the S_4 sequence and are likely volcanic sandstones and mudstones. North of the Tula Fracture Zone, the sequence includes glacial marine sediments. The total thickness of S_4 cannot be estimated from the single channel seismic data. Only the minimum age for S_4 in each segment can be approximated, since, by assumption, S_4 deposition ends as subduction stops.

In terms of source rocks, S_4 should contain marine shales deposited when the climate was temperate, and only areas north of the Anvers Fracture Zone (i.e., areas still actively subducting 10 Ma) would contain subpolar to polar glacial marine sediments and biogenic muds. In conjunction with the high heat flow from the subducting spreading center, S_4 sequences may have generated hydrocarbons. This is based on the occurrence of thermogenic gases and higher hydrocarbons in surface sediments in the Bransfield Basin (Han, 1988). The volcanic sandstones and mudstones of S_4 are poor hydrocarbon reservoir candidates.

Although the S_3-S_4 contact is obscured by the sea bottom multiple on the outer shelf, a conservative estimate of S_3 thickness can be made by extrapolating the dip of the S_3-S_4 contact on the inner shelf. Isopachs of the S_3 interval show that this sequence thins northward (Figure 3).

North of the Tula Fracture Zone, S_3 is so thin (approximately 500 m) and at such a shallow depth that the probability of source rock maturation within this sequence is quite small. The Anvers-Hero Fracture Zone segment is so young that it never coexisted in time with S_3 as a tectonically passive margin.

By the time the Tula-Adelaide and Adelaide-Bisco post-subduction wedges had evolved, the climate had cooled, probably leading to a decline of delta systems and the terrigenous carbon source. The highest hydrocarbon potential lies south of the Tula Fracture Zone where S_3 is about 1500 m thick, and is buried the deepest. Sequence S_3 may be as old as middle Eocene ($\approx$ 40 Ma) in this area. The climate of that period should have been conducive to reservoir and source rock formation, with large delta systems, plus abundant terrigenous and marine carbon supplies.

Even in a best case scenario, using St. John's (1984) estimate of equivalent barrels of oil/volume of sediment, S_3 in the study area would contain only 312 million BOE. However, hydrocarbon potential improves south of the Tula Fracture Zone. St. John (1984) estimates the Graham and Bellingshausen Basins, which are situated south of the study area, to contain in excess of 13 billion BOE. The region south of the Tula Fracture Zone has experienced a tectonic evolution similar to the study area. Consequently, this study provides a model for evaluating the tectonic and climatic controls on sedimentation and stratigraphy in these provinces.

ACKNOWLEDGMENTS

This research was funded by the National Science Foundation-Division of Polar Programs (Grant Number DPP85-16908). We wish to thank Dennis Cassidy of the Antarctic Marine Geology Research Facility for providing us with important literature related to this investigation. We thank Howard Stagg for providing us with a thorough and informative review of the manuscript.

REFERENCES CITED

Anderson, J. B., in press, The Antarctic continental shelf: results from recent marine geologic and geophysical investigations: *in* R. Tingey, ed., Antarctic Geology, London, Oxford Press.

Anderson, J. B. and B. F. Molnia, 1989, Glacial-marine sedimentation: Short Course In Geology, 28th International Geological Congress, American Geophysical Union, Washington, D.C., 127 p.

Anderson, J. B., C. Brake, E. W. Domack, N. C. Myers and R. Wright, 1983, Sedimentary dynamics of the Antarctic continental shelf, *in* R. L. Oliver, P. R. James and J. B. Jago, eds, Antarctic Earth Science: Australian Academy of Sciences, Canberra, Australia, p. 387-389.

Askin, R. A. and R. F. Fleming, 1982, Palynological investigations of Campanian to lower Oligocene sediments on Seymour Island, Antarctic Peninsula: Antarctic Journal of the U.S., v. 17, p. 70-71.

Barker, P. F., 1982, The Cenozoic subduction history of the Pacific margin of the Antarctic Peninsula: ridge crest-

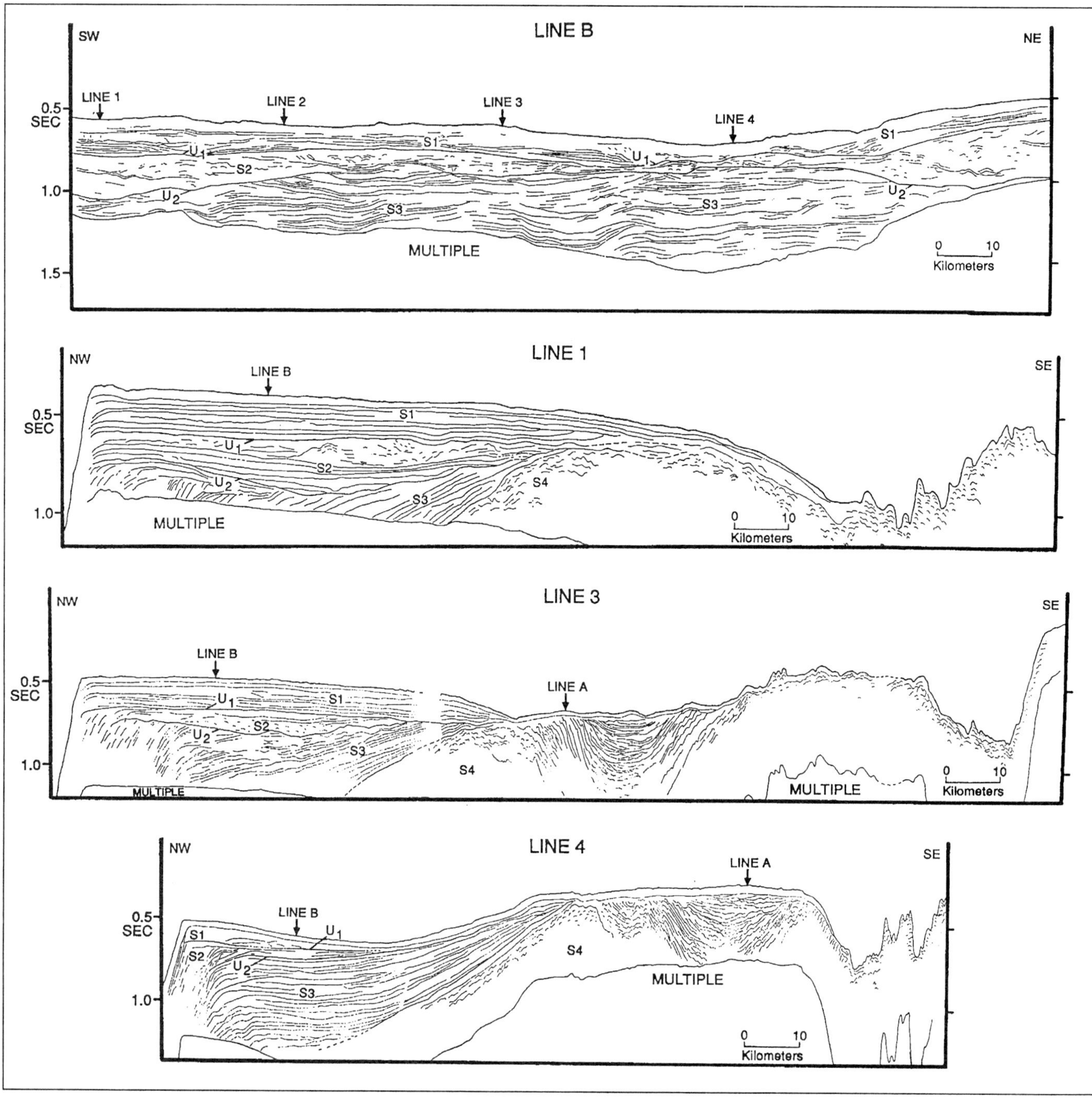

Figure 6—Representative interpreted seismic profiles from the study area illustrating differences in shelf geology within the study area. See Figure 2 for profile locations.

trench interactions: Journal Geological Society of London, v. 139, p. 787-802.

Barker, P. F., I. W. D. Dalziel and B. C. Storey, 1988, Tectonic development of the Scotia arc region, *in* R. Tingey, ed., Antarctic Geology: Oxford University Press, in press.

Biernat, G., K. Birkenmajer and E. Popiel-Barczyk, 1985, Tertiary braciopods from the Moby Dick Group of King George Island (South Shetland Islands, Antarctica): Studia Geologica Polonica, v. 81, p. 109-141, 12 plts.

Birkenmajer, K., 1981a, Geological relations at Lions Rump, King George Island (South Shetland Islands, Antarctica): Studia Geologica Polonica, v. 72, p. 75-87.

Birkenmajer, K., 1981b, Lithostratigraphy of the Point Hennequin Group (Miocene volcanics and sediments) at King George Island (South Shetland Islands, Antarctica): Studia Geologica Polonica, v. 72, p. 59-73.

Birkenmajer, K., 1982a, Late Cenozoic phases of block faulting on King George Island (South Shetland Islands, West Antarctica): Polish Academy of Sciences Bulletin, Earth Sciences, v. 33, p. 21-32.

Birkenmajer, K., 1982b, Pliocene tillite-bearing succession of King George Island (South Shetland Islands, Antarctica):

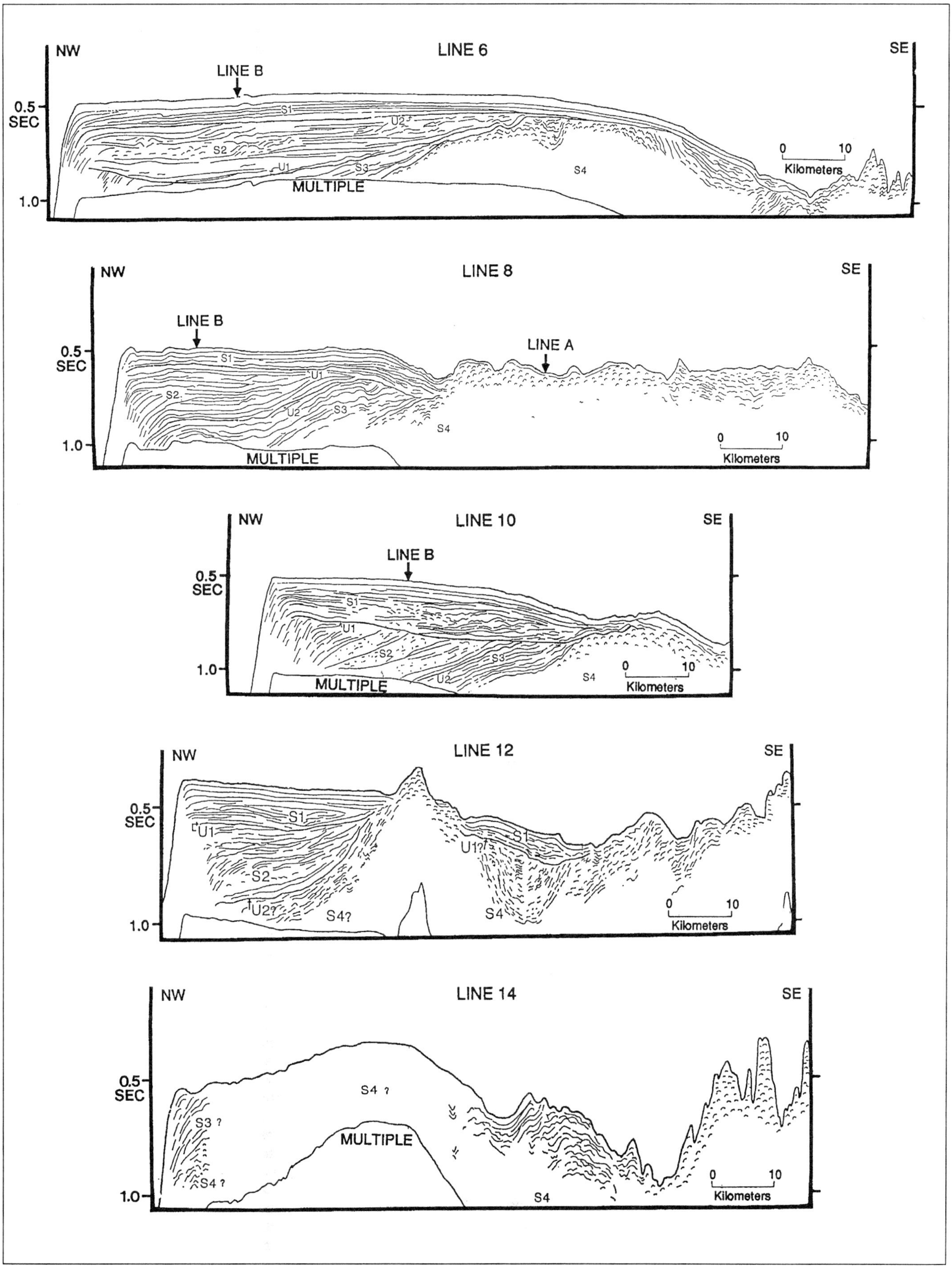
LINE 6
NW
SE
LINE B
0.5
SEC
1.0
S1
U2
S2
U1
S3
S4
MULTIPLE
0 10
Kilometers
LINE 8
NW
SE
LINE B
LINE A
0.5
SEC
1.0
S1
U1
S2
U2
S3
S4
MULTIPLE
0 10
Kilometers
LINE 10
NW
SE
LINE B
0.5
SEC
1.0
S1
U1
S2
S3
U2
S4
MULTIPLE
0 10
Kilometers
LINE 12
NW
SE
0.5
SEC
1.0
S1
U1
S2
U2?
S4?
U1?
S4
0 10
Kilometers
LINE 14
NW
SE
0.5
SEC
1.0
S4 ?
S3 ?
S4 ?
MULTIPLE
S4
0 10
Kilometers

Studia Geologica Polonica, 74, p. 7-72.
Birkenmajer, K., 1984, Geology of the Cape Melville Area, King George Island (South Shetland Islands, Antarctica): pre-Pliocene glaciomarine deposits and their substratum: Studia Geologica Polonica, v. 79, p. 7-36.
Birkenmajer, K., 1985a, K-Ar dating of the Melville Glaciation (Early Miocene) in West Antarctica: Polish Academy of Sciences Bulletin, Earth Sciences, v. 33, p. 15-23.
Birkenmajer, K., 1985b, Onset of Tertiary continental glaciation in the Antarctic Peninsula sector (West Antarctica): Acta Geologica Polonica, v. 35, p.1-31.
Birkenmajer, K., 1987, Oligocene-Miocene glacio-marine sequences of King George Island (South Shetland Islands), Antarctica: Palaeontologia Polonica, No. 49, p. 9-36.
Birkenmajer, K., 1989, Geochronology and climatostratigraphy of Tertiary glacial and interglacial successions on King George Island, South Shetland Islands: International Workshop on Antarctic Geochronology, Munich, Germany.
Birkenmajer, K., A. Gazdzicki, H. Kreuzer and P. Muller, 1985a, K-KAr dating of the Melville Glaciation (Early Miocene) in West Antarctica: Polish Academy of Sciences Bulletin, Earth Sciences, v. 33, p. 15-23.
Birkenmajer, K., A. Paulo and A. K. Tokarski, 1985b, Neogene marine tillite at Magda Nunatak, King George Island (South Shetland Islands, Antarctica): Studia Geologica Polonica, v. 81, p. 99-107.
Birkenmajer, K., E. Soliani and K. Kawashita, 1989, Reliability of Potassium-Argon dating of Cretaceous-Tertiary island-arc volcanic suite of King George Island, South Shetland Islands: International Workshop on Antarctic Geochronology, Munich, Germany.
Dalziel, I. W. D. and D. H. Elliot, 1982, West Antarctica: problem child of Gondwanaland: Tectonics, v. 1, p. 3-19.
Dunbar, R. B., J. B. Anderson, E. W. Domack and S. S. Jacobs, 1985, Oceanographic influences on sedimentation along the Antarctic continental shelf: American Geophysical Union Antarctic Research Series, v. 43, p. 291-312.
Gazdzicki, A. and H. Pugaczewska, 1984, Biota of the "Pecten Conglomerate" (Polonez Cove Formation, Pliocene) of King George Island (South Shetland Islands, Antarctica): Studia Geologica Polonica, v. 79, p. 59-115.
Han, M. W., 1988, Hydrothermalism and hydrocarbon potential in the King George Basin of the Bransfield Basin, Antarctica: Proceedings of the First International Symposium on Antarctic Sciences, Korea Ocean Research and Development Institute, Seoul, Korea, p. 37-55.
Hawkes, D. D., 1981, Tectonic segmentation of the northern Antarctic Peninsula: Geology, v. 9, p. 220-224.
Herron, E. M. and B. E. Tucholke, 1976, Sea floor magnetic patterns and basement structure in the southeastern Pacific, *in* C. D. Hollister and C. Craddock et al., eds., Initial Reports of the Deep Sea Drilling Project, v. 35, Washington, (U.S. Government Printing Office), p. 263-278.
Hollister, C. D. and D. Craddock, 1976, Geologic evolution of the South Pacific Basin, *in* C. D. Hollister and C. Craddock et al., eds. , Initial Reports of the Deep Sea Drilling Program, v. 35, Washington, (U.S. Government Printing Office), p. 723-744.
Jeffers, J. D., J. B. Anderson and L. A. Lawver, in press, Evolution of the Bransfield Basin, Antarctic Peninsula, *in* M. R. A. Thomson, J. A. Crame and J. W. Thomson, eds., Geological Evolution of Antarctica: Cambridge, Cambridge University Press.
King, L. H. and G. B. J. Fader, 1986, Wisconsinan glaciation of the Atlantic continental shelf of southeast Canada: Geological Survey of Canada Bulletin 363, 72 p.
Kimura, K., 1982, Geological and geophysical survey in the Bellingshausen Basin, off Antarctica: Antarctic Record, National Institute of Polar Research, Tokyo, Japan, No. 75, p. 12-24.
LaBrecque, J. L., 1987, The USAC aerosurvey: accelerating exploration of the Antarctic: Lamont-Doherty Geological Observatory Yearbook, Columbia University, New York, p. 52-59.
Larter, R. D. and P. F. Barker, 1989, Seismic stratigraphy of the Antarctic Peninsula Pacific margin: a record of Pliocene-Pleistocene ice volume and paleoclimate: Geology, v. 17, p. 731-734.
Lawver, L. A., J. Y. Royer, D. T. Sandwell and C. R. Scotese, in press, Evolution of the Antarctic continental margins, *in* M. R. A. Thompson, J. A. Crame and J. W. Thompson, eds., Geologic Evolution of Antarctica.
Renner, R. G. B., L. J. S. Sturgeon and S. W. Garrett, 1985, Reconnaissance gravity and aeromagnetic surveys of the Antarctic Peninsula: British Antarctic Survey Scientific Report 110, 54 p.
St. John, B., 1984, Antarctica-Geology and Hydrocarbon Potential, *in* M.T. Halbouty, ed., Future Petroleum Provinces of the World: Tulsa, Oklahoma, AAPG Memoir 40, p. 55-100.
Stuchlik, L., 1981, Tertiary pollen spectrum from the Ezurra Inlet Group at Admiralty Bay, King George Island (South Shetland Islands, Antarctica): Studia Geologica Polonica, v. 72, p. 109-132.
Tucholke, B. E., C. D. Hollister, F. M. Weaver and W. R. Vennum, 1976, Continental rise and abyssal plain sedimentation in the Southeast Pacific Basin-Leg 35 Deep Sea Drilling Project, *in* C. D. Hollister and C. Craddock et al., eds., Initial Reports of the Deep Sea Drilling Project, Washington, (U.S. Government Printing Office), p. 359-400.
Weissel, J. K., D. E. Hayes and E. M. Herron, 1977, Plate tectonics synthesis: the displacements between Australia, New Zealand, and Antarctica since the late Cretaceous: Marine Geology, v. 25, p. 231-277.
Zastawniak, E., R. Wrona, A. Gazdzicki and K. Birkenmajer, 1985, Plant remains from the top part of the Point Hennequin Group (Upper Oligocene), King George Island (South Shetland Islands, Antarctica): Studia Geologica Polonica, v. 81, p. 143-164, 6 plts.
Zinsmeister, W. J., 1982, Review of the Upper Cretaceous-Lower Tertiary sequence on Seymour Island, Antarctica: Journal Geological Society of London, v. 139, p. 779-785.

Chapter 2

Sequence Stratigraphy of the Bransfield Basin, Antarctica: Implications for Tectonic History and Hydrocarbon Potential

John D. Jeffers*
John B. Anderson
Department of Geology and Geophysics
Rice University
Houston, Texas, U.S.A.

*Present address: Mobil Exploration & Producing U.S., Houston, Texas, U.S.A.

ABSTRACT

Application of sequence stratigraphic concepts to seismic reflection profiles from the Bransfield Basin indicates that this modern backarc basin began to form during the waning stages of subduction at the South Shetland Trench at about 4 Ma. Two distinct systems tracts stack to form depositional sequences; organic-rich hemipelagic sediments drape the basin during highstands/interglacial periods, whereas large volumes of glacially eroded terrigenous sediments prograde into the basin during lowstands/glacial maxima. Although the juxtaposition of organic-rich diatomaceous muds with the high heat flow of the backarc spreading system is favorable for the generation of hydrocarbons, reservoir quality sands and suitable traps have yet to be identified.

INTRODUCTION

The Bransfield Basin is an actively spreading backarc[1] basin which separates the South Shetland Islands from the northernmost Antarctic Peninsula (Figure 1). It differs from the well-studied backarc basins of the western Pacific both in its tectonic evolution and its sedimentary character. It is associated with the waning stages of Pacific-Antarctic subduction, rather than with ongoing subduction and arc magmatism. Sedimentation is dominated by glacial-marine processes and their associated lithofacies. Because of its proximity to southern South America, it is one of the best studied and most accessible of the offshore Antarctic basins. Interest in the basin as a potential hydrocarbon prospect was heightened by reports of thermogenic hydrocarbons observed at the sea floor (Whiticar et al., 1985). Although unique in the present, the Bransfield Basin may be analogous to several deformed ancient basins. The Mesozoic Rocas Verdes basin of Patagonia (Dalziel et al., 1974) once occupied a similar tectonic setting, while the backarc basin sequences of South Georgia (Bell et al., 1977; Storey and Macdonald, 1984), the early rifted margins of the Weddell Sea (Barker et al., 1988), and the glaciated margins of many tectonically active small ocean basins of previous glacial episodes may be very similar sedimentologically.

[1]We classify the Bransfield Basin as a backarc basin in the general sense, i.e., an extensional basin behind a B-subduction arc. It is not associated with an active calc-alkaline volcanic arc.

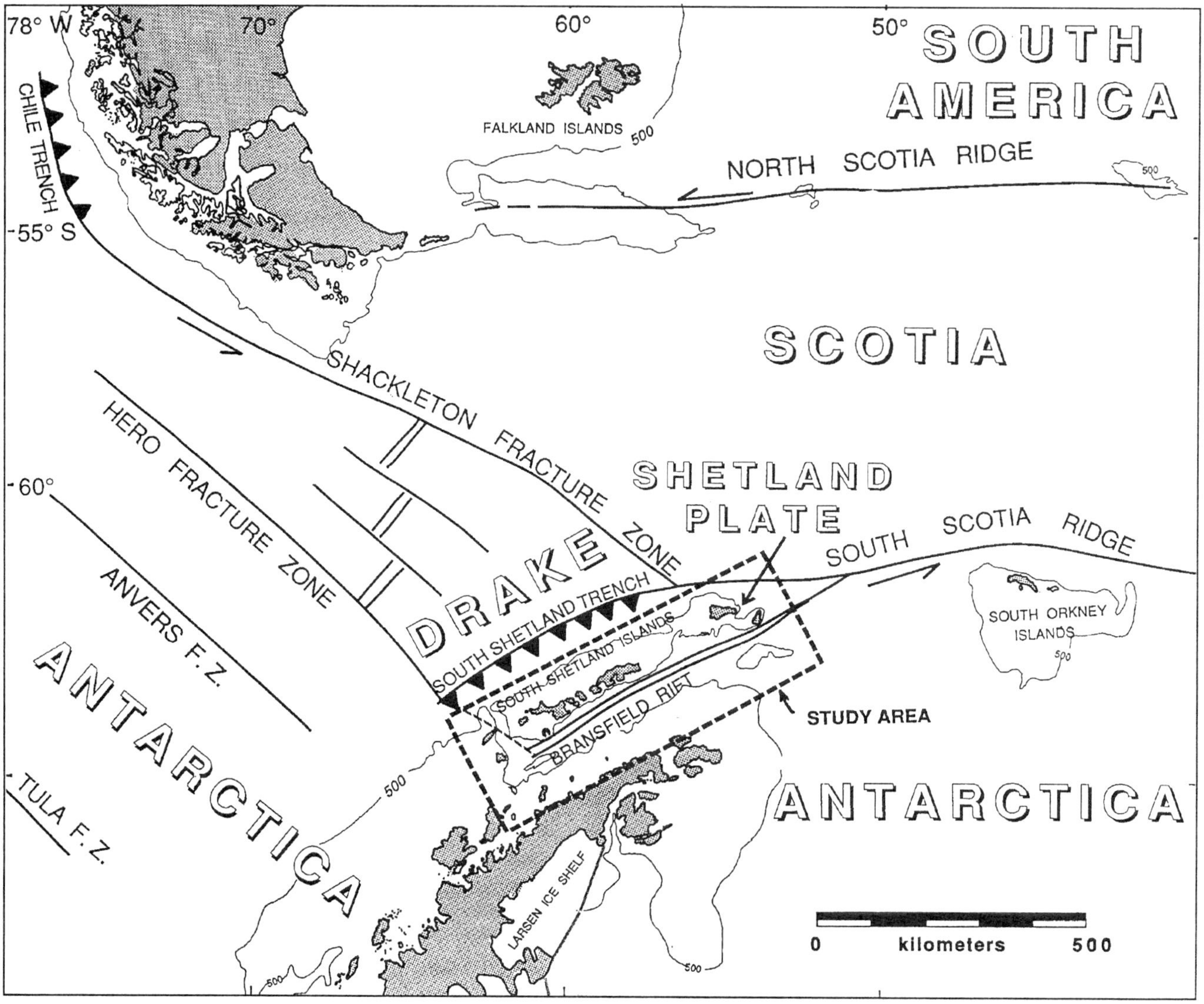

Figure 1—Tectonic map of the Scotia Arc region. Plate names in outlined lettering. Bransfield Basin study area highlighted.

This chapter presents results of five marine geological and geophysical cruises to the Bransfield Basin between 1982 and 1987 as part of the United States Antarctic Program. Approximately 2200 km of single-channel seismic reflection profiles were collected in the Bransfield Basin between 1985 and 1987 from *USCGC Glacier* and R/V *Polar Duke* (Figure 2) using 4.6 kJ sparker arrays and 100 in.^{3} water guns as acoustic sources. In addition, over 100 piston cores and bottom grab samples were collected from the study area (Figure 2).

The sediment samples and high resolution seismic reflection data provide the basis for construction of a simple sequence stratigraphic model, which allows projection of the distribution of sediments in the Bransfield Basin through glacial and sea level cycles. The sequence stratigraphic analysis, when integrated with previously published regional work, allows examination of the tectonostratigraphic evolution of the basin. Finally, we evaluate the potential for hydrocarbon maturation and accumulation in the Bransfield Basin.

BACKGROUND

Regional History

The Bransfield Basin is the latest in a series of convergent margin basins which have developed throughout the history of the Pacific margin of Antarctica and South America. The Scotia Arc connects the Andean Cordillera of South America to the Antarctic Peninsula (Barker et al., 1988) through a complicated system of plate boundaries between South America and Antarctica, in much the same way that the Caribbean Arc connects the North and South American plates (Figure 1).

The Cenozoic tectonic development of the Antarctic Peninsula has been controlled largely by the incremental subduction of the Aluk-Antarctic spreading center, which occurred progressively northeastward along the margin from 50 Ma in the southern Antarctic Peninsula to 4 Ma immediately south of the Hero Fracture Zone (Barker, 1982).

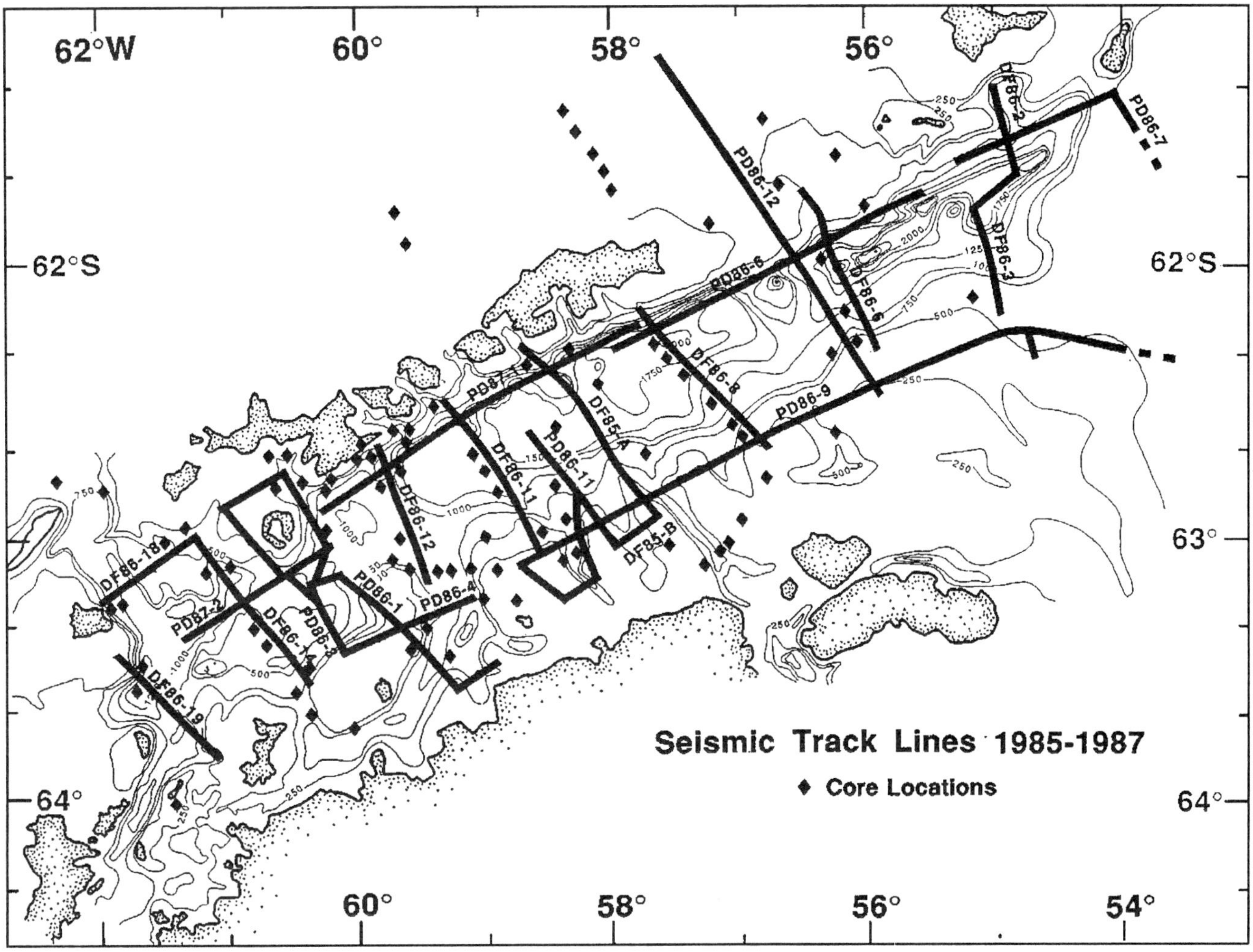

Figure 2—Track lines of seismic reflection profiles and locations of piston cores used in this study.

As segments of the spreading ridge approached the trench, magmatism ceased in the corresponding arc segments, and the arc and forearc experienced uplift and tectonic erosion (Barker, 1982). When the ridge arrived at the trench, both spreading and subduction ceased. The dates for cessation of subduction are determined from the sea-floor magnetic anomaly patterns (Figure 3), which increase in age away from the Antarctic Peninsula. The Drake microplate (Figure 1), now attached to the Antarctic plate, is the last relic of the Aluk plate.

Throughout the Cenozoic, the Antarctic Peninsula subducted Pacific sea floor with fracture zones parallel to the direction of subduction, so that boundaries between segments of different age sea floor stayed at the same point on the margin (Barker et al., 1988). Major fracture zones in the subducted oceanic crust are reflected in the overlying continental crust by "transverse megafractures" (Hawkes, 1981) which divide the northern Antarctic Peninsula into segments having different histories of magmatism, uplift, and erosion. The margin segments remained passive after the ridge-trench collisions occurred, so the relationship between subduction history—recorded in the magnetic anomaly patterns of the surviving flank of the spreading center—and the resulting onshore geologic record is preserved.

The Bransfield Basin, a young extensional basin approximately 100 km wide, occupies the segment of the Antarctic Peninsula between the Hero and Shackleton Fracture Zones which contains the last surviving Aluk-Antarctic spreading ridge segments and the only remaining trench topography (Figure 1). These three ridge segments apparently stopped spreading at 4 Ma, before reaching the trench (Barker, 1982). The Shetland microplate contains the South Shetland Islands; it is bounded to the northwest by the South Shetland Trench and to the southeast by the backarc extensional system in Bransfield Strait. Its northeastern and southwestern boundaries are poorly defined, but they coincide approximately with the landward projections of the Hero and Shackleton Fracture Zones.

Recent volcanicity (Saunders and Tarney, 1982) and seismicity (Forsyth, 1975; Pelayo and Wiens, 1986) along the axis of the Bransfield Basin indicate that extension is continuing presently. The only active subaerial volcanism in the region occurs on Deception Island, although there is evidence of recent activity on Penguin and Bridgeman Islands (Figure 4). In addition, there is considerable submarine volcanism and hydrothermal activity in Bransfield Strait related to backarc extension (Han and Suess, 1987).

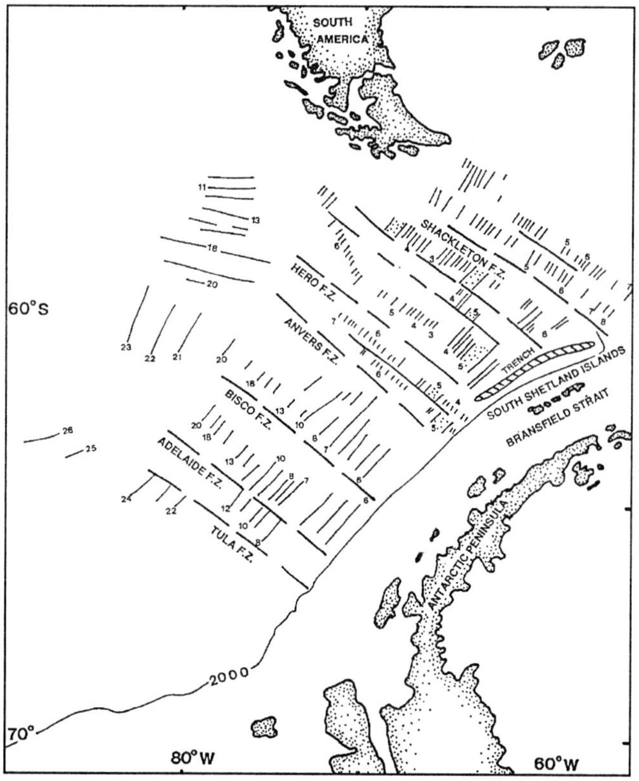

Figure 3—Sea-floor magnetic anomaly patterns of the southeastern Pacific ocean, after Barker and Dalziel (1983).

Bathymetry

The Bransfield Basin is a deep asymmetric trough, trending northeast-southwest, composed of three separate subbasins (Figure 4). The relatively shallow, irregularly shaped western subbasin south and west of Livingston and Deception Islands trends northeast-southwest toward the Gerlache Strait and also branches northwestward through Boyd Strait. The abrupt northeastward deepening of the basin from 900 to 1300 m marks the eastern limit of the western basin. The central subbasin lies south of Robert, Nelson, and King George Islands, and extends northeastward to a bathymetric divide associated with Bridgeman Island. It is bounded to the northwest by the steep slope of the South Shetland Islands, which is incised by troughs associated with the outlets of six bays. Four troughs aligned perpendicular to the axis of the basin cut into the broad shallow shelf of the Trinity Peninsula to a depth of 750 m. The U-shaped cross-sectional profiles of these troughs suggest that they are glacially carved features, but their orientation may be structurally controlled. They appear to connect with glacial drainage systems onshore. The shelf break occurs at about 250 m, and a second platform deepens gradually from 750 m at the outlets of the troughs to about 900 m. A steeper (~9°) slope leads to the basin floor, which deepens from 1300 m at the southwest end of the central subbasin to more than 2000 m southeast of King George Island. The bathymetric highs falling on a line connecting Deception and Bridgeman Islands appear to be submarine volcanoes associated with backarc extension.

The eastern subbasin, extending northeastward from Bridgeman Island past Elephant and Clarence Islands, is narrower than the central subbasin and reaches a depth of 2500 m. Several seamounts may be an extension of the line of submarine volcanoes in the central subbasin. The margins of the eastern basin are free of the large troughs seen farther to the southwest.

Crustal Structure

Seismic refraction data reveal the general crustal structure of the Bransfield Basin (Ashcroft, 1972). The South Shetland Islands sit atop a 20-25 km thick block of igneous and metamorphic continental lithosphere, nearly identical to the adjacent Antarctic Peninsula. The crust beneath the axis of the basin is much thinner, resembling a slightly thickened oceanic section with structure similar to a mid-ocean ridge. Sediments and volcanics overlie basic igneous rocks to a total thickness of 15-20 km. The crust generally thickens to the northeast along strike, both along the South Shetland Islands and within the backarc basin. Northeast-southwest trending normal faulting associated with Pliocene to Recent volcanics exposed on the South Shetland Islands (Barton, 1965; Weaver et al., 1979) also occurs offshore, forming the steep northwestern margin of the Bransfield Basin (Ashcroft, 1972); similar faulting is seen on the Trinity Peninsula Shelf. Sedimentary wedges prograde across the shelves north of the Trinity Peninsula and north of the South Shetland Islands. The similarity of crustal structures led Ashcroft (1972) to propose that the South Shetland Islands and the Antarctic Peninsula were joined originally, before being separated by a Tertiary rifting event.

The Bouguer gravity high over the center of the Bransfield Basin (Davey, 1972) indicates the presence of dense, quasi-oceanic crust beneath a thin sedimentary cover. Simple Airy-Heiskanen isostatic anomalies, ranging from +80 mgal over the South Shetland Islands to -40 mgal over the trench, are smaller in amplitude than isostatic anomalies over other island arcs, and may delineate a formerly active arc which has begun to reequilibrate isostatically.

Aeromagnetic surveys (Renner et al., 1985) defined a series of 50-100 km wide, long-wavelength positive magnetic anomalies exceeding 500 nT in amplitude throughout the Scotia Arc. The largest continuous anomaly, the West Coast Magnetic Anomaly (WCMA), parallels the west coast of the Antarctic Peninsula for a distance of some 1300 km (Garrett and Storey, 1987). Correlation of the magnetic anomalies with associated gravity anomalies, seismic refraction data, and surface geology indicates that the WCMA is due to a group of linear batholiths emplaced during Mesozoic-Cenozoic subduction (Garrett et al., 1987). The WCMA bifurcates at the southern end of Bransfield Strait, suggesting that a subvolcanic batholith was split by the Bransfield rift.

Geology of the South Shetland Islands

The South Shetland Islands between Livingston Island and King George Island are blocks of late Paleozoic and Mesozoic subduction complex accreted to the margin of the Antarctic Peninsula which were later intruded by late Mesozoic-Cenozoic subvolcanic plutons and buried under their associated volcanic rocks (Barker and Dalziel, 1983). Pliocene to Recent volcanics associated with backarc rifting are exposed at several places on King George and Livingston Islands, as well as Deception, Bridgeman, and Penguin Islands (Saunders and Tarney, 1982). Elephant, Clarence, and Smith Islands are distinctly different; they apparently are old (Paleozoic?) blueschist, greenschist, and

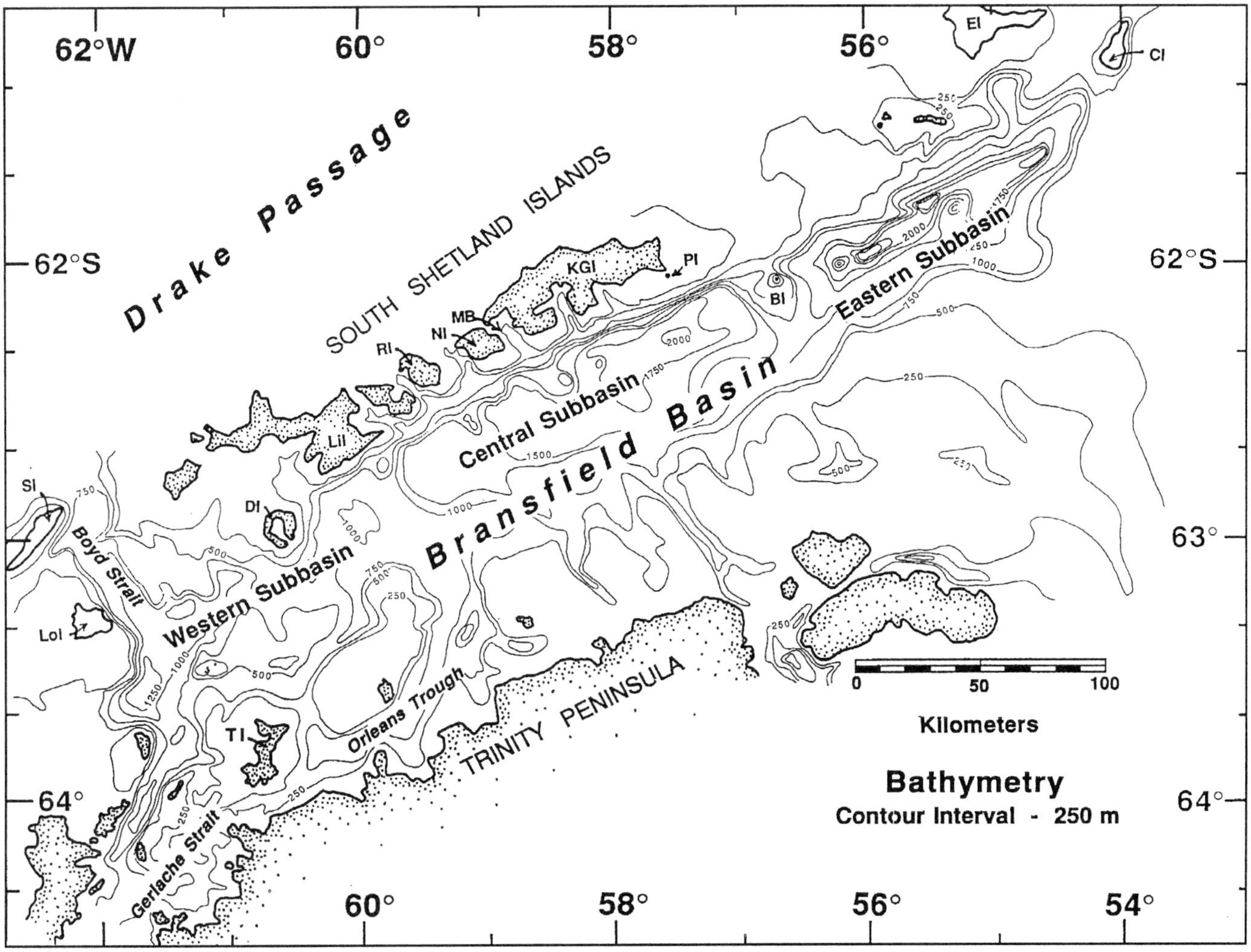

Figure 4—Bathymetry of the Bransfield Basin, showing locations of western, central, and eastern subbasins. Contour interval 250 meters. BI = Bridgeman Island, CI = Clarence Island, DI = Deception Island, EI = Elephant Island, KGI = King George Island, LiI = Livingston Island, LoI = Low Island, MB = Maxwell Bay, NI = Nelson Island, PI = Penguin Island, RI = Robert Island, SI = Smith Island, TI = Trinity Island.

amphibolite facies metamorphics which were uplifted from great depth sometime in the late Cenozoic. Grunow et al. (1987) relate their emplacement to subduction of the Hero and Shackleton Fracture Zones.

Birkenmajer (1982) and Birkenmajer et al. (1986) have helped to reconstruct the chronology of tectonic events in the Bransfield Strait area. They recognize four phases of Late Cretaceous to Cenozoic deformation on King George Island. Northeast-southwest trending right-lateral strike-slip faults and associated folds, mappable the entire length of King George Island, formed in a transpressional regime during Cretaceous to early Miocene time (Birkenmajer et al., 1986). Numerous shorter strike-slip faults, spaced 5-15 km apart, offset the earlier faults and reflect the transition from compression to extension during the late Miocene. Normal faulting related to rifting began in the Pliocene, and continues, along with volcanism, to the present.

Recent subaerial volcanism related to backarc extension in the Bransfield Basin is restricted to Deception, Bridgeman, and Penguin Islands; the seamounts between Bridgeman and Deception Islands are probably submarine equivalents. In general, the Pliocene-Recent volcanic rocks on the three islands are transitional between ocean floor basalts and island arc volcanics (Weaver et al., 1979). Analyses of basalts and basaltic andesites dredged from the seamounts (White and Cheatham, 1987; Keller and Fisk, 1987) show $^{87}Sr/^{86}Sr$ and $^{143}Nd/^{144}Nd$ ratios typical of backarc basin basalts.

The late Cenozoic glacial history of the South Shetland Islands has been reconstructed from geomorphologic and stratigraphic evidence. Curl (1980) identified an earlier Pleistocene (Illinoian or earlier) glaciation, with an ice cap of dimensions 270 x 70 km entirely covering the South Shetland Islands and grounded to a depth of 200 (John and Sugden, 1971) to 375 m (Curl, 1980) on the shelf. The subsequent interglacial, correlated with the Sangamon, was followed by a less extensive advance during the Wisconsinan, when enlarged ice domes covered the individual islands (John and Sugden, 1971). See Anderson et al. (1990) for a detailed discussion of the glacial and climatic history of the region.

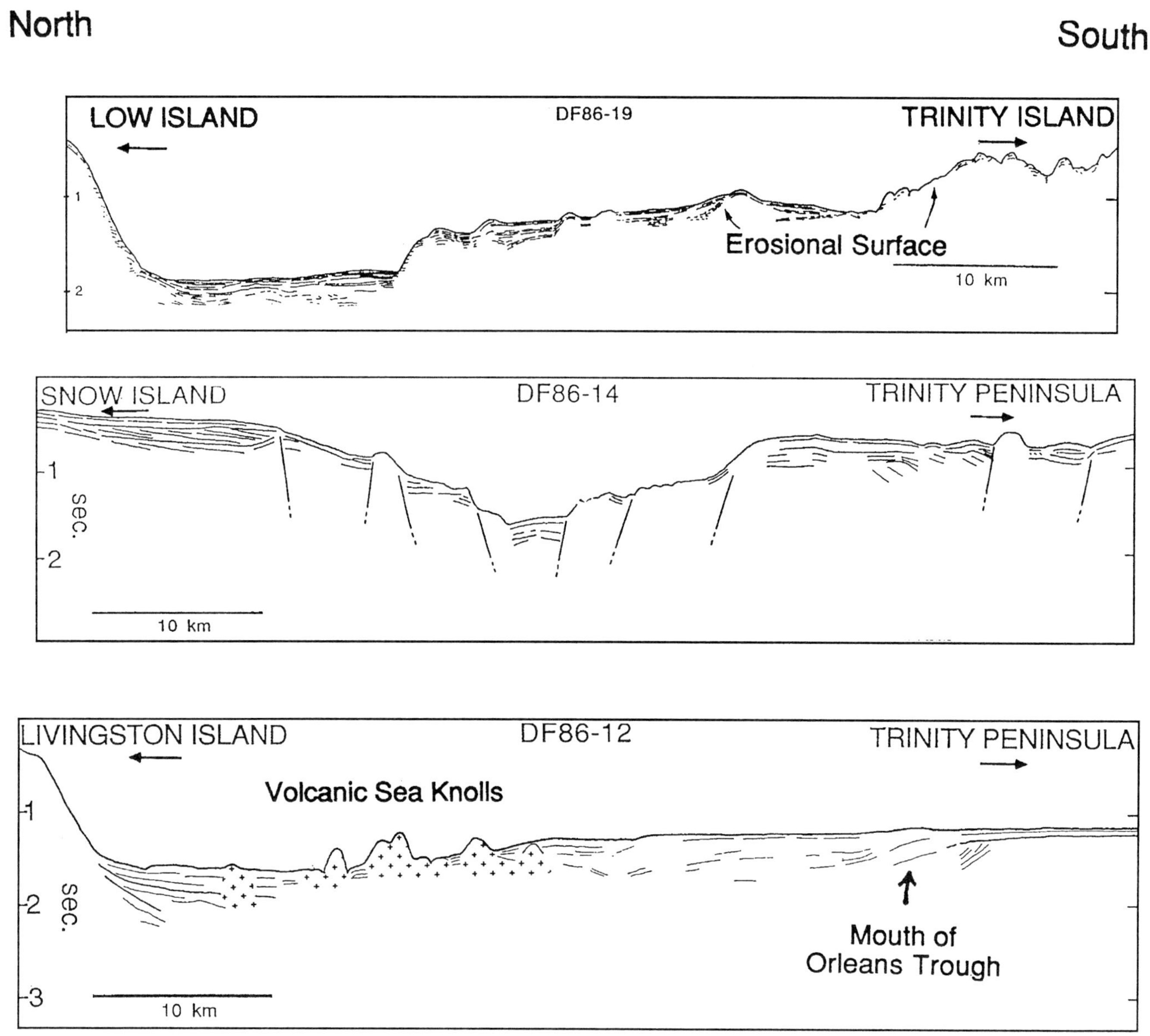

Figure 5—Line drawings of seismic profiles from the western subbasin. See Figure 2 for track lines.

Tectonic Evolution of the Bransfield Basin

Several different mechanisms for the formation of the Bransfield Basin have been proposed. Barker and Dalziel (1983) attributed the rifting to continued sinking of the subducted slab after cessation of Aluk-Antarctic, resulting in northwest migration of the trench and extension in the Bransfield Strait. Sea-floor spreading at the Aluk-Antarctic Ridge in Drake Passage was rapid from Anomaly 6 time (20 Ma) to Anomaly 3 time (4 Ma), and apparently ceased thereafter (Barker and Dalziel, 1983). Bransfield Strait rifting began at about the same time as oceanic sea-floor spreading ended; the initial phase of normal faulting began at about 4 Ma, and later gave way to sea-floor spreading at about 1.3 Ma (Roach, 1978). Their hypothesis depends very much on the coincidence of the *inception* of backarc extension with the *cessation* of oceanic spreading; while the timing of the latter is relatively well constrained by their data, the former is not.

Tokarski (1987) suggested that propagation of stress from the eastern Scotia Sea to the Bransfield Basin along the South Scotia Ridge may have been important in initiating the rifting. He observed a transition from transpressional strike-slip deformation to extension on King George Island from the late Miocene to the present, and a concomitant clockwise rotation of the principal stress axis from east-southeast to west-southwest. Tokarski (1987) attributed the change in stress to both the cessation of subduction at the South Shetland Trench and the onset of east-west sea-floor spreading in the eastern Scotia Sea.

González-Ferrán (1985) viewed the extensional volcanism of the Bransfield Basin within the larger framework of the entire northern Antarctic Peninsula. In addition to Bridgeman, Deception, and Penguin Islands, he considered the Pliocene-Recent extensional volcanic centers of the

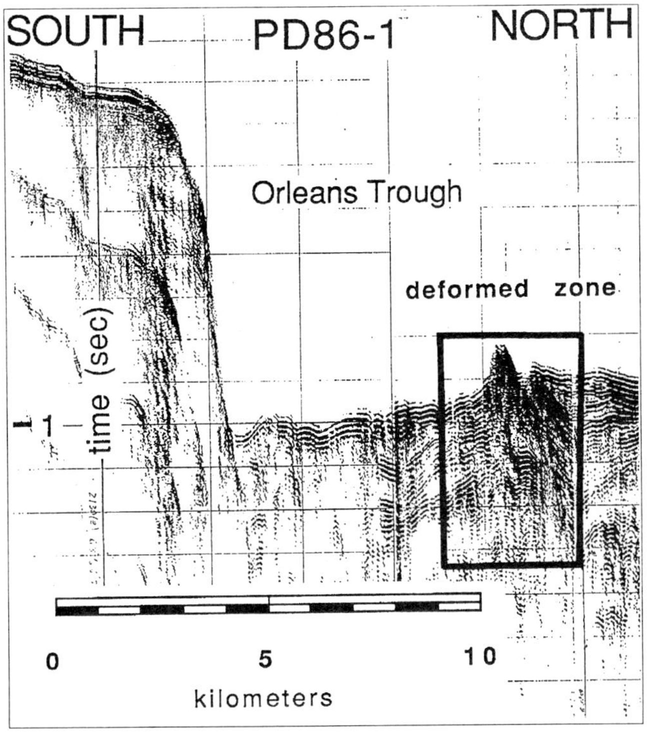

Figure 6—Seismic profile PD86-1 from the western subbasin. See text for discussion.

Prince Gustav Rift (Paulet Islands) and the Larsen Rift (Coley, Argo, and Seal Nunataks) to be part of one large intraplate "fan rift" system.

RESULTS

Western Subbasin

Lines DF86-12, 14, and 19, and PD87-1 and 2 were acquired in the western subbasin (Figure 2). The bottom topography in this portion of the Bransfield Basin is extremely rugged at depths shallower than 500 m; there is little evidence of substantial sediment accumulation at water depths shallower than 800 m. Lines DF86-12, 14, and 19 show that the western subbasin is a narrow, fault-bounded feature (Figure 5). The hummocky slope and incoherent subbottom reflectors on the southern side of the basin indicate that slumping has taken place on this slope. Line PD86-1 (Figure 6) shows the general nature of the troughs and banks of this part of the basin. The banks are generally devoid of acoustically penetrable sediment, whereas the troughs are rough-bottomed and contain at least 100-200 m of sediment. Contorted reflectors show that sediments are being tectonically deformed soon after deposition (Figure 6). Lines DF86-18 and PD87-2 (Figure 7) cross the northward trending branch of the western subbasin (Boyd Strait, Figures 2, 4) and show that a series of downstepping rotated fault blocks forms this physiographic trough.

Line DF86-12 (Figures 2, 5) crosses the western subbasin from the axis of the Orleans Trough to the south end of Livingston Island. Approximately coincident with the mouth of the trough, the sediment package thickens from 100 m to at least 400 m, and a prograding sequence of basinward-dipping reflectors subcrops below the upper draping unit. The prograding package appears to be sediment delivered to the basin through the Orleans Trough, perhaps glacially; it continues basinward as far as the group of 300 m high seaknolls of the submarine volcanic axis.

Central Subbasin

The boundary between the western and central subbasins is marked by an abrupt deepening of the basin floor (Figure 4). Line PD87-1 (Figure 8) crosses this boundary just south of Livingston Island. Lines DF85-A and DF86-11 cross the central subbasin from the Trinity Peninsula to the South Shetland Islands (Figure 9). The central subbasin has received terrigenous sediment both from the Trinity Peninsula and from the South Shetland Islands. Stratal patterns on the Trinity Peninsula margin indicate basinward transport of sediment through deeply incised troughs, forming a prograding complex of coalescing trough-sourced wedges. Although the basinal sequences cannot be tied updip to the sequences of the prograding wedge, turbidites in the deep basin are probably distal equivalents of the prograding complex. When the troughs are not active, the supply of terrigenous sediment to the basin floor is relatively low, and basinal sedimentation is mostly biogenic. The volcanic ridge acts as a barrier to sediment transport across the basin (Figure 8).

The trough-outlet wedges are most prominent in lines DF85-A (Figure 9A) and PD86-11 (Figure 10). The top surface of the wedge complex dips gently from 670 m water depth to approximately 800 m. The internal geometry shows a series of onlapping wedges prograding into the basin. Two major onlapping packages are identified, separated by a substantial basinward shift in onlap (Figure 10). A steep (~9°) slope separates the wedge from the ~1500 m deep basin floor. The basin floor is covered by as much as 700 m of sediment represented by coherent, continuous reflectors divisible into two general types. Strong, continuous reflectors form draping units of nearly constant thickness. Packages of weaker, somewhat discontinuous reflectors onlap and thin toward the slope. The geometry and the character of these reflectors indicate that these are cycles of dominantly pelagic (draping) and dominantly terrigenous (turbidite) deposition. All of the basinal strata eventually lap out against the central volcanic complex, seen as a twin-peaked ridge in lines DF85-A and DF86-11 (Figure 9).

Eastern Bransfield Basin

Line PD86-12 (Figures 2, 11) crosses the northernmost Trinity Peninsula Shelf, the eastern segment of the Bransfield Basin, and the submarine extension of the South Shetland Shelf northeast of King George Island. Two sets of prograding shelf sequences are seen—one on the Trinity Peninsula Shelf which progrades into the Bransfield Basin, and one which progrades northward from the South Shetland Platform over the inner slope of the South Shetland Trench (Figure 11). Four distinct seismic sequences are identified on the South Shetland Shelf and three on the Trinity Peninsula Shelf by basinward shifts in onlap; these represent relative sea level falls, possible glacial advances, and resultant development of lowstand deposits. On the South Shetland Shelf (Figure 11), the youngest sequence (S1) is bounded at the top by the sea bottom, and its lower boundary is marked by a downward shift in onlap. Below it are

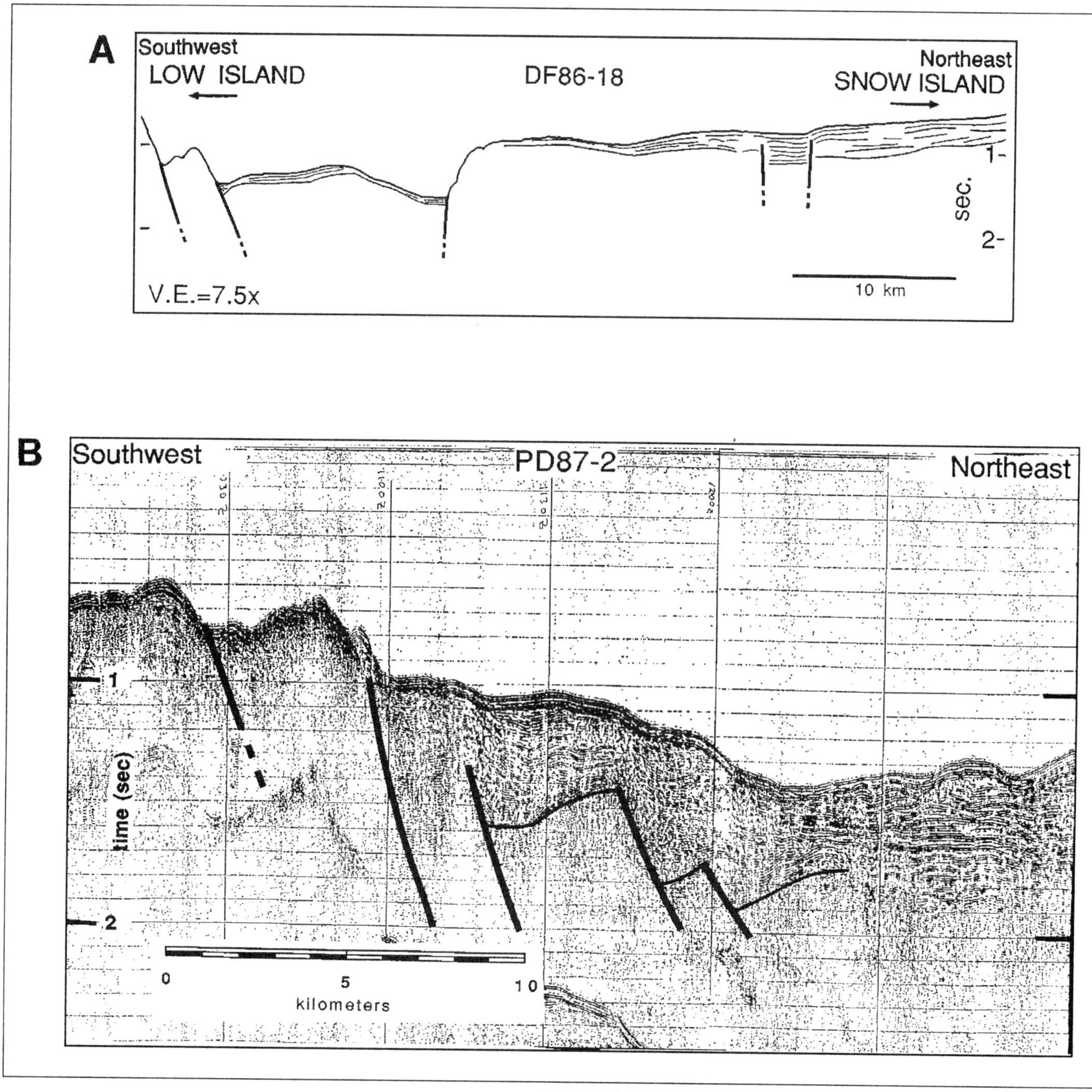

Figure 7—Seismic profiles from the western subbasin. A. Line drawing of line DF86-18, 4.6 kJ sparker. B. Line PD87-2, 100 in.^{3} water gun.

three sequences, identified by downward shifts in onlap, whose upper parts have been truncated erosionally near the sea floor. The basal surface of S4 is a very strong reflector, suggesting that it is the product of prolonged exposure and/or erosion. Strata beneath this surface could be considerably older. The three sequences on the Trinity Peninsula Shelf are similar in geometry to the youngest three sequences on the South Shetland Shelf. S1 is a "wedge" at the shelf margin, clearly showing a downward shift in onlap. Below it, two older sequences onlap onto acoustically opaque basement.

SEQUENCE STRATIGRAPHIC MODEL

Sequence stratigraphy is particularly useful in unraveling the geologic history of frontier areas, like Antarctica, where little or no well control exists, and where biostratigraphy is frequently unreliable or completely unavailable. For the purpose of this study, we developed a sequence stratigraphic model applicable to the Bransfield Basin, and possibly to other portions of the Antarctic margin as well. All of the previous descriptive work and models for sequence stratigraphy have been derived from studies of basins in

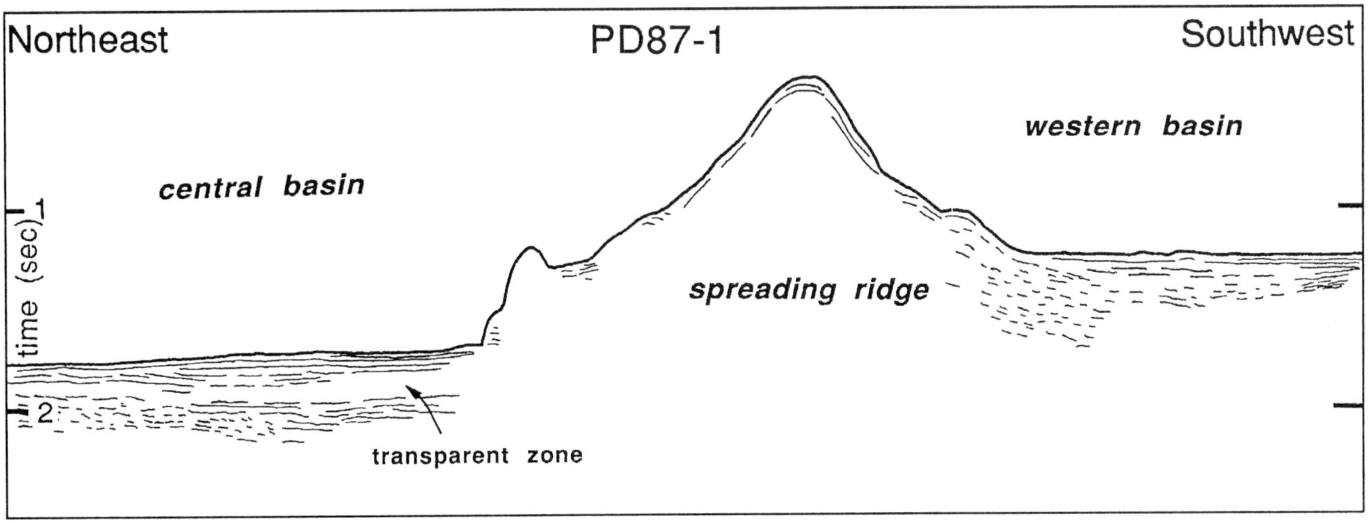

Figure 8—Segment of seismic profile PD87-1. Note the abrupt increase in water depth between the western and central subbasins.

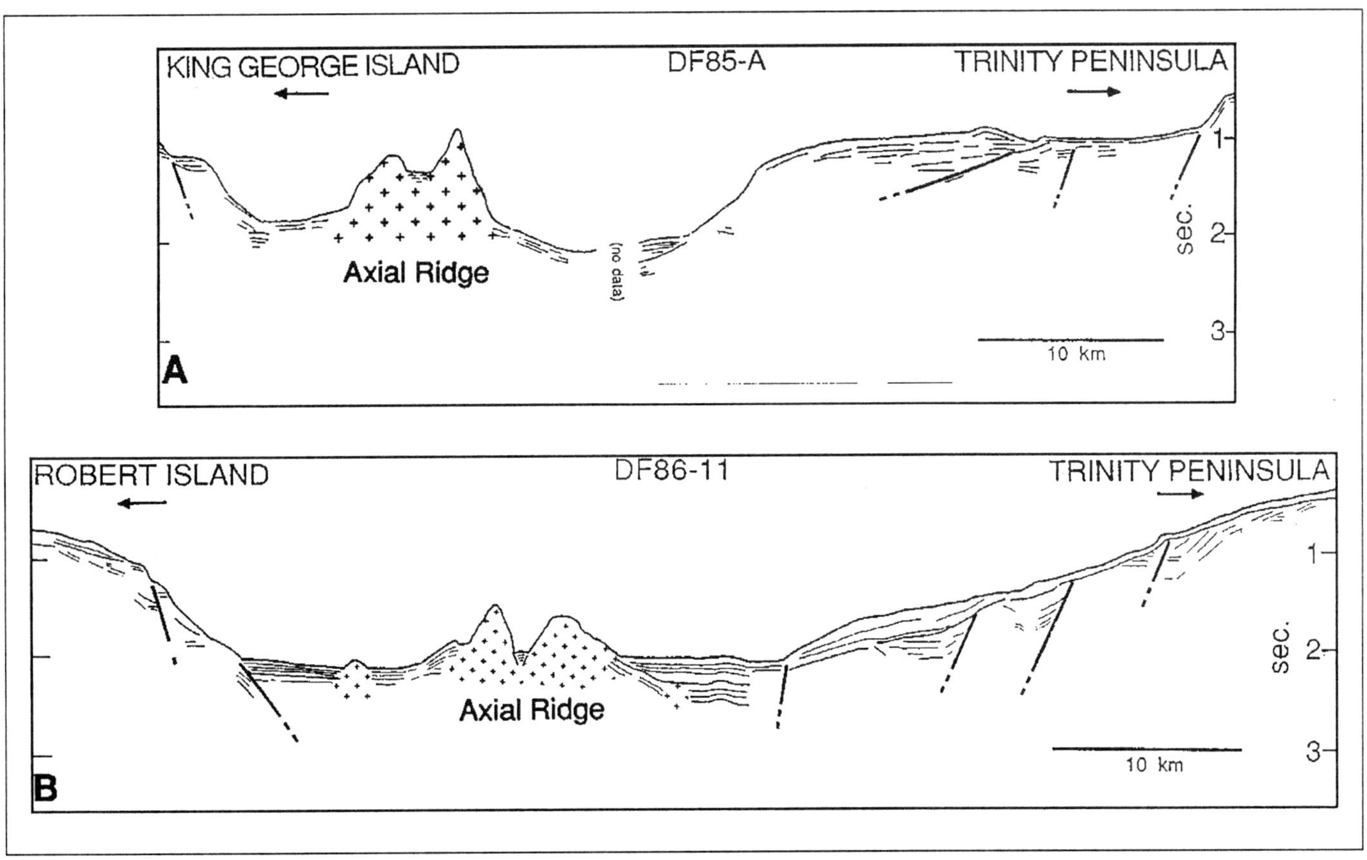

Figure 9—Seismic profiles typical of the central subbasin. A. Line DF85-A. B. Line DF86-11. See text for discussion.

temperate climates where either fluvial-marine, aeolian, evaporite, or carbonate reef processes dominate sedimentation. No rigorous treatment of sequence stratigraphy in glacial-marine settings exists.

In order to interpret the sequence stratigraphy of the Antarctic margin, it is necessary to understand the effect of climate and sediment supply on sedimentation in glacial marine settings and the relationship between glacial/

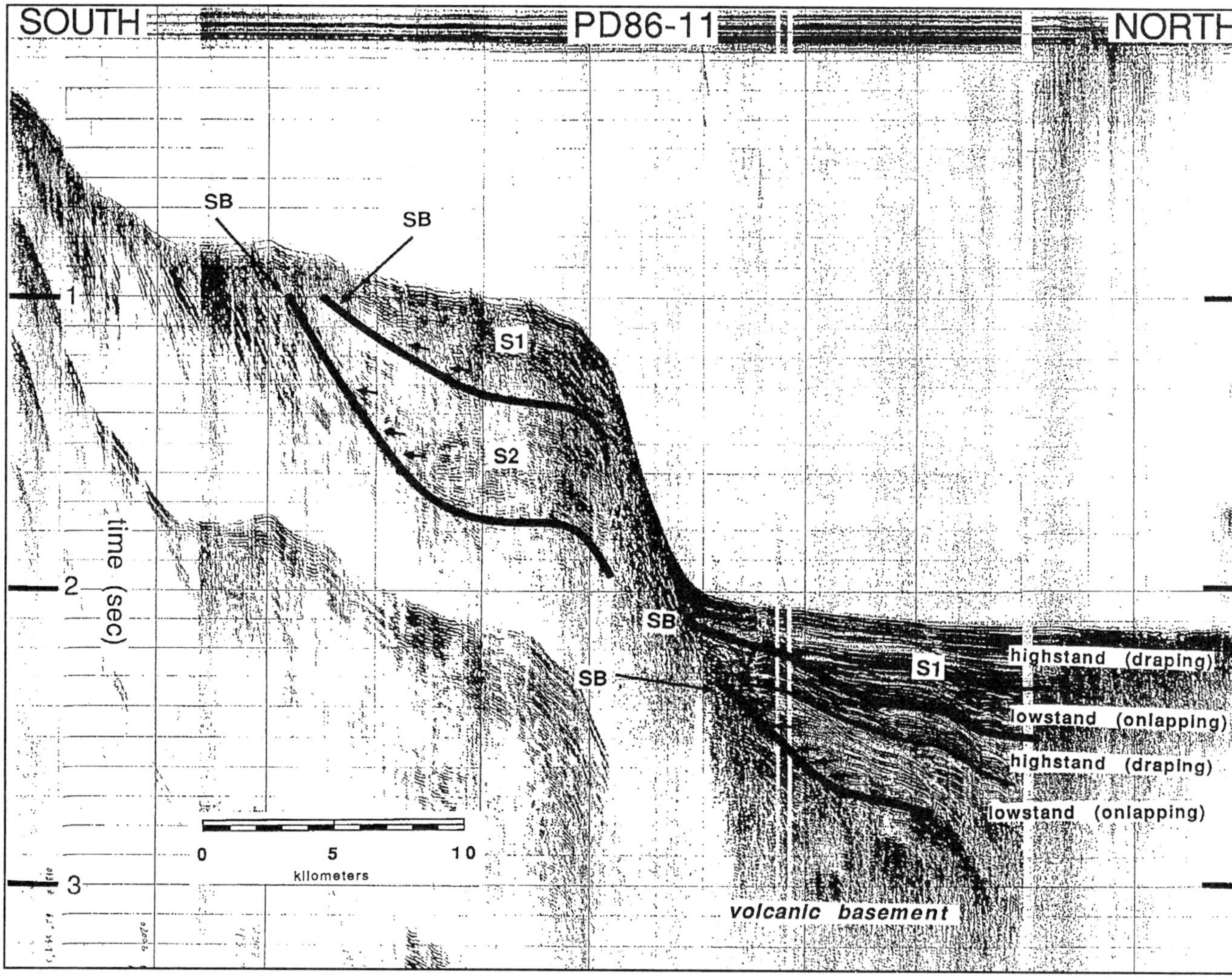

Figure 10—Sequence stratigraphic interpretation of seismic profile PD86-11.

interglacial cycles and eustatic cycles. Many glacial-marine sedimentary environments have been described, and their related lithofacies have been analyzed, both in ancient and modern settings (Anderson et al., 1983). Sediment supply varies greatly and is highly dependent on the glacial regime. In Antarctica, the thermal regime of glaciers controls their ability to erode and transport sediment, and determines the supply of terrigenous sediment to the sea. Warm, wet-based glaciers which move by basal sliding erode and transport sediment much more effectively than do cold, dry-based glaciers which move by internal deformation. Glacial meltwater is an effective sediment transporter, but presently is rare in Antarctica. Anderson et al. (1983) concluded that significant quantities of terrigenous sediment are delivered to the Antarctic sea floor only during major glacial advances, when basal tills and glacial-marine sediments are deposited on the shelf, and during significant warming events when meltwater contributions become substantial; the present climatic regime is intermediate between the two extremes.

The primary factors governing the development of sedimentary sequences, tectonic subsidence, and eustasy are no different on the Antarctic margin than elsewhere in the world. The secondary factors, sediment supply and climate, and the glacial and glacial-marine processes which erode, transport, and ultimately deposit sediments distinguish the Antarctic margin from others. In the area around the Bransfield Basin, glacial cycles are mainly driven by eustacy. The small mountain and valley glaciers of Trinity Peninsula and the South Shetland Islands have small drainage areas and are incapable of advancing out onto the shelf without the help of a sea level fall (see Hollin, 1962; Stuiver et al., 1981). Moreover, we presume that these small ice masses are mostly floating and only lightly grounded, so that they do not cause significant isostatic depression of the shelf when grounded, nor do they have an appreciable effect on global sea level. Changes in the volumes of the large continental ice sheets of the southern and northern hemispheres cause the major eustatic fluctuations, which in turn allow the advance and force the retreat of the tidewater glacial drainage systems. Although the sedimentary packages deposited during a sea level cycle in the Bransfield Basin may not resemble those deposited elsewhere, just as sequences deposited in a siliciclastic province differ from those in a carbonate province, there is no reason to believe that the timing of the sequences and sequence boundaries

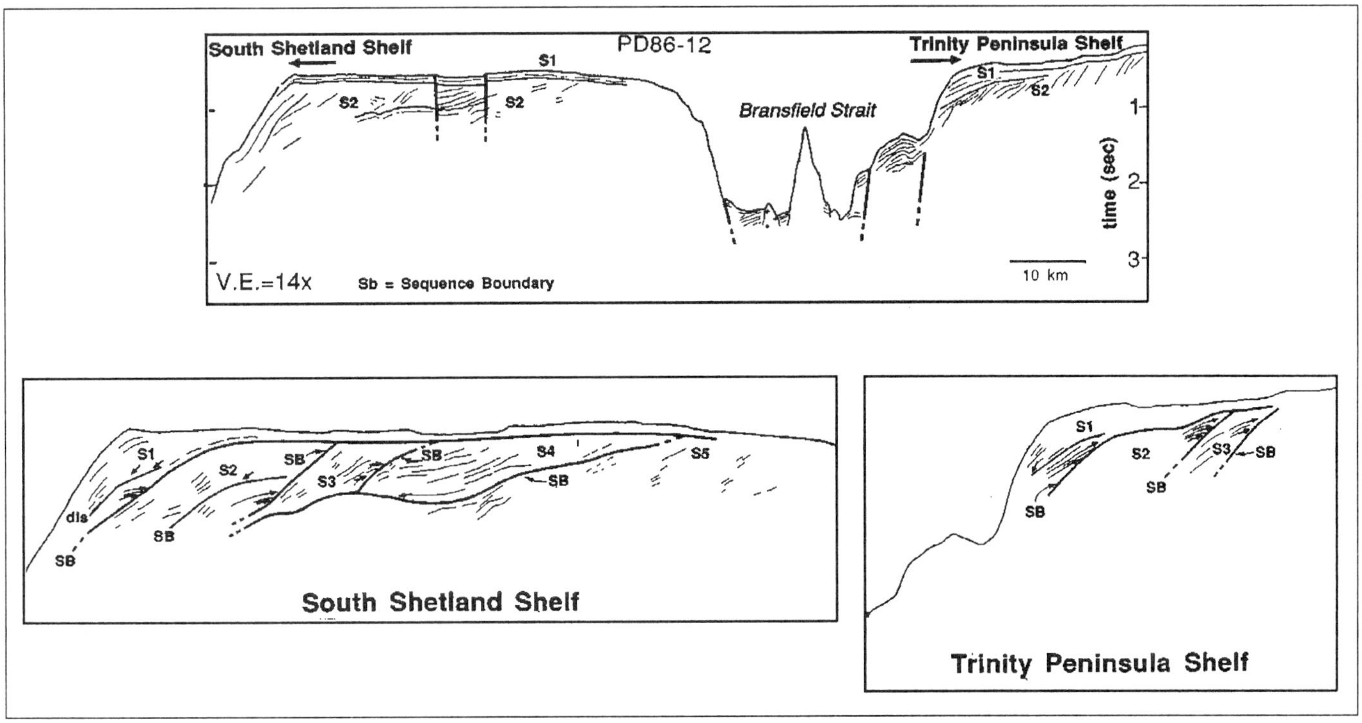

Figure 11—Sequence stratigraphic interpretation of seismic profile PD86-12, showing parallel evolution of Trinity Peninsula Shelf and South Shetland Shelf. Heavy lines marked "SB" indicate sequence boundaries; arrows indicate stratal terminations.

is not approximately synchronous with eustatically controlled depositional sequences in basins elsewhere.

Glacial-Marine Systems Tracts

Two maps (Figures 12, 13) show the sets of depositional systems active at two times during the history of the Bransfield Basin: one which is presently active, and one which was active during glacial maxima and sea level lowstands. Each field on each map represents a depositional system, characterized by a specific environment in the basin, a distinctive set of processes which are inferred to occur within that environment, and a related set of lithofacies which derive from the environment and the process. Since depositional systems alone have no chronostratigraphic significance, each map links contemporaneous depositional systems to form two systems tracts, highstand/interglacial and lowstand/glacial maximum. These form the basis for the simple sequence stratigraphy. The unconformity-bounded sequences observed in the seismic data are then interpreted in terms of the glacial-marine systems tracts, and their time significance is estimated. Finally, the implications of the sequence stratigraphy in terms of the tectonic evolution of the basin are considered.

Highstand/Interglacial Systems Tract

Figure 12 shows the major depositional systems in the modern Bransfield Basin. They have been compiled by combining the surface sediment distribution observed in piston cores and bottom grab samples with the bathymetry of the basin and with some observations from the seismic records. In total, these comprise a representative time slice of the highstand/interglacial systems tract.

Banks (shallower than 250 m) are intensely scoured by marine currents. Singer (1987) observed textural evidence of winnowing by marine currents exceeding 0.11 m/sec, and classified the sediments present as "residual glacial-marine" (Anderson et al., 1983). Lithofacies are generally coarse sand and gravel lags which characteristically limit acoustic penetration. Since fine-grained material is not deposited in these areas, except in topographic depressions, coarse ice-rafted debris and volcanic lapilli are concentrated.

Basinward slopes deeper than 250 m show diminishing effects of currents with depth and are the sites of deposition of the current-derived sand and silt eroded from shallower depths and of material which bypasses the shelf entirely. Deposits are generally muddy sands and sandy muds, with an overall fining as water depth increases. Ice-rafted pebbles are present, but in much smaller proportions than in shallower water sediments. Sediment gravity flow deposits are ubiquitous on steeper slopes, with proximal facies preserved on the slopes. Typical deposits are disorganized conglomerates and crudely graded gravels and coarse sands.

Sediments on the basin floor are composed of three components, in varying proportions. The dominant component is biosiliceous material, mostly diatom tests. The terrigenous component, primarily quartz silt, is most abundant near the base of slopes. The volcanic component includes ash from submarine and subaerial eruptions, and occurs in disseminated fashion throughout the sediments and as discrete ash layers sometimes several centimeters thick. Typical basin floor deposits are ash-bearing diatomaceous muds and oozes; total organic carbon contents may exceed 2% (DeMaster et al., 1987).

The bays of the South Shetland Islands and associated outlet canyons are potential conduits for delivery of terrigenous sediment to fanlike depositional lobes on the basin

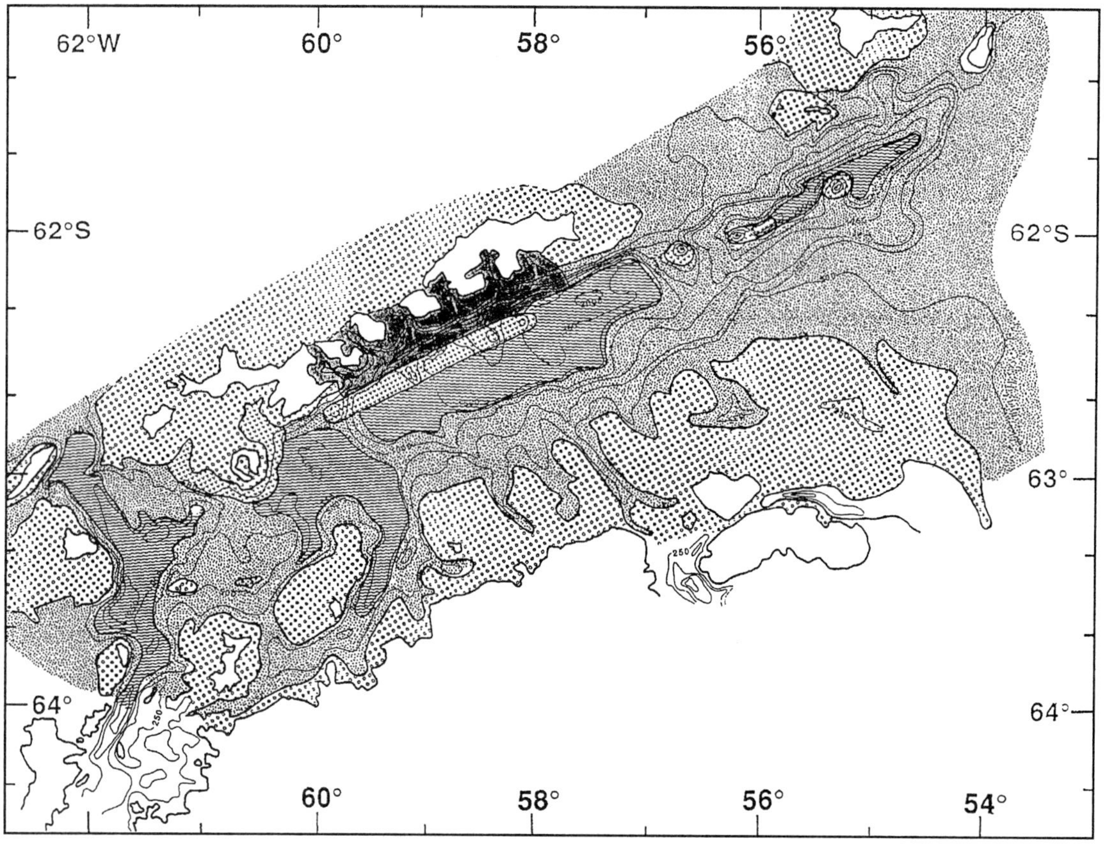

	Sedimentary Environments	Sedimentary Processes	Lithofacies
	Shallow banks (<250 m)	Intensely scoured by marine currents.	Coarse gravel lags (residual glacial marine)
	Slopes	Current-related transport diminishes with depth. Debris flows and slumps are ubiquitous on the steeper slopes.	Muddy sands and sandy muds, with an overall fining offshore. Disorganized conglomerates and crudely graded gravels and sands on steep slopes
	Bays, canyons, and associated fanlike lobes	Glacially eroded sediment from the South Shetland Islands delivered to the bays by glacial meltwater is transported offshore and downslope to depositional lobes on the basin floor.	Terrigenous muds with occasional sandy horizons; basin floor sediments are enriched in biogenic material.
	Central volcanic ridge	Juvenile material is delivered directly by lava flows and by passive settling through the watercolumn. Secondary downslope transport is by sediment gravity flow processes. Pelagic sedimentation dominates between eruptions	Graded volcanic ash units interbedded with diatomaceous muds and oozes.
	Basin floor	Biosiliceous sedimentation with increasing terrigenous influence near the basin margins and volcanic influence near the central ridge.	Ash-bearing diatomaceous muds and oozes. Total organic carbon may exceed 2%.

Figure 12—Depositional systems of the highstand/interglacial systems tract.

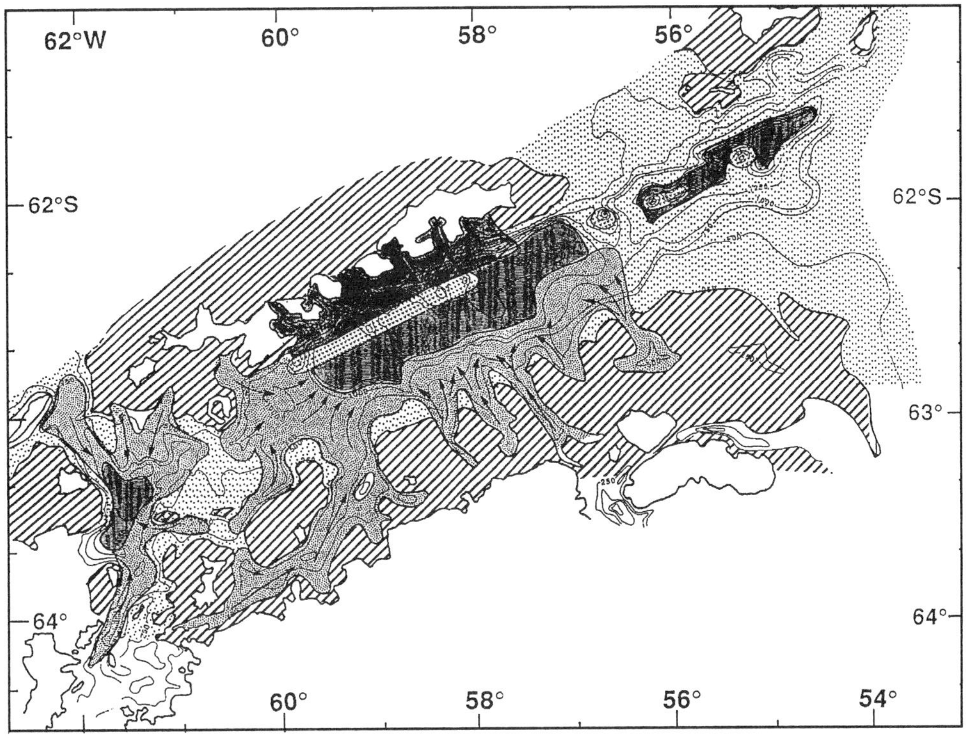

Sedimentary environments	Sedimentary Processes
Shallow banks	Glacial erosion and deposition of basal tills.
Slopes	Sediment eroded from the shelf progrades basinward where canyons or troughs are not present.
Bays, canyons, and associated fan-like lobes	Sediment from the bays is glacially eroded and transported offshore and downslope to depositional lobes on the basin floor.
Central volcanic ridge	Juvenile material is delivered directly by lava flows and by passive settling through the watercolumn. Secondary downslope transport is by sediment gravity flow processes. Pelagic sedimentation dominates between eruptions
Troughs and associated prograding wedges	Material eroded from the continent and shelf is carried through the troughs to prograding wedges on the slope.
Basin floor	Biogenic sedimentation is sharply reduced. Turbidites on the basin floor are sourced from the trough-wedge complex.

Figure 13—Depositional systems of the lowstand/glacial maximum systems tract.

floor. Only Maxwell Bay (Figure 4) has overfilled its bathymetric sill and is capable of delivering large volumes of sediment to the basin. The other bays may deliver some suspended sediment through glacial meltwater plumes. Sediment travels downslope from Maxwell Bay to a fanlike depositional lobe on the basin floor. Evidence of current reworking of these sediments once they reach the basin floor has been observed in seismic profiles. The lithofacies of these lobes are similar to the basin floor sediments, but their seismic character is quite distinctive.

The troughs of the Trinity Peninsula Shelf, by contrast, do not appear to carry a significant amount of terrigenous sediment to the basin. They are filled with diatomaceous muds and oozes identical to those on the basin floor. Slumping of the sidewalls of the canyons may account for the occasional coarse-grained material in cores from the troughs. Glaciers on the Trinity Peninsula are probably not warm enough to produce sediment-laden meltwater in large volumes whereas those of the South Shetland Islands do (Griffith and Anderson, 1989).

The central volcanic ridge is thinly blanketed with pelagic sediment. In many places, juvenile volcanic material, ash and basalt flows are exposed at the sea floor. Secondary downslope transport produces graded ash units. Its elevation above the basin floor means that the ridge is unlikely to receive terrigenous sediment. In addition, it acts as a barrier to cross-basin transport, so the terrigenous components associated with the South Shetland Islands generally are segregated from the components associated with the Trinity Peninsula. Volcanic material from recent subaerial eruptions is abundant near Deception Island.

Glacial Maximum/Lowstand Systems Tract

The modern setting represents only a small portion of a complete sedimentary cycle. During eustatic lowstands and glacial advances, sedimentation in the Bransfield Basin was quite different. Since the processes which occurred then cannot be observed directly now, and since very few pre-Holocene core samples have been recovered, the processes must be inferred from seismic character. Several depositional systems different from those seen in the modern setting are immediately evident (Figure 13).

During glacial advances, portions of the shallow platforms on the coast of the Trinity Peninsula and around the South Shetland Islands would be covered with ice. Reconstructions place the Pleistocene grounding line at 200-375 m depth north of the South Shetland Islands (Curl, 1980; John and Sugden, 1971), whereas ice may have been somewhat thicker on the Trinity Peninsula Shelf since its drainage area is larger. Thin basal tills might be deposited in these net erosional areas. The truncated reflectors seen near the sea floor on seismic profiles north of the South Shetland Islands indicate that glacial erosion has taken place there (Anderson, 1985).

The glacially carved troughs on the Trinity Peninsula Shelf may have been filled entirely with ice. Seismic records show wedges of sediment prograding from the outlets of all of these troughs, which appear to coalesce to form the ~750 m deep bathymetric platform on the Trinity Peninsula Slope. Ice flow apparently diverged at the trough outlets, and eroded sediment was deposited there. Another wedge might have prograded down the axis of the basin from the outlets of the troughs of the western basin and across the boundary between the western and central basins. It seems likely that these depositional systems remained active as long as the shelf was covered with ice and was not available for deposition of significant volumes of sediment. This is analogous to the lowstand systems tracts in fluvio-marine settings where lowstand fans and lowstand wedges develop when sea level drops below the shelf edge, exposing the shelf and making it unavailable for sediment deposition. Thus, it is not necessary to drop sea level below the shelf edge to bypass the shelf in the glacial-marine setting; it is only necessary to drop sea level sufficiently to allow ice to ground at the shelf edge.

The seismic records from the basin floor show acoustically transparent, onlapping units underlying the modern pelagic sequence. These are interpreted as turbidites, deposited as the trough-outlet wedges prograded and oversteepened, initiating slumps, slides, and debris flows. The turbidites on the basin floor are the distal products of this process. At the same time, primary productivity and biogenic sedimentation rates were greatly reduced as the region was covered with multi-year sea ice or an ice shelf.

The modern (highstand) and the lowstand/glacial maximum systems tracts are the only two which have been identified from the seismic data and the sediment samples. However, they represent only a part of the entire sea level/glacial cycle. It is likely that during sea level rises, long highstands, and extensive deglaciations, sufficiently deep water existed on the shelf for deposition of fine-grained sediments below the depth of intense currents. During extensive deglaciations, glacial/fluvial systems may have developed. Since the shelf was deepened by glacial erosion during lowstands, it is unlikely that highstand deposits ever prograded beyond the relict shelf edge, so there was little deposition of terrigenous material in the deep basin during highstands. However, each successive major glacial advance eroded the shelf and redeposited shelf sediments as lowstand turbidites on the basin floor.

DISCUSSION

Tectonic Segmentation of the Bransfield Basin

The division of the Bransfield Basin into three subbasins, clearly reflected in the morphology of the basin, has had a profound effect on sedimentation patterns. The western subbasin is relatively shallow, and is adjacent to landmasses with large glacial drainage systems. During glacial maxima, two large northeasterly trending troughs carried sediment along the axis of the basin to a wedge which prograded across into the central basin. The central subbasin, over 2000 m deep, received significant amounts of terrigenous sediment from the South Shetland Islands and from the Trinity Peninsula, through submarine troughs connected to glacial drainage systems.The eastern basin is still deeper, in places exceeding 2500 m; sedimentation along the shelf here is not controlled by troughs and canyons, but instead, the shelf seems to prograde from a linear shelf edge.

Regional geophysical evidence indicates that the observed segmentation of the basin is a manifestation of its underlying structure. Seismic refraction data (Ashcroft, 1972) show that the variation in depth is not merely due to greater sediment thickness in the shallower basin segments, but also reflects differences in thickness of the underlying crust. Furthermore, numerous discontinuities, possibly related to transverse structural elements, suggest that there may be sharp structural boundaries between the subbasins. Seismicity, too, suggests that the segmentation is tectonic in origin; the eastern and western subbasins have relatively frequent seismic events, while the central subbasin is conspicuously aseismic (Pelayo and Wiens, 1986). This apparent tectonic segmentation of the Bransfield Basin may reflect the influence of South Shetland subduction complex, or it may be unique to the backarc extensional system.

The Age of the Bransfield Basin

In order to attribute the backarc extension to the cessation of ridge-driven subduction, it is vital to confirm the age relationship between the subduction history at the South Shetland Trench and the onset of rifting of the Bransfield Basin. Interpretation of magnetic anomalies as possible indicators of a sea-floor spreading history in the Bransfield Basin has not proven a reliable method for estimating the age of the basin. Barker (1976) interpreted a single magnetic profile across the Bransfield Basin to show no magnetic reversals, and suggested that the sea floor in the Bransfield Basin formed during a single epoch of uniform normal magnetic polarity, but Roach (1978) modeled the same profile by inserting narrow, reversely magnetized strips of sea floor on the flanks of the positively magnetized spreading center, and speculated that sea-floor spreading had been occurring for the past ~1.3 Ma. Because of the short history and the processes which formed the sea floor of the Bransfield Basin, the ages determined from interpretation of the magnetic profiles probably are not reliable. Several factors inhibit the formation, preservation, and recognition of correlatable sea-floor spreading magnetic anomaly patterns. The seismic data shown here and elsewhere (e.g., Barker et al., 1988) show that lavas and sediments are interbedded, and that much of the igneous activity involved injection of dikes and sills into the sedimentary strata, creating thick intercalations of lavas and terrigenous sediment. Magnetic minerals formed in this setting are likely to be much more diffusely distributed and less strongly magnetized than in normal oceanic crust (Lawver and Hawkins, 1978), so typical symmetric sea-floor spreading anomalies may not form.

Interpretation of the significance of seismic sequences observed in several of the seismic profiles of the Bransfield Basin in terms of the glacial-marine systems tracts defined above allows the age of the basin to be estimated independently from the questionable magnetic data. The progression of events seen in the seismic data, subsidence of the forearc, followed by rifting and then sea-floor spreading in the backarc, is consistent with the subduction cessation model summarized by Barker and Dalziel (1983).

In the central subbasin, the lowstand/glacial maxima are characterized by sediment transport through the shelf troughs, construction of a deep prograding wedge of sediment from the outlets of the troughs, and increased turbidite deposition relative to pelagic sedimentation in the deep basin. During highstand/interglacial periods, pelagic sediments drape the basin floor. In line PD86-11 (Figure 10), two downward shifts in onlap are observed in the prograding wedge, in conjunction with two episodes of predominantly turbidite deposition on the basin floor. The inferred processes indicate that these are correlative packages representing two complete glacial/sea level cycles.

The margins of the eastern basin lack troughs, and slope sediments appear to be line-sourced from the shelf edge rather than point-sourced. Since there are no canyons, lowstand wedges prograde from the shelf margin. Downward shifts in onlap mark the sequence boundaries; the wedges are truncated updip by glacial? erosion. Three sequences onlap the Trinity Peninsula margin whereas four are seen on the South Shetland Shelf. The youngest three sequences on both margins are similar in geometry and probably are correlative, representing sedimentation as the two margins evolved in parallel after rifting of the Bransfield Basin. The basal surface in the forearc (base of S4) is probably a tectonically enhanced unconformity, reflecting subsidence of the forearc related either to the slowing of spreading and subduction which took place at 6 Ma, or the cessation of oceanic spreading and ridge-driven subduction at 4 Ma.

The relative ages of the seismic sequences are consistent with the subduction-related hypothesis for origin of the Bransfield Basin, where backarc rifting follows the cessation of Aluk Ridge-driven subduction at the South Shetland Trench.

The absolute duration of the cycles cannot be determined directly. Assuming that cycles here are correlative with global cycles, they may be the third-order cycles of Haq et al. (1987, 1988) which span hundreds of thousands of years, or higher order (fourth-order) cycles which span tens of thousands of years. A reasonable estimate can be made from the thicknesses of the sequences. The youngest pelagic unit in line PD86-11 is 180 m thick; at a velocity of 1.8 km/sec this represents 160 m of sediment. At present accumulation rates (1.8 mm/yr; DeMaster et al., 1987), this unit spans approximately 100,000 years of continuous sedimentation. This calculation should be viewed as an absolute minimum since it ignores compaction and since accumulation rates determined from ^{210}Pb may be erroneously high because of hydrothermal contamination (van Enst, 1987). Furthermore, biogenic productivity falls far below present rates during periods of more extensive ice cover (Dunbar et al., 1985). The top pelagic unit could reflect several hundred thousand years of predominantly pelagic sedimentation and probably represents a significant portion of the last 0.8 Ma third-order cycle; it certainly reflects a period much longer than the Holocene. The dominant signal in the seismic data appears to be third-order cyclicity, with individual glacial events (fourth-order cycles) as short duration events below seismic resolution.

Haq et al. (1987, 1988) identified five third-order eustatic cycles in the Plio-Pleistocene, bracketed by sequence boundaries at 0.8 Ma, 1.6 Ma, 2.4 Ma, 3.0 Ma, and 3.8 Ma. Although the depositional sequences in the Bransfield Basin have not been dated directly, if they are correlative with the Haq et al. (1987, 1988) global cycles, the chronology of the physical stratigraphic record of the Bransfield Basin is consistent with cessation of Aluk-Antarctic spreading at 4 Ma and formation of the backarc basin at 2-3 Ma. Bransfield Basin sequence S4 would span 2.4-3.0 Ma, the period of initial forearc subsidence after ridge-driven subduction stopped; S3, the first sequence deposited on the rifted backarc margin would span 1.6-2.4 Ma. Sequence S2, the first sequence deposited onto quasi-oceanic crust, would be younger than 1.6 Ma. Regardless of their absolute ages, the relative ages of the Bransfield Basin depositional sequences suggest that a period of forearc subsidence, possibly due to the cessation of ridge-driven subduction at the South Shetland Trench, predated the backarc rifting.

Evaluation of Hydrocarbon Potential

The Bransfield Basin is of particular interest as a potential hydrocarbon province because oil shows have been described in near surface sediments (Whiticar et al., 1985). Thus, conditions in the Bransfield Basin are at least somewhat favorable for the generation of hydrocarbons. Potential source material is abundant; organic rich diatomaceous muds accumulate rapidly over large areas of the basin floor during interglacial periods. The high heat flow due to subsea volcanism allows for rapid maturation, and hydrothermal circulation provides a mechanism for flushing hydrocarbons out of source rocks at relatively low overburden pressures. Suess (1987) documented an extensive zone of hydrothermal activity along the axial ridge and fluid advection through its sedimentary cover. In addition to vertical fluid migration, Han and Suess (1987) report widespread lateral pore fluid advection through relatively permeable turbidite layers.

The existence of reservoir-quality sands is less certain. Sediments which might form favorable hydrocarbon reservoirs generally are not seen in the modern (cored) sediments. Our stratigraphic model predicts, however, that terrigenous clastics might ultimately be deposited during glacial advances as shelf-margin and trough-outlet wedges and as basin floor turbidites. The basin floor turbidites are particularly attractive as petroleum reservoirs because of their probable high degree of textural sorting, their juxtaposition with potential source rocks (the interbedded diatomaceous muds), and their proximity to the thermal source. The trough-outlet wedges are also prospective, but might not be as texturally mature. Exploratory or scientific drilling will be necessary to confirm the presence of suitable reservoirs.

Because hydrocarbons are generated at very shallow depths, there may not be adequate reservoirs and traps present when primary migration occurs. In the postrift sediments, very little faulting is observed. Stratigraphic traps may form updip of the basin floor turbidites as porosity decreases due to influx of terrigenous silt. There also may be some potential in the highly faulted underlying prerift sediments.

In summary, conditions in the Bransfield Basin have favored hydrocarbon generation, but probably have not allowed for substantial accumulation. More favorable conditions may have existed in forearc basins farther south. Similar sediments may have been deposited, then heated by the subducted spreading ridge, and migration of hydrocarbons into reservoirs in the overlying passive margin sequences could have occurred (see Anderson et al., 1990).

CONCLUSIONS

1. Sequence stratigraphic concepts can be applied to the Antarctic continental margins if care is taken to account for differences in sedimentary processes between glacial-marine settings and lower latitude regions. Since eustasy is a global effect, and tectonic subsidence rates in the Bransfield Basin are similar to tectonic subsidence rates in rifted basins elsewhere, there is no reason to believe that sequences in the Bransfield Basin are not approximately synchronous with sequences deposited elsewhere. This is especially true where glacial advances and retreats, and their resultant effects on sedimentation, are largely controlled by sea level changes.

2. Sedimentation in the Bransfield Basin varies greatly through glacial-eustatic cycles. During highstand/interglacial periods, organic-rich diatomaceous muds and oozes are deposited pelagically in most of the basin. The shallowest areas presently are scoured free of fine-grained material; however, during extensive deglaciations, fluvial systems may develop and deliver terrigenous sediment to the shelf. During lowstand/glacial maximum periods, wedges of glacially eroded terrigenous sediment prograde from the shelf edge and from the outlets of incised troughs, and turbidites are deposited on the basin floor. The stacking pattern of these highstand/interglacial and lowstand/glacial maximum systems tracts provides the basis for a simple sequence stratigraphy.

3. Relative ages of the sequences deposited on the forearc margin of the South Shetland Islands and the rifted backarc margin of the Bransfield Basin suggest that subsidence of the forearc, possibly due to slowing or cessation of ridge-driven subduction, predated backarc rifting.

4. Estimation of the duration of the glacial-eustatic cycles observed in seismic reflection data from the Bransfield Basin, based on extrapolation from modern pelagic sedimentation rates and assessment of thermal subsidence rates, suggests that if they correlate with global cycles, these sequences are the products of the ~0.8 Ma third-order global cycles of Haq et al. (1987).

5. The Bransfield Basin is segmented longitudinally into three subbasins which differ in width, depth, and structural style. This segmentation exerts a strong influence on the distribution of principal sedimentary environments and lithofacies. Geophysical evidence, including magnetics, seismic refraction, and earthquake seismology, suggests that the segmentation is tectonically controlled and does not merely reflect differences in the surficial morphology of the basin.

6. The juxtaposition of organic-rich muds of the highstand-interglacial systems tract with the heat of the backarc spreading system is favorable for maturation of hydrocarbons. However, the existence of suitable hydrocarbon traps and reservoir-quality sandstones has not been proven. Our stratigraphic model predicts that updip stratigraphic traps in lowstand basin floor turbidites are the most likely site of hydrocarbon accumulation.

ACKNOWLEDGMENTS

This chapter was extracted from J.D. Jeffers' Masters thesis at Rice University. Portions were first presented at the Fifth International Symposium on Antarctic Earth Sciences in 1987. Research was funded by the National Science Foundation under Grant No. DPP 88-18523 to J.B. Anderson. Critical reviews of the manuscript by D. Macdonald, S. Eittreim, and L. Lawver greatly improved the clarity of the text.

REFERENCES

Anderson, J. B., 1985, Preliminary results from the USCGC *Glacier* 1985 cruise: Antarctic Journal of the United States, 1985 Review Issue, p. 81-85.

Anderson, J. B., P. G. Pope and M. A. Thomas, 1990, Evolution and hydrocarbon potential of the northern Antarctic Peninsula Continental shelf, this volume.

Anderson, J. B., C. F. Brake, E. W. Domack, N. C. Myers and R. Wright, 1983, Development of a polar glacial-marine sedimentation model from Antarctic Quaternary deposits and glaciological information, *in* B. Molnia, ed., Glacial Marine Sedimentology: Plenum Press, p. 233-264.

Ashcroft, W. A., 1972, Crustal structure of the South Shetland Islands and Bransfield Strait: British Antarctic Survey, Scientific Report No. 66, 43 p.

Barker, P. F., 1976, The tectonic framework of Cenozoic volcanism in the Scotia Sea Region, A review, *in* O. González-Ferrán, ed., Andean and Antarctic Volcanology Problems: International Association of Volcanology and Chemistry of the Earth's Interior, Rome, Italy, p. 330-346.

Barker, P. F., 1982, The Cenozoic subduction history of the Pacific margin of the Antarctic Peninsula: ridge crest-trench interactions: Journal Geological Society of London, v. 139, p. 787-802.

Barker, P. F. and I. W. D. Dalziel, 1983, Progress in geodynamics in the Scotia Arc region: American Geophysical Union Geodynamic Series 9, p. 137-170.

Barker, P. F., I. W. D. Dalziel and B. C. Storey, 1988, Tectonic development of the Scotia arc region, *in* R. Tingey, ed., Antarctic Geology: Oxford University Press.

Barton, C. M., 1965, The geology of the South Shetland

Islands III, The stratigraphy of King George Island: British Antarctic Survey, Scientific Report No. 44, 33 p.

Bell, C. M., B. F. Mair and B. C. Storey, 1977, The geology of part of an island-arc marginal basin system in southern South Georgia: British Antarctic Survey, Scientific Report No. 46, 109 p.

Birkenmajer, K., 1982, Late Cenozoic phases of block faulting on King George Island (South Shetland Islands, West Antarctica): Academy of Polish Science Bulletin, Earth Sciences, v. 30 (1-2), p. 21-32.

Birkenmajer, K., M. C. Delitala, W. Narebski, M. Nicoletti and C. Petrucciani, 1986, Geochronology of Tertiary island arc volcanics and glacigenic deposits, King George Island, South Shetland Islands (West Antarctica): Academy of Polish Science Bulletin, Earth Sciences, v. 34, p. 257-273.

Curl, J. E., 1980, A glacial history of the South Shetland Islands, Antarctica: Institute for Polar Studies Report, The Ohio State University, Columbus, Ohio, 129 p.

Dalziel, I. W. D., M. S. de Wit and K. F. Palmer, 1974, A fossil marginal basin in the southern Andes: Nature 250, p. 291-294.

Davey, F. J., 1972, Marine gravity measurements in Bransfield Strait and adjacent areas, *in* R. J. Adie, ed., Antarctic Geology and Geophysics: Oslo, Universitetsforlaget, p. 39-46.

DeMaster, D. J., T. M. Nelson, C. A. Nittrouer and S. L. Harden, 1987, Biogenic silica and organic carbon accumulation in modern Bransfield Strait sediments: Antarctic Journal of the United States, v. 22. 1987 Review Issue, p. 108-110.

Dunbar, R. B., J. B. Anderson, E. W. Domack and S. S. Jacobs, 1985, Oceanographic influences on sedimentation along the Antarctic continental shelf: American Geophysical Union Antarctic Research Series, v. 43, p. 291-312.

Forsyth, D. W., 1975, Fault plane solutions and tectonics of the South Atlantic and Scotia Sea, Journal of Geophysical Research, v. 80, p. 1429-1443.

Garrett, S. W. and B. C. Storey, 1987, Lithospheric extension on the Antarctic Peninsula during Cenozoic subduction, *in* M. P. Coward, J. F. Dewey and P.L. Hancock, eds., Continental extensional tectonics: Geological Society of London, Special Publication No. 28, p. 419-431.

Garrett, S. W., R. G. B. Renner, J. A. Jones and K. J. McGibbon, 1987, Continental magnetic anomalies and the evolution of the Scotia arc: Earth Planetary Science Letters, v. 81, p. 273-281.

González-Ferrán, O., 1985, Volcanic and tectonic evolution of the northern Antarctic Peninsula— Late Cenozoic to Recent, *in* E. S. Husebye, G. L. Johnson and Y. Kristoffersen, eds., Geophysics of the Polar Regions: Tectonophysics, 114, p. 389-409.

Griffith, T. W. and J. B. Anderson, 1989, Climatic control on sedimentation in bays and fjords of the northern Antarctic Peninsula: Marine Geology, v. 85, p. 181-204.

Grunow, A. M., I. W. D. Dalziel and T. M. Harrison, 1987, Tectonic significance of new $^{40}Ar/^{39}Ar$ ages from the comparatively high P/low T metamorphic rocks of the Scotia arc (abstract): Fifth International Symposium on Antarctic Earth Sciences, Cambridge, p. 57.

Han, M. W. and E. Suess, 1987, Lateral migration of pore fluids through sediments of an active back-arc basin, Bransfield Strait, Antarctica (abstract): American Geophysical Union EOS Transactions, 68(50), p. 1769.

Haq, B. U., J. Hardenbol and P. R. Vail, 1987, The new chronostratigraphic basis of Cenozoic and Mesozoic sea level changes: Cushman Foundation for Foraminiferal Research, Special Publication No. 24, p. 7-13.

Haq, B. U., J. Hardenbol and P. R. Vail, 1988, Mesozoic and Cenozoic chronostratigraphy and cycles of sea-level change, *in* C. K. Wilgus et al., eds., Sea Level Changes: An Integrated Approach: SEPM Special Publication No. 42, p. 71-108.

Hawkes, D. D., 1981, Tectonic segmentation of the northern Antarctic Peninsula: Geology, v. 9, p. 220-224.

Hollin, J. T., 1962, On the glacial history of Antarctica: Journal of Glaciology, 4, p. 173-195.

John, B. S. and D. E. Sugden, 1971, Raised marine features and phases of glaciation in the South Shetland Islands: British Antarctic Survey Bulletin, 24, p. 45-111.

Keller, R. A. and M. R. Fisk, 1987, Magmatism associated with the initial stages of backarc rifting, Bransfield Strait, Antarctica (abstract): American Geophysical Union EOS Transactions, 68(44), p. 1533.

Lawver, L. A. and J. W. Hawkins, 1978, Diffuse magnetic anomalies in marginal basins: their possible tectonic and petrologic significance: Tectonophysics, 45, p, 323-339.

Pelayo, A. M. and D. A. Wiens, 1986, Seismotectonics of the western Scotia Sea region (abstract): American Geophysical Union EOS Transactions, 67 (44), p. 1239.

Renner, R. G. B., L. J. S. Sturgeon and S. W. Garrett, 1985, Reconnaissance gravity and aeromagnetic surveys of the Antarctic Peninsula: British Antarctic Research, Scientific Report, 110, 54 p.

Roach, P. J., 1978, The nature of back-arc extension in Bransfield Strait: Geophysical Journal of Research, Astronomical Society, v. 53, p. 165.

Saunders, A. D. and J. Tarney, 1982, Igneous activity in the southern Andes and northern Antarctic Peninsula: a review: Journal Geological Society of London, v. 139, p. 691-700.

Singer, J. K., 1987, Terrigenous, biogenic, and volcaniclastic sedimentation patterns of the Bransfield Strait and bays of the northern Antarctic Peninsula: Implications for Quaternary glacial history: Unpublished Ph.D. dissertation, Rice University, Houston, Texas.

Storey, B. C. and D. I .M. Macdonald, 1984, Processes of formation and filling of a Mesozoic back-arc basin on the island of South Georgia, *in* B. P. Kokelaar and M. F. Howells, eds., Marginal Basin Geology: Volcanic and associated sedimentary and tectonic processes in modern and ancient marginal basins: Oxford, Blackwell Scientific Publications, p. 207-218.

Stuiver, M., G. H. Denton, T. J. Hughes and J. L. Fastook, 1981, History of the marine ice sheet in West Antarctica during the last glaciation: a working hypothesis, *in* G. H. Denton and T. J. Hughes, eds., The Last Great Ice Sheets: New York, Wiley-Interscience, p. 319-436.

Suess, E., 1987, Hydrothermalism in the Bransfield Strait, Antarctica: an overview (abstract): American Geophysical Union EOS Transactions, 68(50), p. 1768.

Tokarski, A. K., 1987, Late Cretaceous-Cenozoic structural history of King George Island (South Shetland Islands) (abstract): Fifth International Symposium on Antarctic Earth Sciences, Cambridge p. 130.

Van Enst, J. W. A., 1987, 210Pb activities in the Bransfield Strait, additional proof of hydrothermal activity (abstract): Fifth International Symposium on Antarctic Earth Sciences, Cambridge, p. 166.

Weaver, S. D., A. D. Saunders, R. J. Pankhurst and J. Tarney, 1979, A geochemical study of magmatism associated with the initial stages of back-arc spreading: the Quaternary volcanics of Bransfield Strait from South Shetland Islands: Contributions to Mineralogy and Petrology, v. 68, p. 151-169.

White, W. M. and M. Cheatham, 1987, Geochemistry of back-arc basin volcanics from Bransfield Strait, Antarctic Peninsula: American Geophysical Union EOS Transactions, 68(44), p. 1520.

Whiticar, M. J., E. Suess and H. Wehner, 1985, Thermogenic hydrocarbons in surface sediments of the Bransfield Strait, Antarctic Peninsula: Nature, v. 314, p. 87-90.

Chapter 3

Hazards to Antarctic Exploration and Production

David E. Reid
Amoco Production Company
Houston, Texas, U.S.A.

John B. Anderson
Department of Geology and Geophysics
Rice University
Houston, Texas, U.S.A.

ABSTRACT

Antarctica's continental shelf averages 500 m in depth and exhibits a landward slope, due to the combined effects of isostatic loading and glacial erosion. These effects are more pronounced near the continent. The highly rugged topography of the shelf typifies high latitude continental shelves.

Antarctica is the coldest, driest, windiest place on earth and the extremely hostile climate represents a formidable obstacle to the exploration for hydrocarbons. Sea ice covers the entire continental shelf during most of the year and presents another serious threat to the explorationist. The distribution and movement of sea ice on the continental shelf are hard to predict and have historically been responsible for the demise of several research vessels. Even less predictable is iceberg movement. Individual icebergs within the same area may drift at different speeds and in different directions because their size and draft determines to what extent winds and currents affect them. Drift speeds up to 3 km/hr and drafts exceeding 400 m have been reported.

Rugged topography and interstratification of stiff glacial deposits with water-saturated glacial-marine deposits combine to make the sea floor of the Antarctic continental shelf and slope unstable. Evidence for this exists in the form of abundant sediment gravity flow deposits on the shelf and slope. To date, shallow gas has been observed only in the Bransfield Basin. Significant earthquake activity is virtually nonexistent.

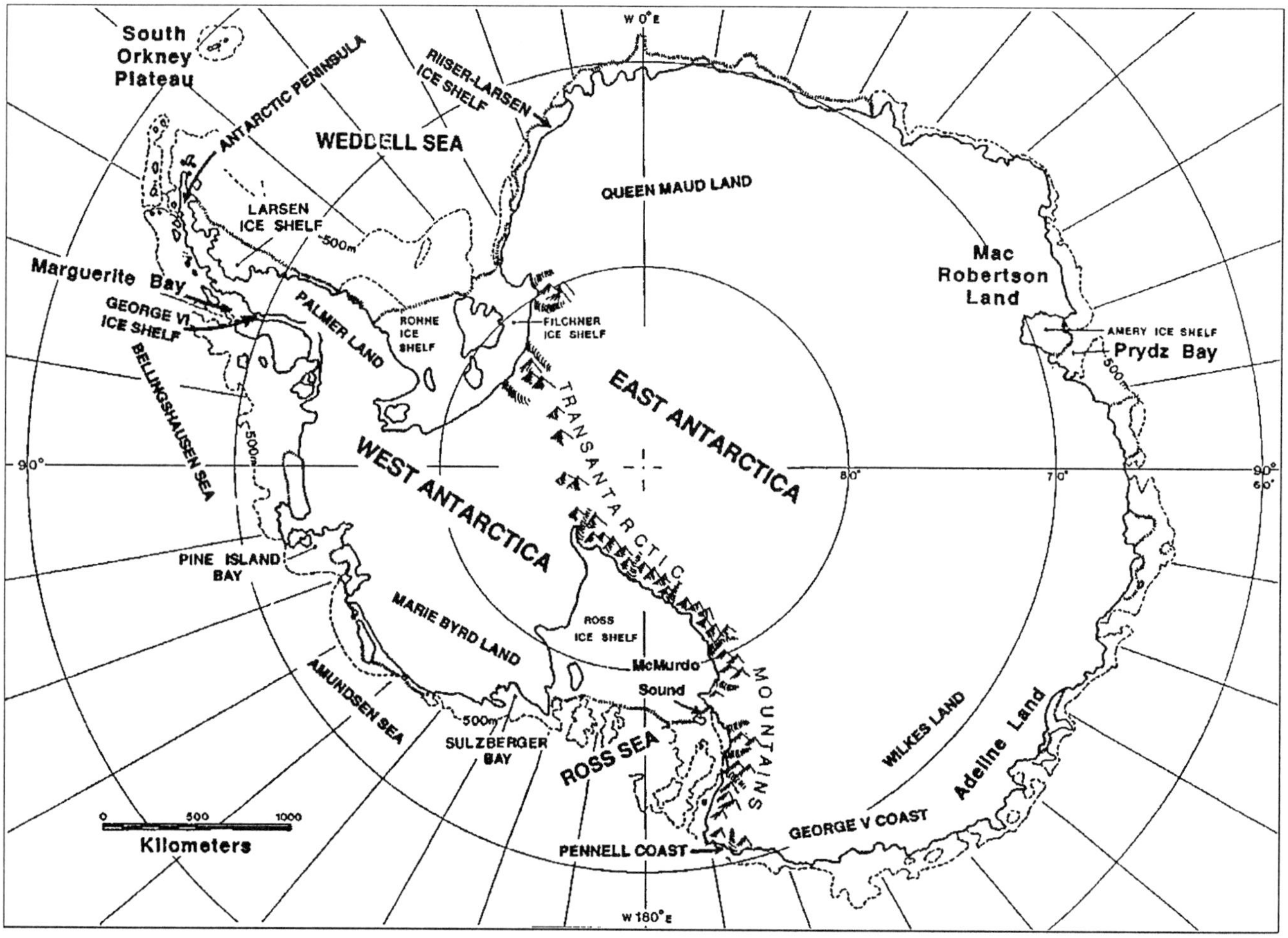

Figure 1—Geographic map of Antarctica.

INTRODUCTION

The Antarctic continental shelf is one of the most unique shelves on earth and poses an immense challenge to those who wish to exploit its oil and gas reserves. This chapter offers a brief description of the shelf and describes some of those unique properties. Also included is a description of the features of the Antarctic continental shelf that pose the greatest threat to exploration and production. Figure 1 is a geographic map which shows areas and features mentioned in the text. For a more detailed discussion of the geology of the Antarctic continental shelf see Anderson (in press, and other chapters in this volume).

SHELF PHYSIOGRAPHY

Among the unique physiographic features of the Antarctic continental shelf are its great depth, rugged topography, and landward gradient (Figure 2). The average shelf depth, roughly 500 m, is approximately eight times the world average. Typically, the inner shelf is the deepest part of the continental shelf (Figure 2), reflecting the combined effects of glacial erosion and isostasy. Glacial erosion is focused primarily on the inner shelf and close to larger glacial drainage systems. Seismic records from the Ross Sea shelf (Houtz and Davey, 1973; Wong and Christoffell, 1981; Hinz and Block, 1983), the Weddell Sea shelf (Fossum et al., 1982; Elverhoi and Maisey, 1983), Prydz Bay and the Mac Robertson Land shelf (Stagg, 1985), Wilkes Land shelf (Wannesson et al., 1985), and the continental shelf of the Antarctic Peninsula (Kimura, 1982; Anderson et al., 1990) all reveal direct evidence of glacial erosion of several hundred meters of strata from the continental margin. A seismic reflection profile from the Antarctic Peninsula shelf provides an example of the foredeepened topography and multiple glacial erosional surfaces typically found on the Antarctic shelf (Figure 2). Note that the entire sediment cover of the inner shelf has been eroded, leaving a deep, rugged sea floor where acoustic basement is widely exposed.

The extent to which the foredeepened gradient of the Antarctic shelf reflects isostatic loading by the ice sheet remains controversial. Most models for isostatic loading suggest the existence of an extensive isostatically depressed region situated adjacent to the terminus of the ice sheet. For a full-bodied ice sheet, such as the East Antarctic Ice Sheet, this area (the proglacial isostatic depression) approaches 200 km in width and is up to 300 m deep near the ice sheet terminus (Walcott, 1970). It is important to recognize that these isostatic models are theoretical and rely on assumptions about bedrock geology that often are unrealistic for continental margins. Still, these models imply that most of the continental shelf experienced isostatic loading and rebound,

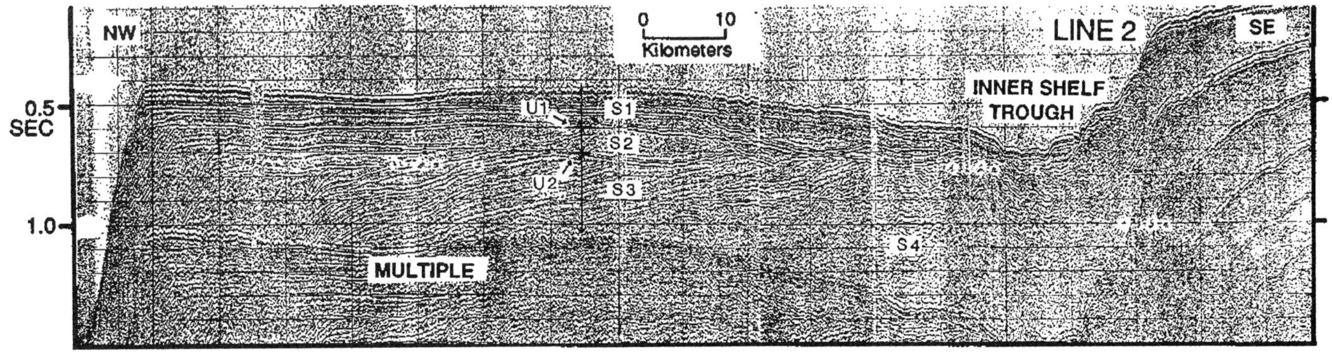

Figure 2—USAP 88 single channel seismic profiles from the Bellinghausen continental shelf showing typical landward gradient of shelf resulting from the combined effects of glacial erosion and isostacy. U_1 and U_2 are glacial erosion surfaces.

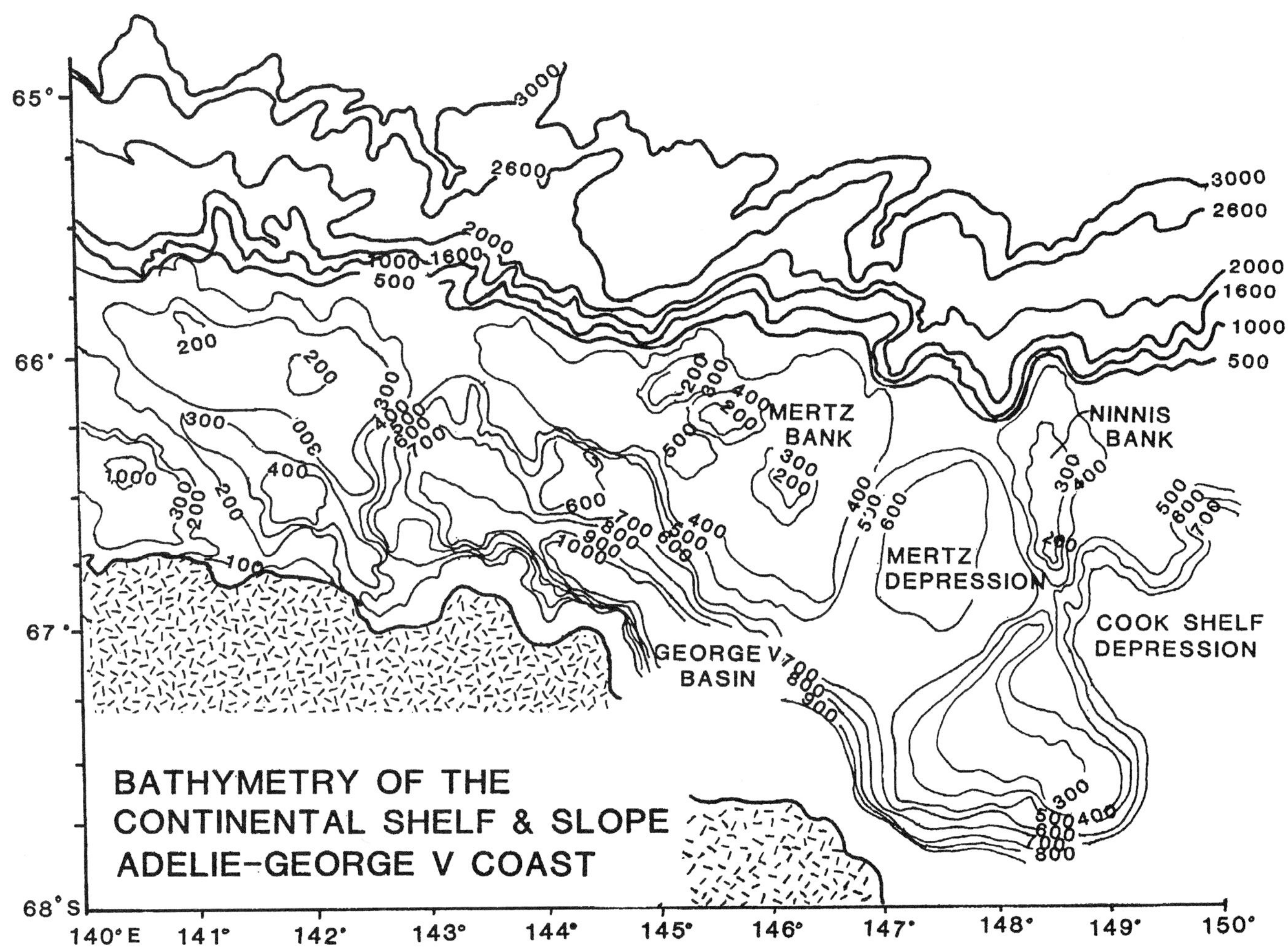

Figure 3—Bathymetric map of continental shelf off George V Land and Adelie Land. The bathymetry of this region is typical of the East Antarctic margin.

and the extent of isostatic loading on the shelf was much greater during glacial maxima when the ice sheet extended well onto the continental shelf. It is generally felt that the continental shelf is presently in isostatic equilibrium (Behrendt, 1962).

The width of the Antarctic continental shelf varies considerably, but in general the West Antarctic shelf is much wider (typically > 200 km) than the East Antarctic shelf. In general, the outer shelf is shallower, with an average depth of 400 m, and flatter than the inner shelf (Figure 2).

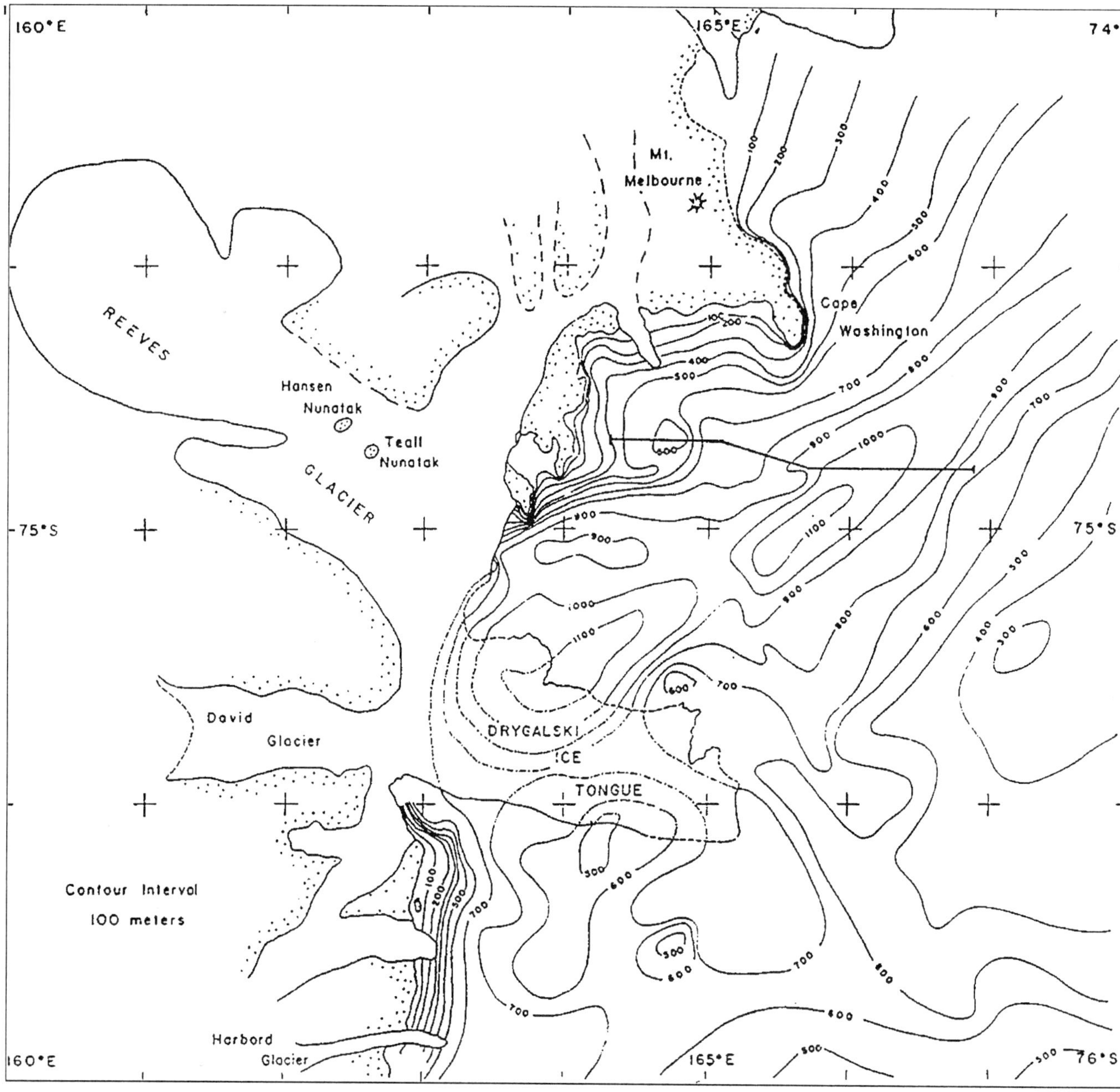

Figure 4—Bathymetric map of the west-central Ross Sea showing the Drygalski Trough. The trough was formed during glacial advances onto the continental shelf by the David Ice Stream.

Deep troughs, aligned subparallel to the coast, occur in a more or less en echelon fashion around East Antarctica. These longitudinal troughs (Holtedahl, 1958) comprise the most pervasive topographic features of the shelf. The probable mechanism for formation of these longitudinal troughs is preferential erosion along contacts between crystalline basement rocks and less resistant sedimentary strata (Holtedahl, 1970; Johnson et al., 1982). The troughs probably coincide with major tectonic features associated with continental rifting. One of the most thoroughly surveyed portions of the East Antarctic continental shelf is the George V-Adelie shelf. The detailed bathymetric map of the George V Basin, which is a longitudinal trough, provides a good example of typical East Antarctic shelf topography (Figure 3).

The primary topographic features of the West Antarctic continental shelf are transverse troughs (aligned transverse to the coast). The heads of the transverse troughs are connected to large ice streams that drain the West Antarctic Ice Sheet. They typically exceed 1000 m water depth and deepen toward the continent where they connect with subglacial troughs (Drewry, 1983). An excellent example is Drygalski Trough in the Ross Sea (Figure 4). Drygalski Trough was carved by David Glacier, a large ice stream that flows into the Ross Sea from East Antarctica.

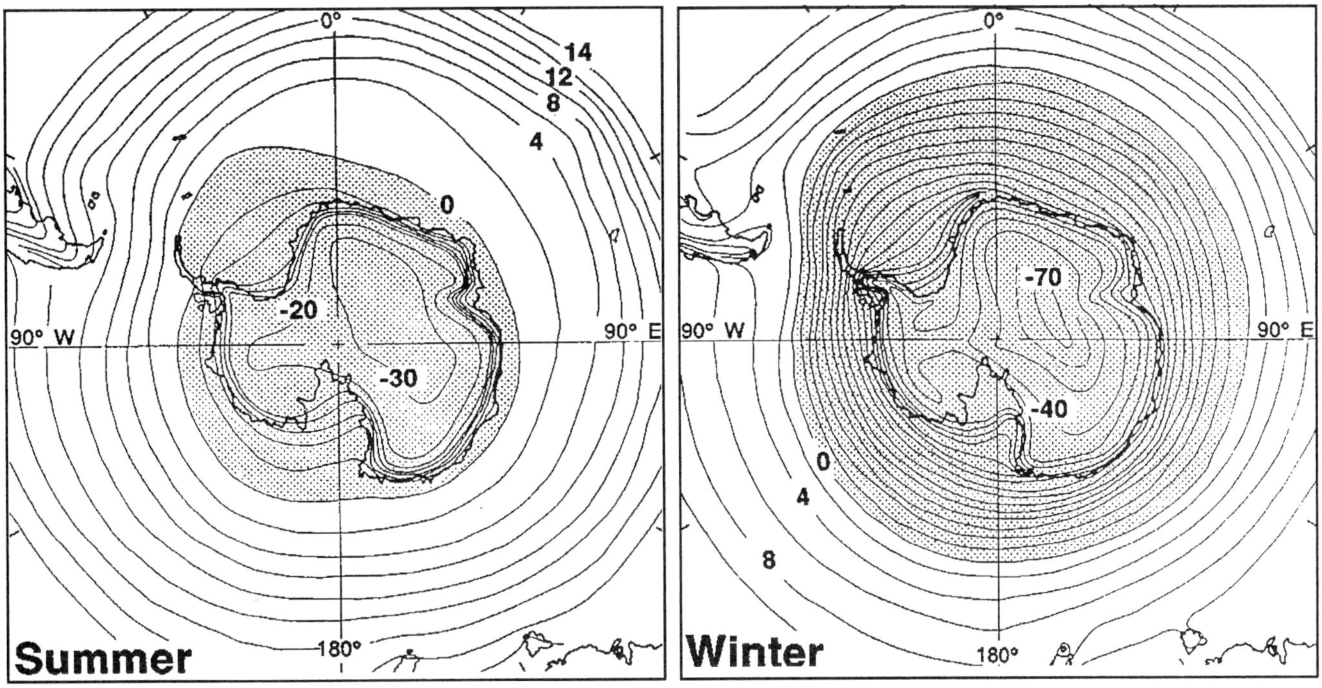

Figure 5—Mean summer and winter isotherms for the Antarctic region. Note the climatic amelioration of the coastal zone during the summer (modified from Dudeney, 1987).

CLIMATE

Antarctica is the coldest, driest, and windiest place upon the earth (Schwerdtfeger, 1984). The climate is very severe, especially in the interior of the continent with the coastal regions experiencing somewhat more moderate conditions. Temperatures along the coast average from just below freezing during the austral summer to about -20° to -30° C during the winter, with a mean annual temperature of about -15° C (Figure 5).

In addition to the extreme cold, Antarctic and Subantarctic storms are the most violent on earth. Researchers in the Antarctic often experience severe storms such as "Hurricane Schepel-Sturm" recently encountered by the German Antarctic North Victoria Land Expedition in the Pennell Coast region (Figure 1). The storm developed to hurricane strength within an hour and lasted four days with measured wind speeds exceeding 180 km/hr. The suddenness and severity of the storm resulted in heavy damage to both shipboard and land-based equipment and at least one casualty (Kothe et al., 1981).

In addition to synoptic events, much of the coastal region of the Antarctic experiences strong katabatic winds. The gravity driven katabatic winds form by radiation cooling on the high interior ice plateau and are characterized by very sudden intense flow (Tauber, 1960). Katabatic winds are often channeled down glacier valleys which may result in an increase in both intensity and duration. In addition to normal katabatic flow, cyclonic storms and katabatic winds can act in concert. As a storm moves inland, the relatively warm and moist air mass associated with the storm acts as a barrier to normal downslope katabatic flow. The passing of the storm removes the barrier and results in a violent avalanche of katabatic winds down the ice slope.

The Cape Denison-Commonwealth Bay region of Adelie Land provides an extreme example of the potential power of katabatic flow. This area is the windiest spot on the earth with an annual mean wind speed of 80 km/hr and maximum measured wind velocities exceeding 320 km/hr.

Strong winds create rough seas for which Antarctica is notorious (Figure 6A). Storms in the Antarctic and Subantarctic are frequent and sea states can change suddenly and unpredictably. Rough seas represent a particular hazard to the collection of geophysical data on the Antarctic margin.

ICE COVER ON THE CONTINENTAL SHELF

"I will not say it was impossible anywhere to get in among this Ice, but I will assert that the bare attempting of it would be a very dangerous enterprise and what I believe no man in my situation would have thought of. I whose ambition leads me not only farther than any other man has been before me, but as far as I think possible for man to go, was not sorry at meeting with this interruption, as it in some measure relieved us from the dangers and hardships, inseparable with the Navigation of the Southern Polar regions."—Captain James Cook, Journals, Voyage of the Resolution and Adventure (1774).

The broad band of sea ice surrounding the Antarctic continent for most of the year poses one of the greatest hazards to petroleum exploration and production on the continental margin (Figures 6B, 7). Sea ice covers virtually all of the continental shelf for much of the year with large areas of the Weddell Sea, Bellinghausen Sea, and eastern Ross Sea under perennial ice cover (Figure 8). The Antarctic ice pack differs in physical and behavioral characteristics from its Arctic counterpart. Arctic sea ice is generally confined to the land-locked Arctic basin. This favors the formation of pressure ridges and multi-year sea ice (Wadhams, 1986). In contrast,

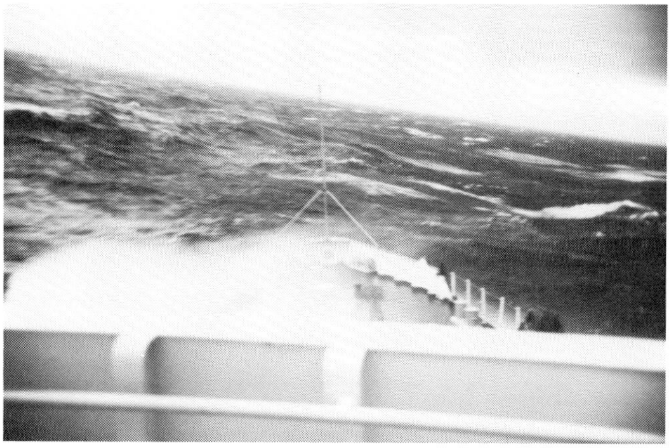

A

Figure 6A—Antarctica is notorious for its frequent storms and rough seas.

B

Figure 6B—The ice pack in the Antarctic is able to move quickly and often unpredictably, presenting an acute hazard to exploration on the continental shelf.

the ice cover surrounding Antarctica is everywhere divergent in its general behavior (Wadhams, 1986). Approximately 85% of the ice pack melts each year, which means that most of the ice pack that forms the following season is composed predominantly of first year ice. Because most of the ice cover is first year ice, Antarctic sea ice is relatively uniform in thickness, ranging from 1 to 3 m and averaging 1.5 m. However, fast ice, which is permanently attached to the coast, may reach several meters in thickness. The divergent forces of winds and currents tend to disperse the ice pack and open up new leads and polynyas. This makes the ice pack, in general, more navigable to ice-strengthened vessels, but these same factors also present the greatest hazard to shipping. The pack is able to move quickly, often in an unpredictable fashion, under the stress of winds and surface currents. Several ships (*Deutschland, Endurance, Aurora, Antarctic,* and *West Wind*) that have ventured through leads within the sea ice have become entrapped and often crushed when the ice shifted.

Another important characteristic of the Antarctic sea ice cover is its large degree of variability, both on a seasonal and yearly basis (Figures 7, 8). The Navy-NOAA Joint Ice Center (JIC) has prepared weekly ice-coverage maps from satellite imagery (Fleet Weather Facility Suitland Antarctic Ice Charts) since 1973. Analyses of interannual and seasonal variation of the areal extent of ice cover, both spatially and temporally, have been performed utilizing this data (Lemke et al., 1980; Chiu, 1983; Ropelewski, 1983; Lemke, 1986).

One of the most striking features of the Antarctic sea ice cover is its large seasonal cycle (Figure 8). Ropelewski (1983) identified six distinct subregions of sea ice which displayed coherent spatial and temporal behavior (Figure 9). The ratio of mean maximum to mean minimum extent of ice cover is approximately 4.9; however, this ratio varies between different sectors of the continental margin from 3 to 30 (Ropelewski, 1983). The largest annual fluctuations occur in the Ross and Weddell Seas, which are considered the ice factories of Antarctica (Lemke et al., 1980; Ropelewski, 1983). Unfortunately, these regions also contain several of the basins considered most prospective for hydrocarbon accumulations (Anderson et al., 1990). The smallest seasonal changes are found in the Bellinghausen Sea, the South Indian Basin, and around the Antarctic Peninsula (Lemke et al., 1980; Ropelewski, 1983). Interannual fluctuations during maximum sea ice extent are on the order of 20%, while the total sea ice extent minima can vary by more than 100% (Ropelewski, 1983). In general, ice anomaly drift appears to correlate with the prevailing surface currents and winds, indicating that advection of sea ice plays a major factor in large-scale sea ice dynamics around Antarctica (Lemke et al., 1980).

The seasonal extremes tend to occur at the same time each year with the annual minimum occurring in February (8 of 9 years analyzed) and the maximum occurring in either August or September (6 of 9 and 3 of 9 years, respectively) (Ropelewski, 1983). The timing of the sea ice minima between the individual subregions shows little variation. All of the sectors exhibit sea ice extent minima in February, except for the South Indian Basin and Southeast Pacific regions which show January minima. The temporal variation of the maximum extent of sea ice between the subregions displays more variability than the minimum. The Weddell and Southeast Pacific sectors typically exhibit maxima in August, the South Atlantic and South Indian basins reach their maxima in September, and the Indian Ocean region displays an October maxima. The Ross Sea sector displays a bimodal maxima with a primary maxima occurring in July and a secondary maxima in October. The Indian Ocean, South Indian Basin, and Southeast Pacific sectors also demonstrate secondary maxima for individual years. The secondary maxima may result from the transportation of sea ice to the north, in sufficient concentrations to produce a secondary maxima, following breakup of the primary maxima.

In addition to the sea ice cover, ice shelves cover large segments of the Antarctic continental margin (Figure 1). Much of the ice drainage from the Antarctic continent flows into these ice shelves. In fact, the three largest ice shelves, the Ross, Ronne-Filchner, and Amery Ice Shelves receive approximately 53% of the total ice drained from the continent while occupying only 10% of the coastline (Anderson et al., 1984).

Icebergs calved from ice shelves are generally large tabular bodies commonly hundreds of square kilometers in area (Figure 10). A good example of the massive size of icebergs calved from ice shelves is one that calved from the Ross Ice Shelf in November 1987, designated B-9 on JIC Ice Charts

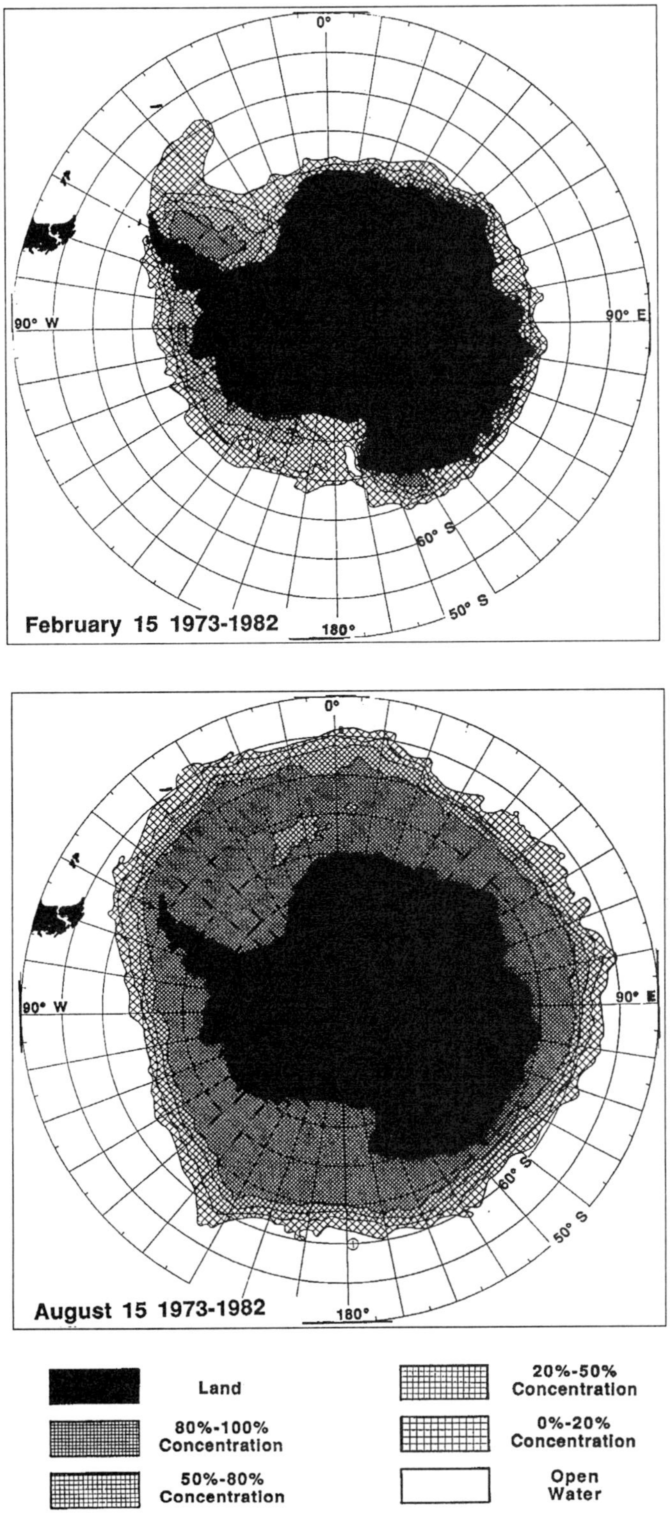

Figure 7—Mean winter and summer sea ice concentration surrounding Antarctica. Note the large seasonal cycle (taken from Department of the Navy, 1985).

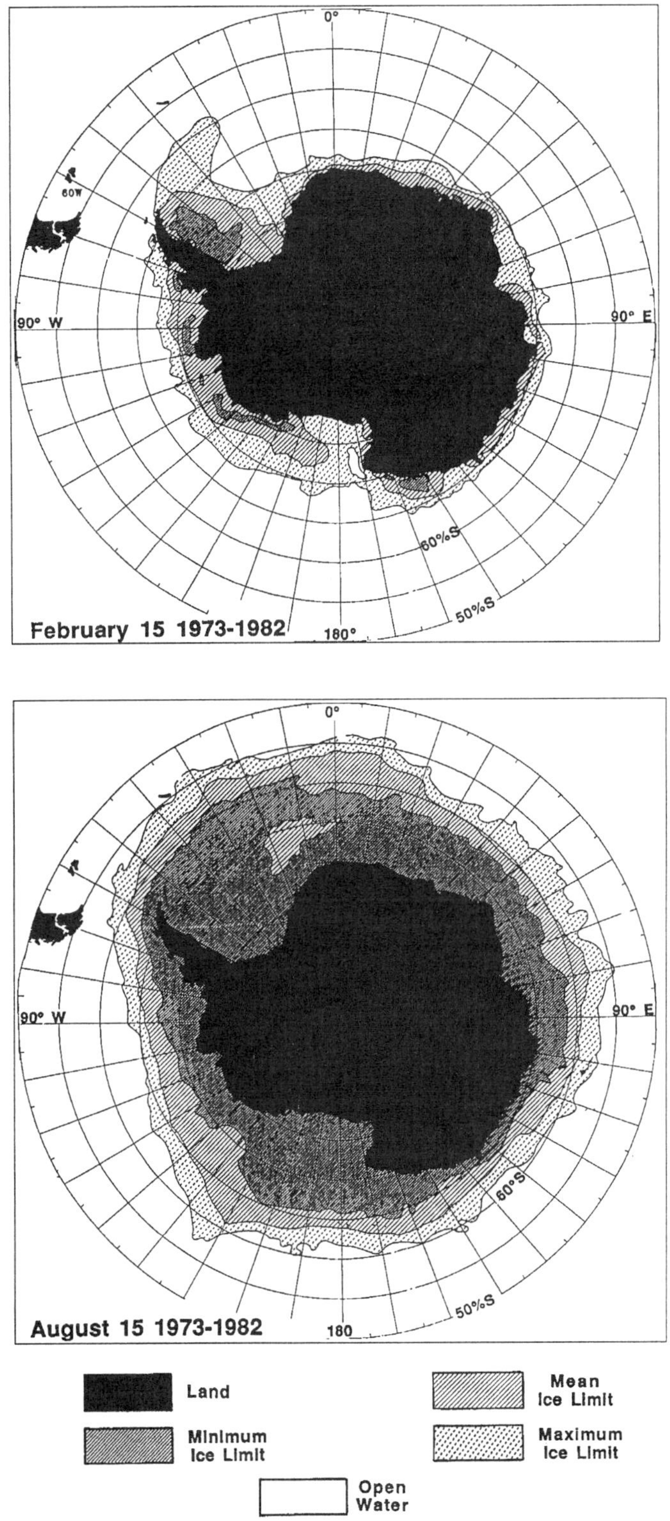

Figure 8—Summer and winter maximum, minimum, and mean sea ice extent. Note the large interannual variation in the summer sea ice cover (taken from Department of the Navy, 1985).

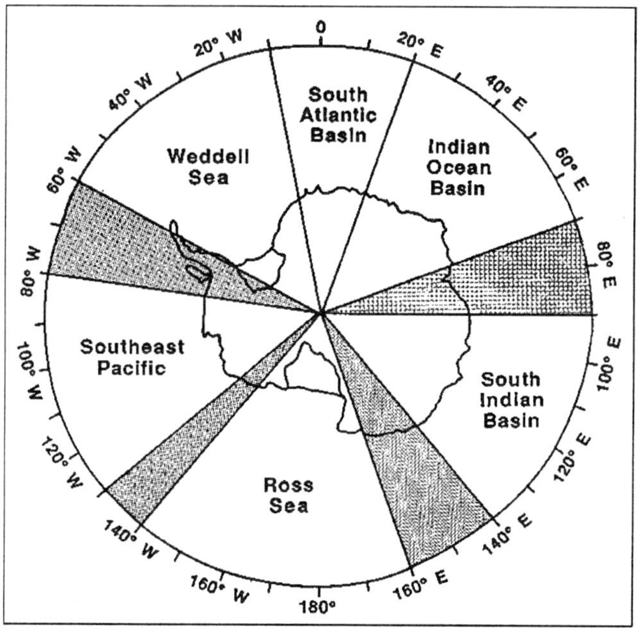

Figure 9—Sea ice subregions displaying coherent spatial and temporal behavior (from Ropelewski, 1983).

Figure 10—Photograph of a large tabular iceberg in the northwestern Ross Sea. Tabular icebergs, usually derived from ice shelves, are commonly hundreds of square kilometers in area.

(Houston Post, December 11, 1987). The iceberg was approximately 40 km by 157 km or roughly twice the size of the state of Rhode Island. In addition to ice shelves, icebergs are derived from ice tongues, outlet glaciers, and tidewater glaciers. These sources combine to produce an estimated annual iceberg flux of 1.3×10^{12} m^3 (Morgan and Budd, 1978).

Studies on the distribution and long-term motion of icebergs in the Southern Ocean have been conducted by Swithinbank et al. (1977), Morgan and Budd (1978), Tchernia and Jeannin (1984), and Wadhams (1988). Analysis of the data shows that the major geostrophic currents operating in the Southern Ocean appear to control long-term iceberg drift patterns (Tchernia and Jeannin, 1984). The data and our experience reveal a high concentration of icebergs in a narrow band surrounding the coast protected by the ice pack and under the general influence of the East Wind Drift.

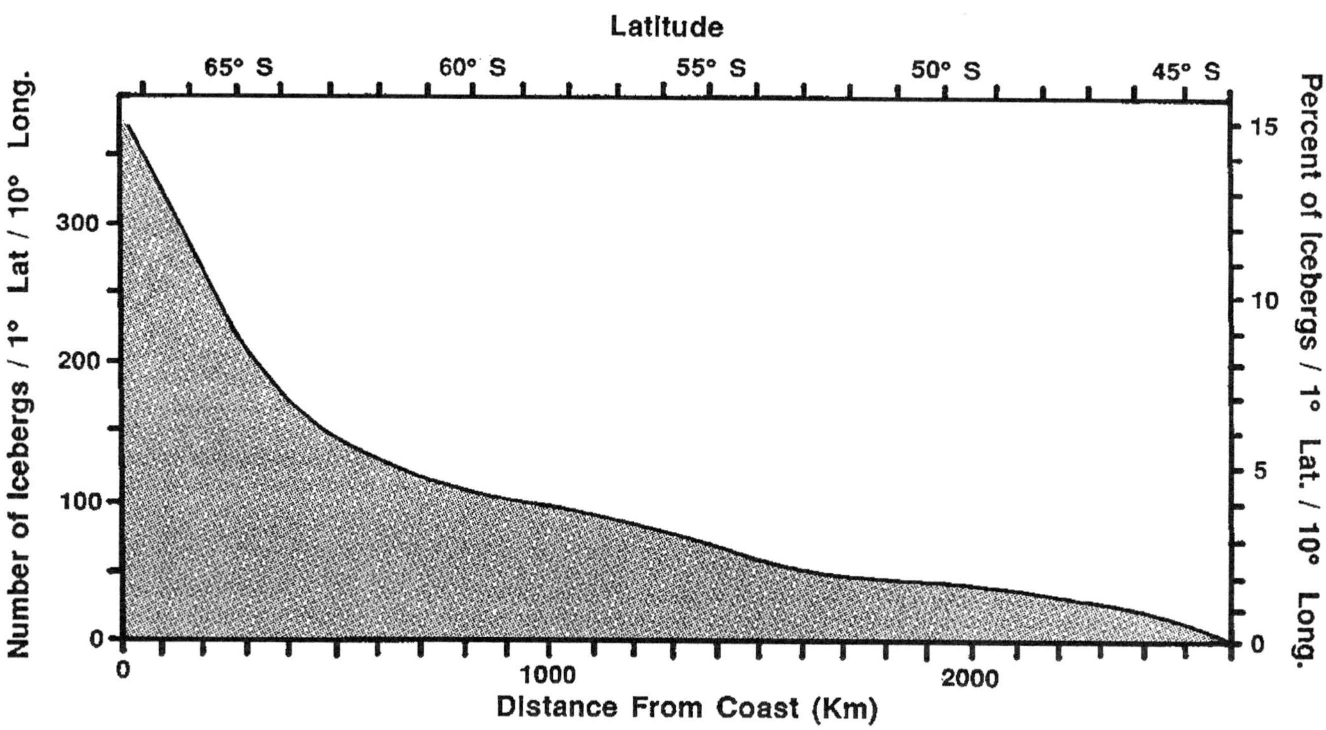

Figure 11—Graph of iceberg population versus degree of latitude (from Morgan and Budd, 1978).

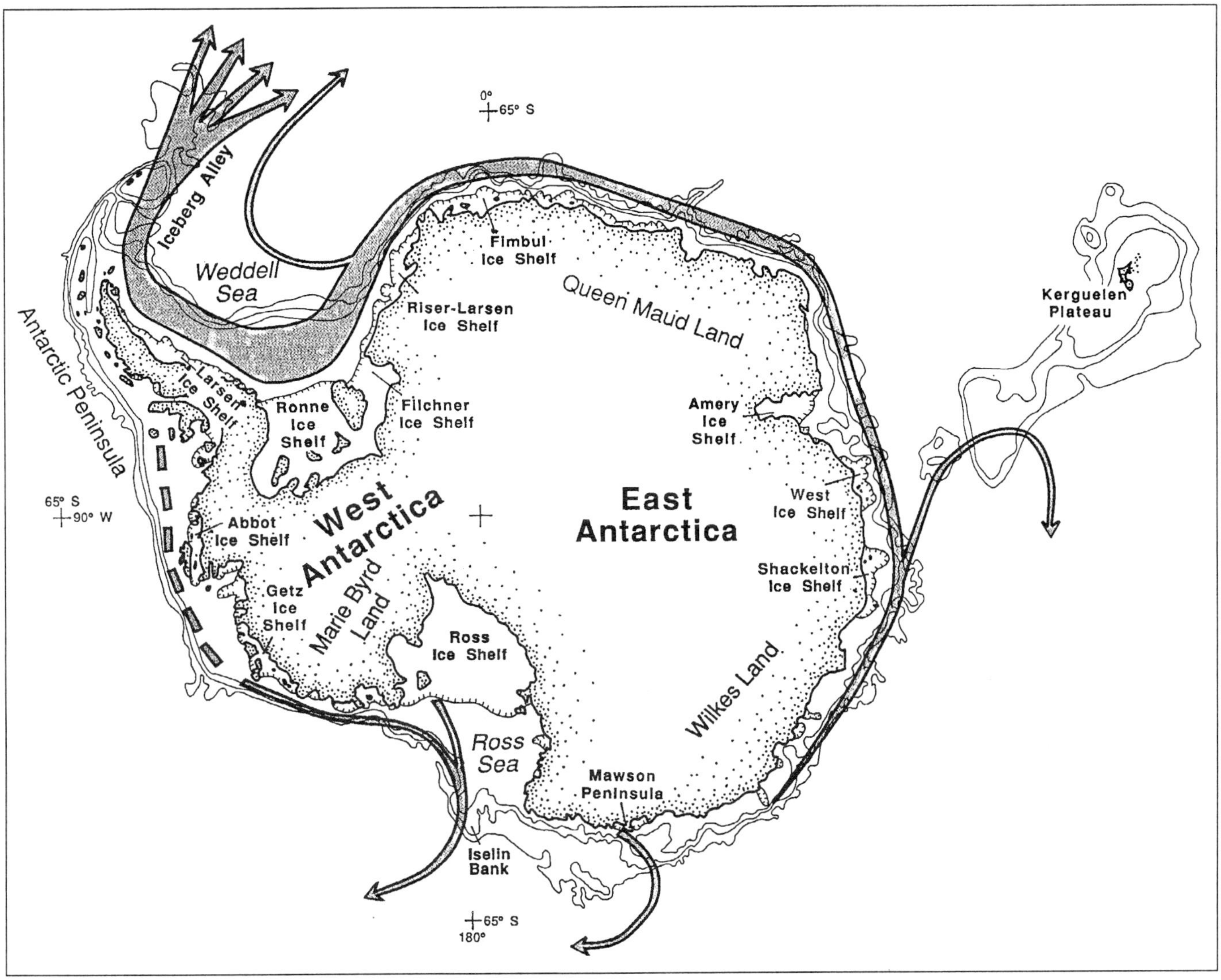

Figure 12—Generalized long-term iceberg drift trajectory. Icebergs from East Antarctica tend to be concentrated by the East Wind Drift in the Weddell Sea.

Moving to the north, away from the coast, there is a rapid decrease in the iceberg population density near 65° S. This is succeeded by a gradual decrease to a near zero population near 45° S, just north of the Antarctic Divergence (Figure 11) (Morgan and Budd, 1978). Concurrent with the northward decrease in the iceberg population is a dramatic decrease in average iceberg length north of 65° S. The average length decreases by a factor of 2.5 with only a minor decrease in average height (Morgan and Budd, 1978).

Superimposed on this latitudinal zonation are longitudinal variations imposed primarily by irregularities in the shape and bathymetry of the Antarctic coast, especially by the Antarctic Peninsula in the Weddell Sea (Figure 12). Figure 12 shows the general, long-term iceberg drift track pattern, based on the visual tracking of large (greater than 15 nm) icebergs by satellite (JIC 1973-1989) and 22 icebergs equipped with radio transponders (Tchernia and Jeannin, 1984). Most icebergs calved from East Antarctica west of about 150° E longitude appear to move under the influence of the East Wind Drift, hugging the coast and moving persistently to the west (Figure 12). Some of the icebergs are deflected to the north near 90° E longitude due to a small gyre formed by the Kerguelen Plateau bathymetric high (Tchernia and Jeannin, 1984). Many of the large icebergs eventually make their way into a large well-developed clockwise gyre located in the Weddell Sea. There icebergs calved from the large ice shelves which ring the landward margin of the Weddell Sea join the older, well-traveled icebergs. Often the icebergs appear to wander aimlessly for some time in the vicinity of Crary Trough near the confluence of the Ronne and Filchner Ice Shelves (Figure 13). Since there is no bathymetric high on which to ground, icebergs probably are trapped in this area by the eastern edge of the dense ice pack that tends to persist in the western Weddell Sea (Figure 8). Eventually, most of the icebergs penetrate the ice pack and make their way in a narrow band along the coast of the Antarctic Peninsula. The concentration of icebergs in the Weddell Sea forms an "iceberg alley" along the east coast of the Antarctic Peninsula and over the Scotia Ridge, greatly increasing the potential iceberg hazard in this region (Figures 12, 13).

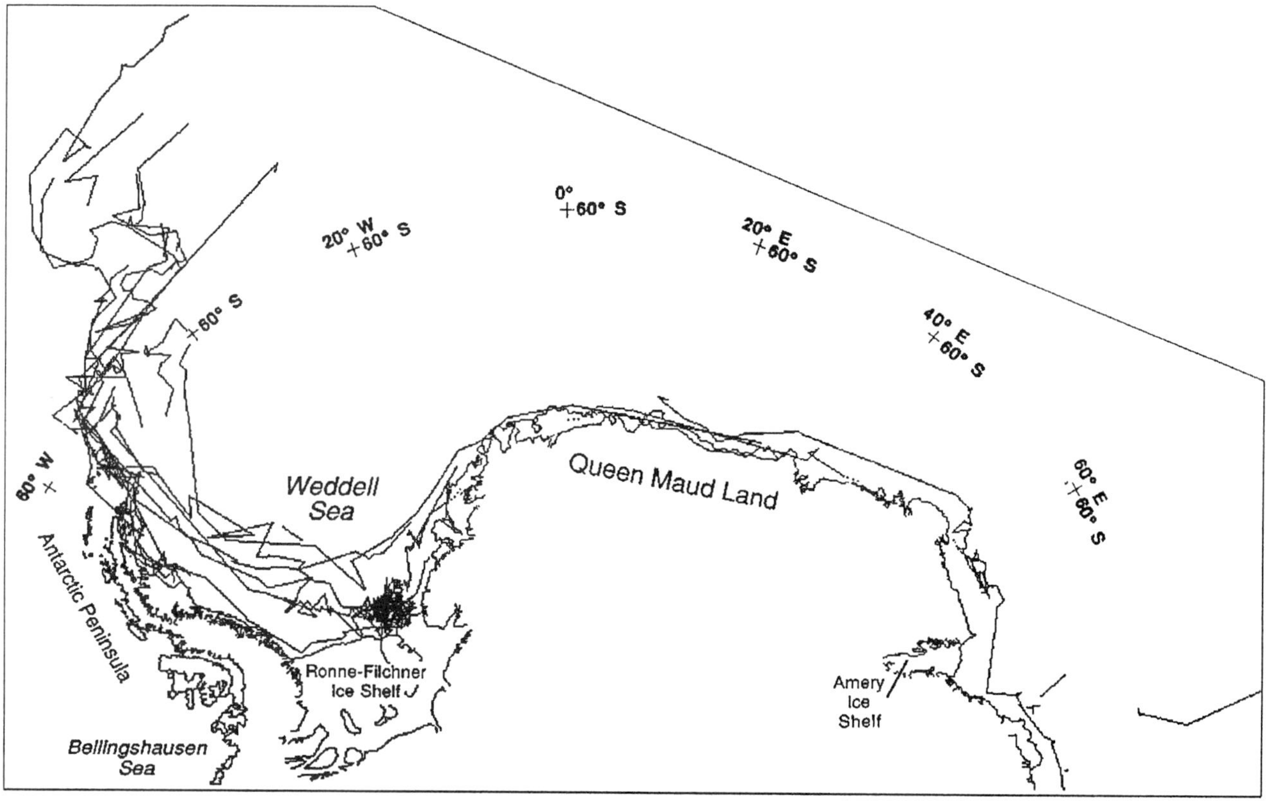

Figure 13—Digitized iceberg drift tracks illustrating: 1) the concentration of icebergs in the Weddell Sea by the East Wind Drift; 2) the relatively long residence time for many icebergs near the Ronne-Filchner Ice Shelf; and 3) the formation of "iceberg alley" along the eastern coast of the Antarctic Peninsula.

Icebergs calved in the Bellinghausen and Amundsen Seas along the West Antarctic margin, also appear to move predominantly under the influence of the East Wind Drift, but data from these segments of the Antarctic coast are relatively scarce. After moving along the coast of Marie Byrd Land, a clockwise gyre located in the Ross Sea and a bathymetric high near 180° W longitude (Iselin Bank) deflect icebergs to the north (Figure 12). Icebergs are also deflected to the north by a gyre located near 150° E longitude along the eastern Wilkes Land margin. The formation of the gyre is probably related to an increase in width of the continental shelf in this region.

At a local scale, the movement of icebergs is fairly unpredictable, especially on a short-term basis. This is because icebergs of different size are not affected in the same way by winds and surface currents, and icebergs with different drafts are pushed along by currents at different levels within the surface water column. Crepon et al. (1988) show that medium to low velocity winds have only a limited effect on the motion of deep draft icebergs with keels below the thermocline. Most Antarctic icebergs have drafts deeper than the mixed water layer (approximately 40 m). In addition, the effect on iceberg motion by moderate winds is further diminished by increasing horizontal scale, which may be particularly significant for the large tabular icebergs typically calved from Antarctic ice shelves.

The unpredictable nature of Antarctic iceberg motion can be illustrated best by observations acquired during Deep Freeze 86 in the South Orkney Plateau area (Figure 1). During the course of several days we used the ship's radar to track the drift of icebergs passing within 50 km of the ship. Several dozen icebergs were tracked and the result is shown in Figure 14. These data illustrate the apparently random drift track of icebergs in this area; the drift direction and speed of icebergs did not correlate well with wind direction and speed. Also note that the measured speeds of several icebergs exceeded 2 knots.

Another source of concern for those considering exploiting the resources of the Antarctic continental shelf is the potentially deep draft of icebergs. The large ice shelves, responsible for most of the large tabular icebergs found in the Antarctic, have thicknesses of about 200 to 250 m near the calving margin and tabular icebergs with drafts up to 330 m have been reported (Orheim, 1980). However, side-scan sonar records show abundant iceberg furrow and gouge marks (Figure 15) to depths of about 400 m, ranging down to 500 m (Lien, 1981; Barnes, 1987; Barnes and Lien, 1988; Dr. Peter Barker, personal communication, 1987), but it is not clear whether these represent modern or relict features. Typically, ice gouge features are most common and best developed on the sides of troughs, subdued on bank tops and absent in deep troughs. Studies show that by tilting or overturning, icebergs can increase their draft by as much as 50% (Keys, 1983) and that the majority of Antarctic icebergs are nontabular or tilted (Keys, 1984) (Figure 16).

The bathymetry of areas where icebergs are frequently grounded also provides information on the draft of Antarctic icebergs. Over the past several years we have observed that inner shelf areas and banks that are shallower than 250 m tend to be the resting grounds for large icebergs which are grounded.

Clearly, more systematic studies are required for establishing the maximum depth of iceberg grounding on the Antarctic continental shelf, but shallower portions of the shelf (<250 m) are subject to frequent grounding events.

SEA-FLOOR STABILITY

The Antarctic continental shelf has been cored rather extensively during the past decade, resulting in a reasonable understanding of the properties and distribution patterns of surface deposits (Chriss and Frakes, 1972; Anderson, 1985; Anderson et al., 1979, 1980, 1983, 1984; Domack, 1982). Few studies addressed the geotechnical properties of Antarctic marine sediments (Kurtz et al., 1979; Edwards et al., 1987). High resolution seismic data, collected in the Ross Sea (Houtz and Davey, 1973; Wong and Christoffel, 1981; Anderson et al., 1987), Weddell Sea (Elverhoi and Maisey, 1983), the Antarctic Peninsula area (Anderson, 1985; Anderson et al., 1990; Jeffers and Anderson, 1990), and a few other scattered areas of the shelf, allow some evaluation of the substrate conditions on the shelf and evidence for such hazards as mass movement, faulting, and gas seeps.

The general stratigraphy of piston cores from the Antarctic continental shelf reveals water-saturated glacial-marine sediments resting directly on cohesive diamictons of glacial and glacial-marine origin (Kurtz et al., 1979; Anderson et al., 1980, 1984). This stratigraphic association marks the boundary between glacial sediments deposited on the shelf during the last glacial maximum (late Wisconsin) and Holocene deposits. Drill cores obtained from the Ross Sea (Hayes and Frakes, 1975; Hambrey et al., 1989) and Prydz Bay (Barron et al., 1988) indicate that interbedded deposits similar to those penetrated in piston cores comprise the upper few hundred meters of the stratigraphic column.

Glacial and glacial-marine sequences rest on a sea floor characterized by very rugged topography. Local sea floor gradients typically exceed 2°. This situation is conducive to sediment mass movement, particularly when stress is applied to these sediments. Evidence for sea-floor instability occurs all around the continent in the form of sediment gravity flow deposits (slumps, debris flows, etc.) commonly seen in seismic records (Figure 17) and in piston cores (Kurtz and Anderson, 1979; Anderson et al., 1979; Wright et al.,1983).

In general, there is little indication of surface faulting or gas seeps on the Antarctic sea floor, with the exception of Bransfield Basin. This basin comprises the only large basin subject to seismic activity (the smaller McMurdo Sound in the Ross Sea is the only other basin, Figure 1). Seismic records from Bransfield Basin show features that imply gas seeps and/or the occurrence of shallow gas (Figure 18). Reports of thermogenic gases and higher hydrocarbons recovered from the area support the seismic interpretation of gas-related features (Han, 1988). The high flux of organic carbon (from surface productivity) to the sea floor in conjunction with high heat flow within this backarc basin may provide a suitable environment for hydrocarbon generation (Suess, 1988). Similar conditions of organic-rich sediments and high heat flow occur in the western Ross Sea, but high resolution seismic records from this area display no evidence of near-surface gas.

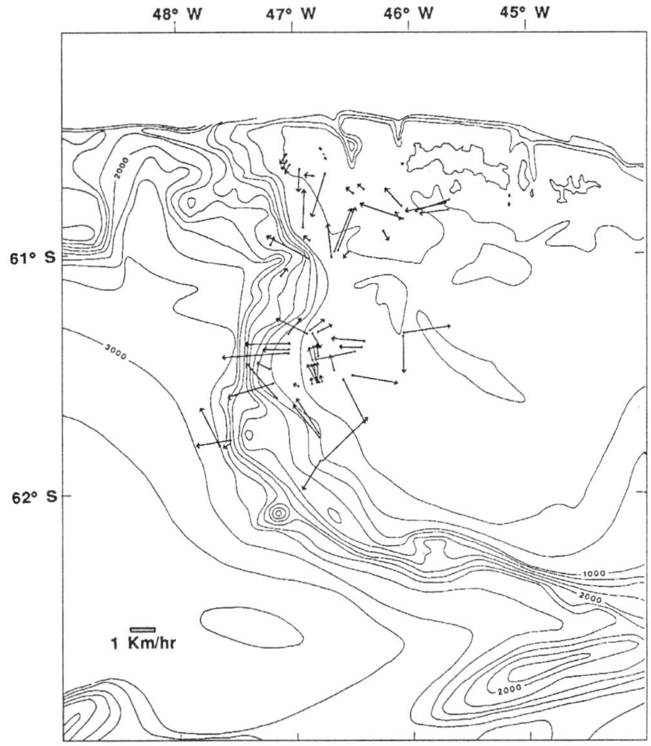

Figure 14—Iceberg drift tracks on the South Orkney Plateau which were recorded using ship's radar.

SUMMARY

1. The Antarctic continental shelf is extremely deep and foredeepened. Rugged topography and a virtual absence of sediment cover characterize the inner shelf. Glacial and glacial marine sediments blanket the outer shelf. These sediments display abrupt changes in geotechnical properties with depth and are subject to failure under stress. Abundant evidence for sediment mass movement is seen in high resolution seismic records and in piston cores from the continental shelf and slope.

2. Antarctica has the most severe climate on Earth. Hurricane force storms are common and unpredictable.

3. Sea ice covers virtually all of the continental shelf during much of the year with large segments of the shelf under perennial ice cover. The Antarctic ice pack is very mobile, and therefore represents a tremendous hazard to exploration on the Antarctic margin. Analysis of sea ice data allows some prediction of sea ice distribution on a regional scale.

4. Iceberg drift tracks show a fairly consistent pattern on a regional scale, but movement is unpredictable on a local scale. Iceberg drafts up to 330 m have been reported, but side-scan sonograph records reveal iceberg furrows at depths up to 500 m. Shallow portions of the shelf (<300 m) are furrowed heavily and frequently constitute the resting grounds for large icebergs.

5. Seismic activity occurs only in isolated portions of the Antarctic margin (Bransfield Basin and McMurdo Sound). Surface faulting is restricted to these areas. The shelf is thought to be in isostatic equilibrium.

6. Evidence for shallow gas hazards has been observed only in the Bransfield Basin area.

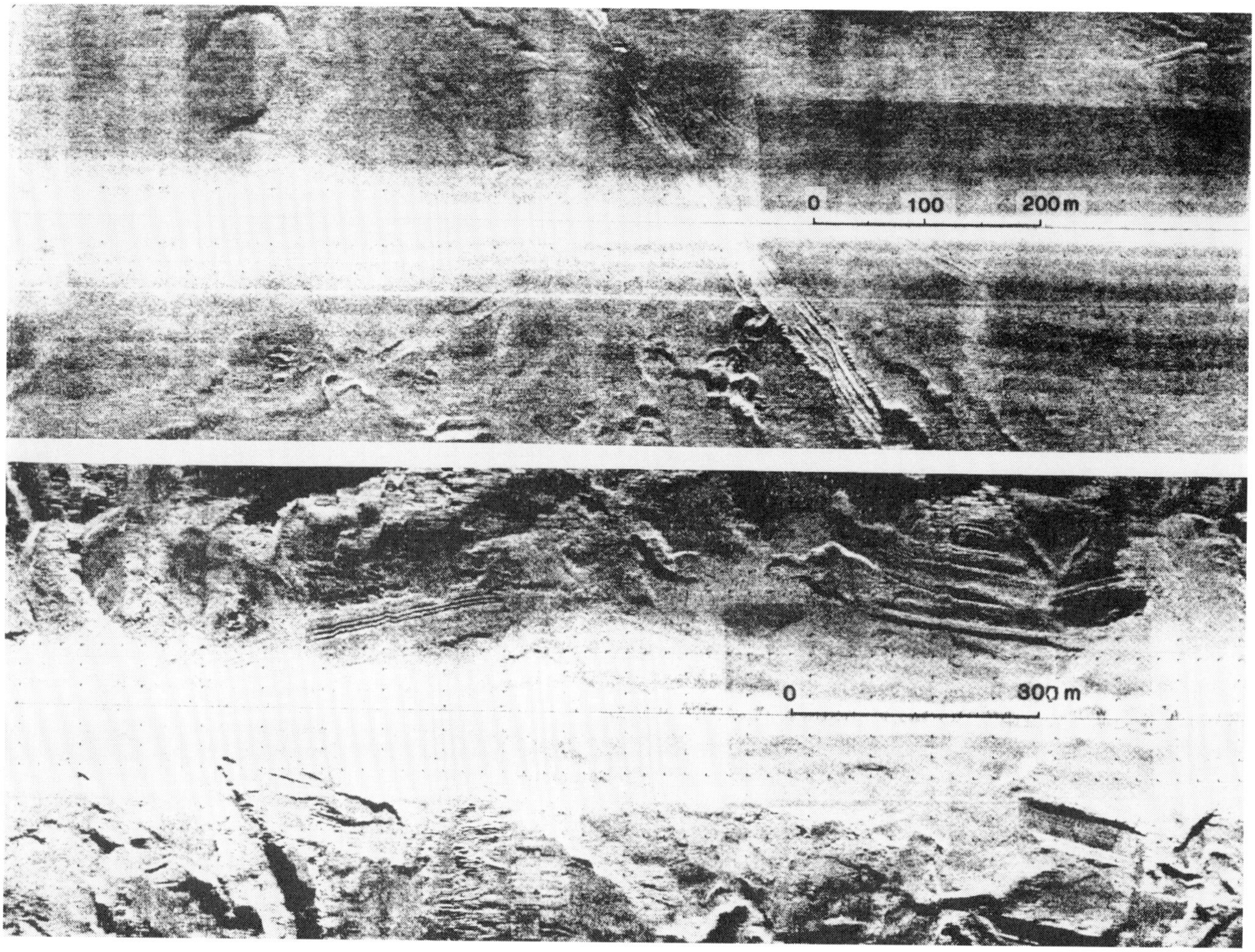

Figure 15—Iceberg furrow and gouge marks on the sea floor along the Wilkes Land margin (from Barnes, 1987).

REFERENCES

Anderson, J. B., 1985, Antarctic glacial marine sedimentation: a core workshop: Workshop Notes, GSA Annual Meeting, Orlando, Florida, 66 p.

Anderson, J. B., in press, Geology and hydrocarbon potential of the Antarctic continental margin: J. Splettstoesser and G. Dreschhoff, eds., American Geophysical Union Antarctic Research Series, Washington, D.C.

Anderson, J. B., D. D. Kurtz and F. M. Weaver, 1979, Sedimentation on the Antarctic continental slope, *in* L. J. Doyle and O.H. Pilkey, eds., Geology of Continental Slopes: SEPM Special Publication No. 27, p. 265-283.

Anderson, J. B., C. Brake and N. Myers, 1984, Sedimentation on the Ross Sea Continental shelf, Antarctica: Marine Geology, v. 57, p. 295-333.

Anderson, J. B., L. R. Bartek and D. E. Reid, 1987, Preliminary results of 1986-87 austral summer marine geological survey of the western Ross Sea: Antarctic Journal of the United States, 1987 Review, v. 22, No. 5, p. 120-122.

Anderson, J. B., P. G. Pope and M. A. Thomas, 1990, Evolution and hydrocarbon potential of the northern Antarctic Peninsula continental shelf, this volume.

Anderson, J. B., D. D. Kurtz, E. W. Domack and K. M. Balshaw, 1980, Glacial and glacial marine sediments of the Antarctic continental shelf: Journal of Geology, v. 88, p. 399-414.

Anderson, J. B., B. A. Andrews, L. R. Bartek and E. M. Truswell, in press, Petrology and palynology of Weddell Sea glacial sediments: implications for subglacial geology, *in* M. R. Thomas, ed., Antarctic Earth Science, Cambridge University Press.

Anderson, J. B., C. Brake, E. W. Domack, N. Myers and J. Singer, 1983, Sedimentary dynamics of the Antarctic continental shelf, *in* R. L. Oliver, P. R. James and J. B. Jago, eds., Antarctic Earth Science, Proceedings of the Fourth International Symposium, Antarctic Earth Sciences, Australian Academy of Science, Canberra, p. 387-389.

Barnes, P. W., 1987, Morphologic studies of the Wilkes Land continental shelf, Antarctica-glacial and iceberg effects, *in* S. L. Eittreim and M. A. Hampton, eds., The Antarctic Continental Margin: Geology and Geophysics of Offshore Wilkes Land: Circum-Pacific Council Energy and Mineral Resources, Houston, Texas, p. 175-194.

Barnes, P. W. and R. Lien, 1988, Icebergs rework shelf sediments to 500 m off Antarctica: Geology, v. 16(12), p. 1130-1133.

Barron, J., B. Larsen and Shipboard Scientific Party, 1988, Leg 119 Kerguelen Plateau and Prydz Bay, Antarctica: Proceedings of the Ocean Drilling Program Initial Reports, Texas A&M University, College Station, TX, v. 119.

Figure 16—Nontabular iceberg. Studies show that tabular icebergs can increase their draft significantly by tilting or rolling.

Behrendt, J. C.,1962, Geophysical and glaciological study in the Filchner Ice Shelf area of Antarctica: Journal of Geophysical Research, v. 67, p. 221-234.

Chiu, L. S., 1983, Variation of antarctic sea ice: An update: Monthly Weather Review, v. 111, p. 578-580.

Chriss, T. and L. A. Frakes, 1972, Glacial marine sedimentation in the Ross Sea, *in* R. Adie, ed., Antarctic Geology and Geophysics, Oslo, Commission on Antarctic Research, p. 747-762.

Cook, J., 1774, Journals, Voyage of the *Resolution* and *Adventure.*

Crepon, M., M. N. Houssais and B. Saint Guily, 1988, Drift of icebergs under wind action, Journal of Geophysical Research, v. 93, p. 3608-3612.

Department of the Navy, 1985, Sea Ice Climatic Atlas: Volume I, Antarctic: (NAVAIR 50-1C-540), 131 p.

Domack, E. W., 1982, Sedimentology of glacial and glacial-marine deposits on the George V-Adelie continental shelf, East Antarctica: Boreas, v. 11, p. 79-97.

Domack, E. W. and J. B. Anderson, 1983, Marine geology of the George V continental margin: Combined results of Deep Freeze 1979 and the 1911-1914 Australasian Expedition, *in* R. L. Oliver, P. R. James and J. B. Jago, eds., Antarctic Earth Science, Australian Academy of Science, Canberra, p. 402-406.

Drewry, D. J., 1983, Antarctica: Glaciological and Geophysical Folio, Scott Polar Research Institute, Cambridge, England.

Dudeney, J. R., 1987, The Antarctic Atmosphere, *in* D.W.H. Walton, ed., Antarctic Science: Cambridge, Cambridge University Press, p. 210.

Edwards, B. D., H. J. Lee, H. A. Karl, E. Reimnitz and L. A. Timothy, 1987, Geology and physical properties of Ross Sea, Antarctica, continental shelf sediment, *in* A. K. Cooper and F. J. Davey, eds., The Antarctic Continental Margin: Geology and Geophysics of the Western Ross Sea, Circum-Pacific Council Energy and Mineral Resources Earth Science Series, v. 5B, Houston, p. 191-216.

Elverhoi, A., 1984, Glacigenic and associated marine sediments in the Weddell Sea, Fjords of Spitsbergen and the Barents Sea: a review: Marine Geology, v. 57, p. 53-88.

Elverhoi, A. and G. Maisey, 1983, Glacial erosion and morphology of the eastern and southern Weddell Sea shelf, *in* R. L. Oliver, P. R. James and J. B. Jago, eds., Antarctic Earth Science: Australian Academy of Science, Canberra, p. 483-487.

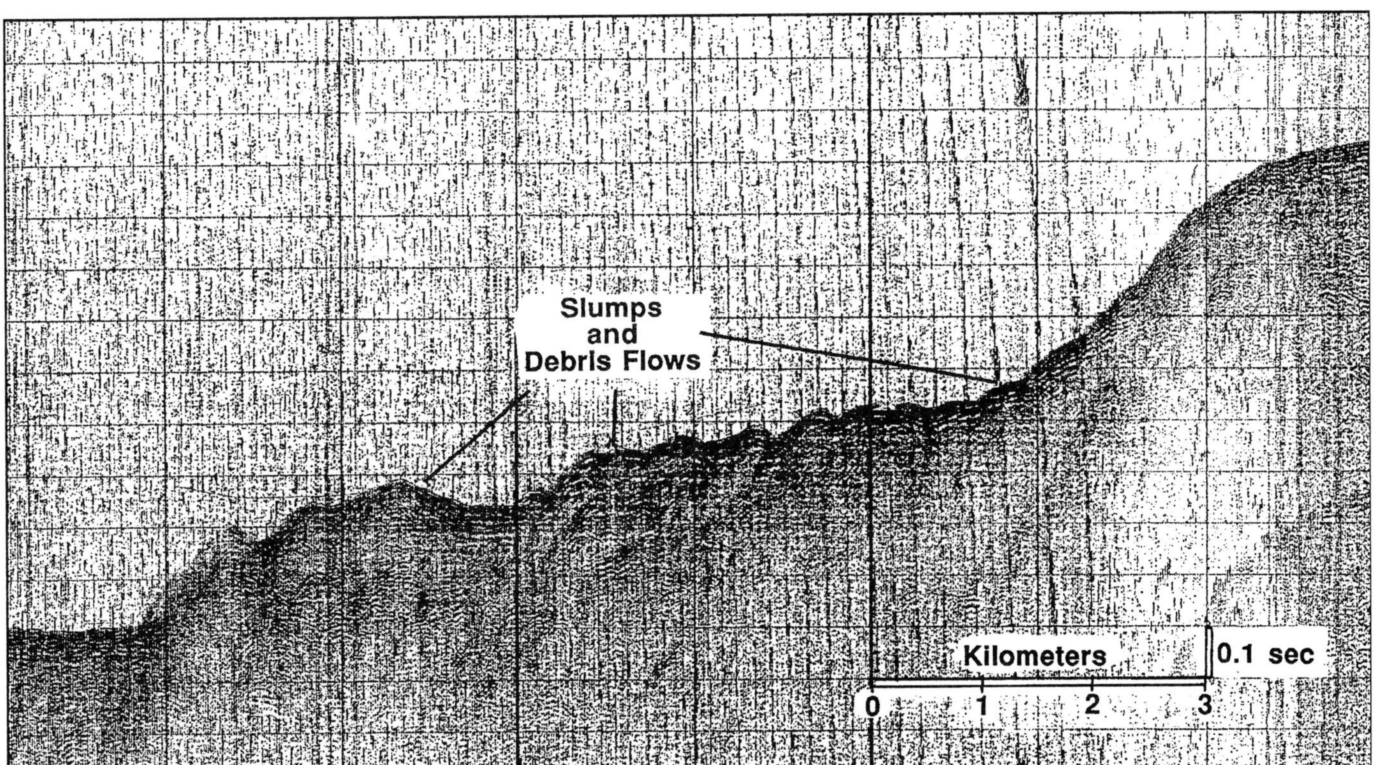

Figure 17—Seismic record from the Bransfield Basin with numerous slumps. This illustrates the unstable nature of much of the sea floor on the Antarctic margin.

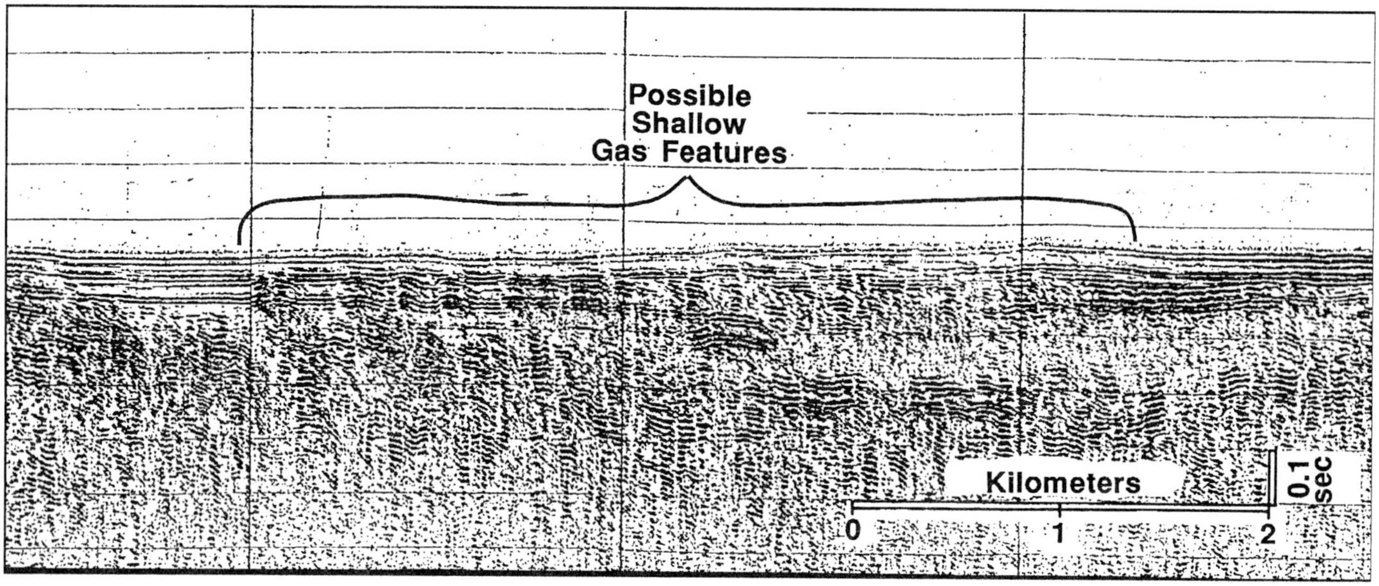

Figure 18—Seismic record from the Bransfield Basin displaying evidence of a potential shallow gas hazard.

Fossum, B. A., G. H. Maisey and H. O. Torsen, 1982, Marine geophysical research in the Weddell Sea during the Norwegian Antarctic Research Expedition of 1976-77, *in* C. Craddock, ed., Antarctic Geoscience: Madison, University of Wisconsin Press, p. 397-404.

Hambrey M. J., P. J. Barrett and P. H. Robinson, 1989, Stratigraphy, *in* P. J. Barrett, ed., Antarctic Cenozoic History from the CIROS-1 Drill Hole, McMurdo Sound: New Zealand Department of Scientific and Industrial Research Bulletin 245, Wellington, p. 23-48.

Han, M. W., 1988, Hydrothermalism and hydrocarbon potential in the King George Basin of the Bransfield Strait, Antarctica, *in* H. T. Huh, B. Park and S. Lee, eds., Antarctic science: Geology and Biology: Proceedings of the First International Symposium on Antarctic Science, Seoul, p. 37-55.

Hayes, D. E. and L. A. Frakes, 1975, General Synthesis: Deep Sea Drilling Project 28, *in* D. E. Hayes and L. A. Frakes, eds.: Initial Reports of the Deep Sea Drilling Project: Washington (U.S. Government Printing Office), v. 28, p. 919-942.

Hinz, K., 1983, Results of geophysical investigations in the Weddell Sea and in the Ross Sea, Antarctica: Proceedings of the Eleventh World Petroleum Congress, London, England, p. 1-13.

Hinz, K. and M. Block, 1983, Results of geophysical investigations in the Weddell Sea and in the Ross Sea, Antarctica, Proceedings of the 11th World Petroleum Congress (London), PD2, Westchester, New York, Wiley and Sons, Ltd., p. 79-91.

Holtedahl, O., 1958, Some remarks on the geomorphology of continental shelves off Norway, Labrador, and Southeast Alaska, Journal of Geology, v. 66, no. 4, p. 461-471.

Holtedahl, O., 1970, On the geomorphology of the West Greenland Shelf, with general remarks on the "marginal channel problem," Marine Geology, v. 8, p. 155-172.

Houtz, R. E. and F. J. Davey, 1973, Seismic profiler and sonic buoy measurements in the Ross Sea, Antarctica: Journal of Geophysical Research, v. 78, p. 3448-3468.

Jeffers, J. D. and J. B. Anderson, 1990, Sequence stratigraphy of the Bransfield Basin, Antarctica: Implications for tectonic history and hydrocarbon potential, this volume.

Johnson, G. L., J. R. Vanney and D. Hayes, 1982, The Antarctic continental shelf, *in* C. Craddock, ed.: Antarctic Geoscience, University of Wisconsin Press, p. 995-1002.

Keys, J. R., 1983, Iceberg quantities, shapes and sizes in Ross and D'Urville Seas: Antarctica Journal of United States, v. 18, p. 125-127.

Keys, J. R., 1984, Antarctic marine environments and offshore oil: Commission for the Environment, Wellington, New Zealand, p. 1-168.

Kimura, K., 1982, Geological and geophysical survey in the Bellingshausen Basin off Antarctica: Antarctic Record, v. 75, p. 12-24.

Kothe, J., F. Tessensohn, W. Thonhauser and R. Wendebourg, 1981, The expedition and its logistics: Geologisches Jahrbuch, Reihe B Heft 41, p. 3-30.

Kurtz, D. D. and J. B. Anderson, 1979, Recognition and sedimentologic description of recent debris flow deposits from the Ross and Weddell Seas, Antarctica: Journal Sedimentary Petrology, v. 94, p. 1159-1170.

Kurtz, D. D., J. B. Anderson, K. M. Balshaw and M. L. Cole, 1979, Glacial marine sedimentation: relationship to the distribution of sediment geotechnical properties on high latitude continental margins: Proceedings of the 11th Annual Offshore Technology Conference, Houston, Texas, p. 683-688.

Lemke, P., 1986, Stochastic description of atmosphere - sea ice - ocean interaction, *in* N. Untersteiner, ed., The Geophysics of Sea Ice, NATO ASI Series B: Physics V. 146, New York, Plenum Press, p. 785-823.

Lemke, P., E. W. Trinkl and K. Hasselman, 1980, Stochastic dynamic analysis of polar sea ice variability: Journal of Geophysical Research, v. 10, p. 2100-2120.

Lien, R., 1981, Sea bed features in the Blaaenga Area, Weddell Sea, Antarctica: Proceedings from the Sixth International Conference on Port and Ocean Engineering Under Arctic Conditions, Quebec, Canada, p. 706-716.

Morgan, V. I. and W. F. Budd, 1978, The distribution, movement and melt rate of antarctic icebergs, *in* A. A. Husseiny, ed., Iceberg Utilization: Proceedings First International Conference on Iceberg Utilization, New York, Pergamon Press, p. 220-228.

O'Lenic, E. A., 1978, U.S. Navy global ice analysis and forecasting, *in* A. A. Husseiny, ed., Iceberg Utilization: Proceedings First International Conference on Iceberg Utilization, New York, Pergamon Press, p. 123-130.

Orheim, O., 1980, Physical characteristics and life expectancy of tabular Antarctic icebergs: Annals of Glaciology, v.2, p. 123-128.

Ropelewski, C. F., 1983, Spatial and temporal variations in antarctic sea-ice (1973-82): Journal of Applied Meteorology, v.22, p. 470-473.

Schwerdtfeger, W., 1970, The climate of the Antarctic, *in* S. Orvig, ed., World Survey of Climatology: Amsterdam, Elsevier, p. 253-355.

Schwerdtfeger, W., 1984, Developments in Atmospheric Sciences 15, New York, Elsevier, 261 p.

Shepard, F. P., 1863, Submarine Geology, New York, Harper and Row, 2nd ed., p. 408.

Stagg, H. M. J., 1985, The structure and origin of Prydz Bay and Mac Robertson shelf, East Antarctica: Tectonophysics, v. 114, p. 315-340.

St. John, B., 1984, Antarctica-geology and hydrocarbon potential *in* M. T. Halbouty, ed., Future Petroleum Provinces of the World: AAPG Memoir 40, Tulsa, p. 55-100.

Suess, E., 1988, Hydrothermalism in the Bransfield Strait: American Geophysical Union EOS, Transactions, v. 68, p. 1769.

Swithinbank, C., P. McClain and P. Little, 1977, Drift tracks of antarctic icebergs: Polar Record, v. 18, p. 495-501.

Tauber, G.M., 1960, Characteristics of katabatic winds, *in* Antarctic Meteorology: Proceedings of Symposium, Australian Academy of Sciences, New York, Pergamon Press, p. 52-64.

Tchernia, P. and P .F. Jeannin, 1984, Circulation in Antarctic waters as revealed by iceberg tracks 1972-1983: Polar Record, v. 22(138), p. 263-269.

Wadhams, P., 1986, The seasonal ice zone, *in* N. Untersteiner, ed., The Geophysics of Sea Ice, NATO ASI Series B: Physics, v. 146, New York, Plenum Press, p. 825-991.

Wadhams, P., 1988, Winter observations of iceberg frequencies and sizes in the South Atlantic Ocean: Journal of Geophysical Research, v. 93, p. 3583-3590.

Walcott, R. I., 1970, Isostatic response to loading of the crust in Canada: Canadian Journal of Earth Science, v. 7, p. 716-727.

Wannesson, J., M. Pelras, B. Petitperrin, M. Perret and J. Segoufin, 1985, A geophysical transect of the Adelie margin, East Antarctica: Marine and Petroleum Geology, v. 2, p. 192-201.

Wong, H. K. and D. A. Christoffel, 1981, A reconnaissance seismic survey of McMurdo Sound and Terra Nova Bay, Ross Sea, *in* L. D. McGinnis, ed., Dirt Valley Drilling Project. Antarctic Research Series, v. 33, p. 37-62.

Wright, R., J. B. Anderson and P. P. Fisco, 1983, Distribution and association of sediment gravity flow deposits and glacial/glacial marine sediments around the continental margin of Antarctica, *in* B. F. Molnia, ed., Glacial-Marine Sedimentation: New York, Plenum Press, N.Y., p. 265-300.

Chapter 4

Geology and Hydrocarbon Potential of the Ross Sea, Antarctica

Alan K. Cooper
U.S. Geological Survey
Menlo Park, California, U.S.A.

Frederick J. Davey
Geophysics Division (D.S.I.R.)
Wellington, New Zealand

Karl Hinz
Bundesanstalt für Geowissenschaften und Rohstoffe
Hannover, Federal Republic of Germany

ABSTRACT

The Ross Sea contains three major depocenters, each underlain by a sediment-filled rift graben and an overlying glacial sedimentary sequence. The sedimentary sections are up to 14 km thick with up to 8 km in the rift grabens and up to 6 km in the presumed-glacial sequences. The rift grabens were downfaulted and filled probably during the late Mesozoic to early Cenozoic continental breakup of Gondwana; their early history may be analogous to coeval rift basins of southeast Australia, Tasmania, and New Zealand. The rift-graben sediments are unconformably overlain by glacial-marine sequences deposited since middle Eocene(?) to early Oligocene time. Renewed down-faulting has occurred along the west and east margins of the Ross Sea probably since Eocene time.

The hydrocarbon potential of the Ross Sea is poorly known because only post-Eocene glacial rocks have been sampled offshore. The age and type of rocks filling the rift grabens, below the glacial sequence, is unknown. Source beds do not occur in the glacial sequence, but may exist in the rift grabens. Structural and stratigraphic traps are likely near basement structures and unconformities, which are common, and near large sedimentary structures found only in the Victoria Land Basin and along margins of the Eastern Basin. Reservoir rocks are unknown but sands could occur throughout the glacial and rift sequences. Lopatin-Waples models indicate that hydrocarbons could be generated presently at depths of 2.5 to 4.0 km, if source beds exist. Migration is likely in dipping strata along rift-graben flanks and in late-rift fault zones of the Terror Rift.

Hydrocarbon seeps and accumulations are unknown in the Ross Sea. The preglacial strata that are deeply buried within the early-rift grabens have the best hydrocarbon potential; however, a definitive assessment awaits sampling of these deep rift deposits.

INTRODUCTION

The Ross Sea lies along the Pacific sector of Antarctica, and covers more than 750,000 km^2, an area larger than the state of Texas. The Ross Sea is the northern end of an 800-km-wide and 1500-km-long marine embayment, the Ross Embayment, which lies between the Transantarctic Mountains in the west, Marie Byrd Land in the east, and the Siple Coast in the south (Figure 1). The southern end of the embayment is covered by the vast (530,000 km^2) floating Ross Ice Shelf. Water depths in the Ross Sea range from 200 to 1100 m and, like most Antarctic shelves, they deepen landward toward the major ice sheets.

The Ross Sea is one of the best known areas of the Antarctic continental margin, principally because large areas of the sea are ice free during the austral summer. Offshore geologic and geophysical studies have been conducted since the early 1970s (Houtz and Davey, 1973; Hayes and Davey, 1975; see Cooper et al., 1987a for summary), and shallow research drilling has been done from ship (DSDP Leg 28, Hayes and Frakes, 1975), and, in the McMurdo Sound area, from nearshore sea ice (MSSTS, CIROS, Barrett, 1986; Barrett, 1989) and from onshore coastal areas (DVDP; McGinnis, 1981). Multichannel seismic reflection data have been collected in the Ross Sea since 1980 by West Germany (Hinz and Block, 1983; Hinz and Kristoffersen, 1987), Japan (Sato et al., 1984), United States (Kim et al., 1986; Cooper et al., 1987a), France (Wannesson, personal communication, 1987), Soviet Union (Zayatz et al., 1990), and Italy (Brancolini, 1990). However, only the United States multichannel data are publicly available (Cooper et al., 1986). The offshore studies show that the Ross Sea contains three major depocenters (Victoria Land Basin, Central Trough, and Eastern Basin, Figure 1). Each depocenter is underlain by a large basement graben that is filled with 3-8 km of rift-related sediments (e.g., early-rift graben). These isolated early-rift grabens are buried beneath a widespread glacial-sedimentary section 2-6 km thick. The widths of the large early-rift grabens define the widths of the Victoria Land Basin and the Central Trough, however, the large early-rift graben underlying the Eastern Basin covers only a small (15-20%) part of the basin. The remainder of the Eastern Basin principally comprises glacial sedimentary rocks.

The early-rift grabens are part of a complex rift zone that formed initially during the Mesozoic breakup of Gondwana. This continental rift zone once extended several thousand kilometers from Australia/New Zealand to and across Antarctica (through the Ross Embayment toward the Weddell Sea along the Transantarctic Mountains). The Ross Sea rift grabens may thus have their structural analogs in formerly adjacent areas of southern Australia and New Zealand, and possibly in the Weddell Sea.

Major Antarctic-block movements southeast of the Ross Embayment in late Mesozoic time have altered the size and orientation of this continental rift zone (Dalziel, 1982; Lawver et al., in press). Still, the zone of Cenozoic extension between East and West Antarctica, which includes the Ross Embayment, is comparable in size to the East African Rift and Great Basin of the western United States (Tessensohn, in press). In the Ross Embayment region, active rifting, as delineated by near-sea-floor faulting and volcanism, is presently identified only in the Victoria Land Basin (Terror Rift), in Marie Byrd Land, and possibly along the Central High.

The petroleum-geologic implications of regional studies in the Ross Sea are discussed by Cameron (1981), Behrendt (1983), Hinz and Block (1983), Cook and Davey (1984), Davey (1985), St. John (1986), Hinz and Kristoffersen (1987), Elliot (1988), Behrendt (1989), and Collen and Barrett (1990). In this chapter, we summarize previous studies and use multichannel seismic reflection surveys to examine the sedimentary depocenters and early-rift grabens of the Ross Sea. The discussion relates also to adjacent areas beneath the Ross Ice Shelf (south) and continental slope (north), where early-rift grabens are postulated, but have not yet been fully delineated by seismic reflection data (Cooper et al., in press).

The hydrocarbon potential of the Ross Sea may be moderately good, but our assessment is highly speculative, principally because few subsurface rock samples have been collected from the Ross Sea.

REGIONAL FRAMEWORK

Onshore Geology

The geology of the Ross Sea region (Figure 2; Table 1) is known from limited outcrops in the Transantarctic Mountains in the west and Marie Byrd Land in the east (see summaries by Grindley, 1981; Barrett, 1981; Webb, 1981; LeMasurier and Rex, 1982; Bradshaw and Laird, 1983; Davey, 1987). In the Transantarctic Mountains, a large suite of Precambrian and early Paleozoic granitic and metamorphic rocks comprise basement. These basement rocks are unconformably overlain by widespread exposures of nearly flat-lying mainly nonmarine sedimentary rocks of Devonian or older to Jurassic age (Beacon Supergroup), which are intruded by extensive Jurassic tholeiitic dolerite sills (Ferrar Group) and are covered, in places, by Jurassic flood basalts. The Jurassic (about 175 Ma) mafic and volcanic rocks are thought to be related to initial breakup of Gondwana (Elliot, 1985). Rocks younger than Jurassic are uncommon, and are mostly late Cenozoic volcanic and glacial-sedimentary rocks.

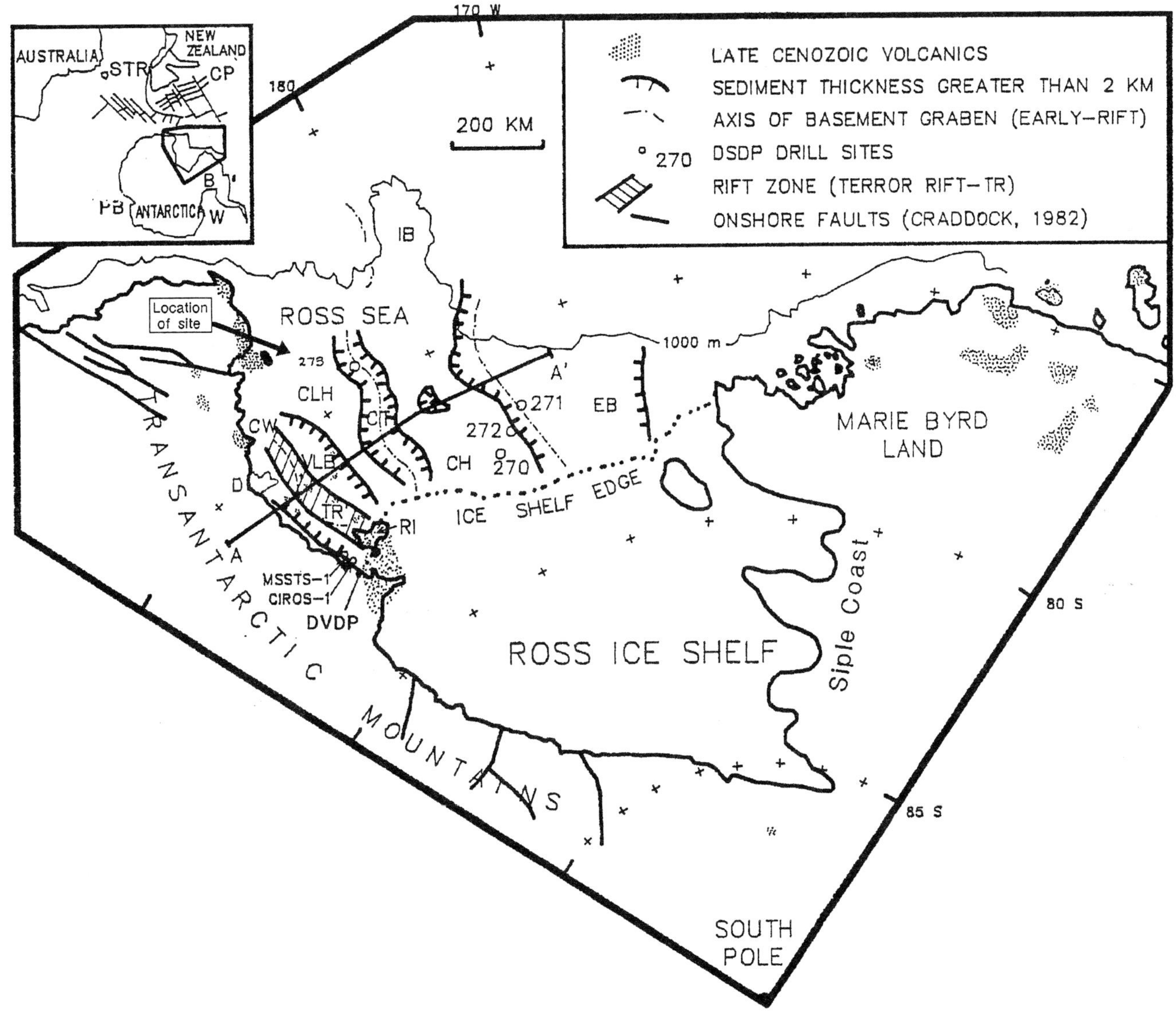

Figure 1—Index map of the Ross Sea showing major sedimentary depocenters, early-rift basement-grabens, and the active Terror Rift. Depocenter limits from Hinz and Block (1983) and Cooper et al. (1987a). The Ross Embayment includes the Ross Sea and Ross Ice Shelf. B-Byrd subglacial basin, CH-Central High, CLH-Coulman High, CP-Campbell Plateau, CT-Central Trough, CW-Cape Washington, EB-Eastern Basin, IB-Iselin Bank, M-McMurdo Sound, PB-Prydz Bay, TR-Terror Rift, RI-Ross Island, STR-South Tasman Rise, VLB-Victoria Land Basin, W-Weddell Sea. Polar stereographic projection.

In Marie Byrd Land, the geology is also dominated by the Precambrian and early Paleozoic igneous and metasedimentary basement rocks that have been intruded by late Paleozoic to early Mesozoic granites and by Middle to Late Cretaceous granites and mafic dikes (Davey, 1987). These pre-Cenozoic rocks have been regionally eroded and covered by hyaloclastic volcanic rocks of late Cenozoic age (younger than 28 Ma). The Cretaceous and late Cenozoic dikes and volcanic rocks are believed to be related to Mesozoic and younger rifting in the Ross Sea region.

Significantly, sedimentary rocks of Mesozoic to Paleogene age (about 160 Ma to 30 Ma) have not been found onshore adjacent to the Ross Embayment.

Offshore Geology

The offshore geology is known from drilling in the central (DSDP) and western (CIROS, MSSTS) Ross Sea (Figure 1) and is inferred from recycled and glacial-erratic materials found within the Ross Sea region (see summaries in Hayes and Frakes, 1975; Webb, 1981; Truswell, 1983; Barrett, 1986; Davey, 1987; Barrett et al., 1987; Barrett, 1989). Glacial-marine sedimentary rocks of early Oligocene and younger age comprise the upper part of the thick offshore sedimentary sections. Unconformities occur throughout the sedimentary section, with major events in the late Oligocene and late Miocene. Mesozoic and lower Paleogene rocks have not

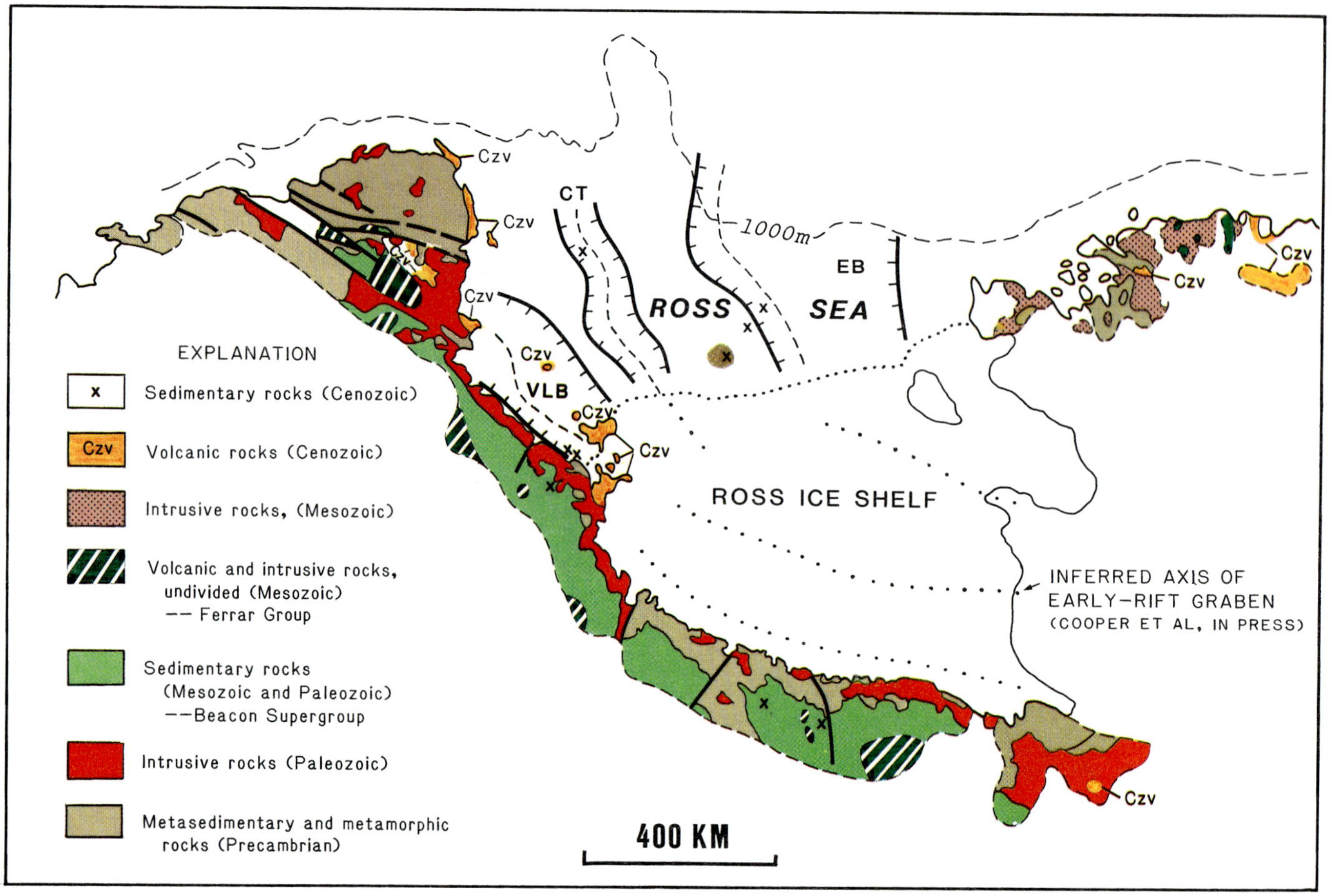

Figure 2—Generalized geologic map of the Ross Embayment (Ross Sea and Ross Ice Shelf) region modified from Davey (1987). See Figure 1 caption for explanation of offshore depocenters.

been sampled, but are suspected within the deeper sedimentary section based on recycled microfossils in DSDP cores, coastal glacial erratics, and sea-floor cores in the eastern Ross Sea. Basement rocks have been recovered only at DSDP Site 270 on a basement high, and are composed of early Paleozoic(?) gneiss.

Crustal Structure

The structure of the Ross Embayment is known from many seismic reflection/refraction surveys in the Ross Sea, from extensive gravity, magnetic, and water depth measurements throughout the Ross Embayment, and from limited seismic reflection data on the Ross Ice Shelf (see summaries by Hayes and Davey, 1975; Jankowski and Drewry, 1981; Davey et al., 1982; Bentley, 1983; Drewry, 1983; Jankowski et al., 1983; McGinnis et al., 1985; Cooper et al., 1987b; Davey and Cooper, 1987; Hinz and Kristoffersen, 1987; Cooper et al., in press).

The Ross Embayment is a broadly depressed area of thinned continental crust (19-27 km thick) that lies between the thick (40 km) cratonic rocks of the Transantarctic Mountains and the intermediate-thickness crust of Marie Byrd Land. In the Ross Sea, the large early-rift grabens are underlain by locally thinner crust (19-21 km, Figure 3). The sea floor within the Ross Embayment has been eroded glacially into a series of linear ridges and troughs that, in the Ross Sea, have a different orientation than the underlying early-rift grabens (Cooper et al., in press). The thinned crust and structural grabens are primary evidence for regional extension within the embayment.

The early-rift grabens of the Ross Sea are separated by broad, eroded basement ridges that have crustal thicknesses of 21-27 km. One ridge (Central High) continues north and west into the Southern Ocean as Iselin Bank. Prior to Gondwana breakup, Iselin Bank may have been adjacent to the South Tasman Rise and/or Campbell Plateau. Early-rift grabens of the Eastern Basin and Central Trough can be traced from near the continental shelf edge to the edge of the Ross Ice Shelf. The Victoria Land Basin, however, terminates at its northern end against the Transantarctic Mountains midway across the Ross Sea (Figure 1). All early-rift grabens may continue south beneath the Ross Ice Shelf (Cooper et al., in press).

The Ross Sea depocenters have up to 14 km of layered (sedimentary?) strata that are thickest in the Victoria Land Basin (Figures 3, 4). Faulting, folding, and tilting of the layered strata occurs principally in the Victoria Land Basin, although some strata along the eastern flank of the Central High and deep within the early-rift grabens are also deformed. In the Victoria Land Basin, the sedimentary section and underlying basement rocks are disrupted by subvolcanic intrusions and faulting along the basin axis, at the

Table 1. Selected References for Mesozoic and Cenozoic Sedimentary Rocks of the Ross Sea and Surrounding Areas[1].

	ROSS SEA REGION				NEW ZEALAND-AUSTRALIA REGION				
	TAM	MBL	ERS	WRS	WLK	CP	TR	GOBC	SO
Neogene	GVFP1,12	GV2	GM3	GMV1,4	GM	M5	M6	M7	M3
Oligocene	FP12	V2	GM3	GMV4	-	M5	M6	M7	M3
Eocene	FP12	-	FP8	F9	-	M5	NM6	NM7	M3
Paleocene	FP12	-	P8	F9	-	NM5,10	NM6	NM7	-
L.Cretaceous[2]	FP12	-	P8,9	F9	-	NM5,10	NM6	NM7	-
E.Cretaceous[3]	-	-	P8	-	N11	-	NM6	NM7	-
Jurassic[4]	V	-	P8	-	-	-	NV6	NV7	-
Triassic	N1	-	-	-	N13	-	N6	N7	-

1 Letters refer to rock types and numbers to reference citations
2 First sea-floor spreading between ANT-NZ and ANT-AUS and NZ-AUS
3 Major downfaulting of basins of SE. AUS, NZ, TAS
4 Initial breakup of Gondwana

ANT - Antarctica
AUS - Australia
CP - Campbell Plateau
ERS - Eastern Ross Sea
GOBC- Gippsland/Bass/Otway/Ceduna Basins
MBL - Marie Byrd Land
NZ - New Zealand
SE. AUS - Offshore SE Australia
SO - Southern Ocean (near Ross Sea)
TAM - Transantarctic Mountains
TAS - Tasmania
TR - Tasman Rise
WLK - Wilkes Basin
WRS - Western Ross Sea

In-Situ Rocks
N - nonmarine
M - marine
G - glacial
V - volcanic

Recycled or Allochthonous Rocks
P - nonmarine (from pollen/spore)
F - marine (from micro-fossils)

1 - Davey, 1987
2 - LeMasurier and Rex, 1982
3 - Hayes and Frakes, 1975
4 - Barrett, 1986, 1989
5 - a) Anderton et al., 1982
b) Shirley, 1983
c) Sandford, 1980
6 - Hinz et al., 1986
7 - a) Williamson et al., 1987
b) Etheridge et al., 1985
c) Fraser and Tillbury, 1979
8 - Truswell, 1983
9 - Webb, 1981
10 - Kennett and Houtz, 1975
11 - Domack et al., 1980
12 - Webb et al., 1984
13 - Davey, 1985

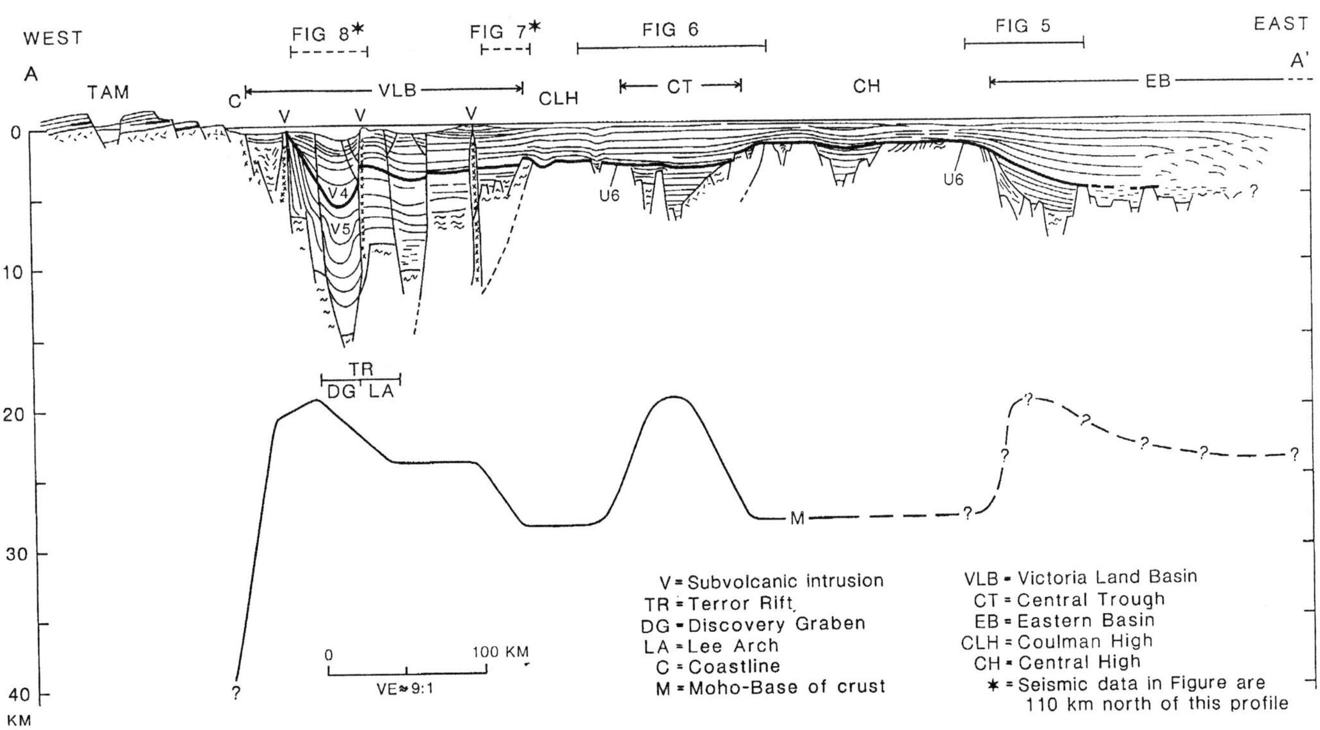

Figure 3—Generalized profile across the Ross Sea based, in part, on offshore multichannel seismic reflection data (Hinz and Block, 1983; Cooper et al., 1987a) and gravity-model studies (Davey and Cooper, 1987). Early-rift grabens lie beneath a buried regional unconformity (U6, heavy line), and are delineated by thin crust. Late-rift faults and intrusive structures deform the Victoria Land Basin and some small basement grabens. Figure modified from Cooper et al. (in press). See Figure 1 for location.

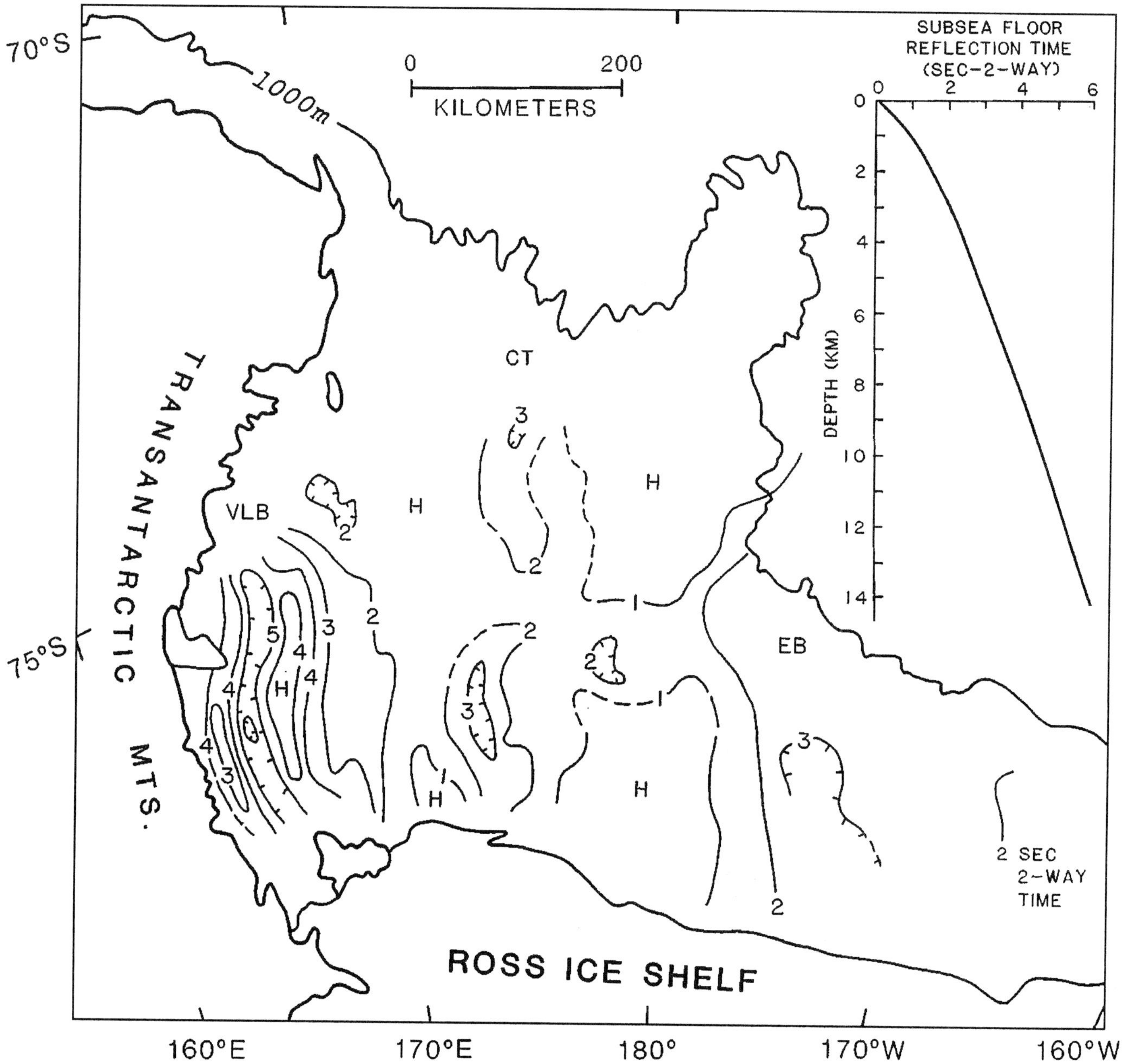

Figure 4—Map of reflection time (2-way, in seconds) between sea floor and acoustic basement for the Eastern Basin (EB), Central Trough (CT), and Victoria Land Basin (VLB) based on Hinz and Block (1983) and Cooper et al. (1987a). The reflection-time versus depth curve, derived for the Victoria Land Basin, is from Cooper et al. (1987a).

Terror Rift, and along the basin edges. Major Neogene volcanic centers lie at the northern (Cape Washington) and southern (Ross Island) ends of the Terror Rift.

Seismic Stratigraphy

Detailed acoustic stratigraphy of the Ross Sea, based on multichannel seismic reflection data, has been described by Hinz and Block (1983), Sato et al. (1984), Cooper et al. (1987a), Cooper et al. (in press). These authors identify up to seven acoustics units and unconformities within the sedimentary section, which is thought to be of late Mesozoic and younger age. Acoustic basement may be composed of sedimentary and/or igneous units of Precambrian to late Mesozoic age.

The uppermost part of the sedimentary section is cut regionally by the Ross Sea unconformity, which lies 2-42 m below the sea floor (Karl et al., 1987) and spans the interval 14.7 to 4.0 Ma at DSDP Site 273 (Savage and Ciesielski, 1983). Another major unconformity, which cuts the sedimentary sections and tops of basement ridges throughout the Ross Sea, lies at 3-6 km depth (U6, Figure 3). Hinz and Block (1983) name this unconformity U6 in the Eastern Basin and Central Trough, and Cooper et al. (1987a) refer to it as the unconformity between acoustic units V4 and V5 in the

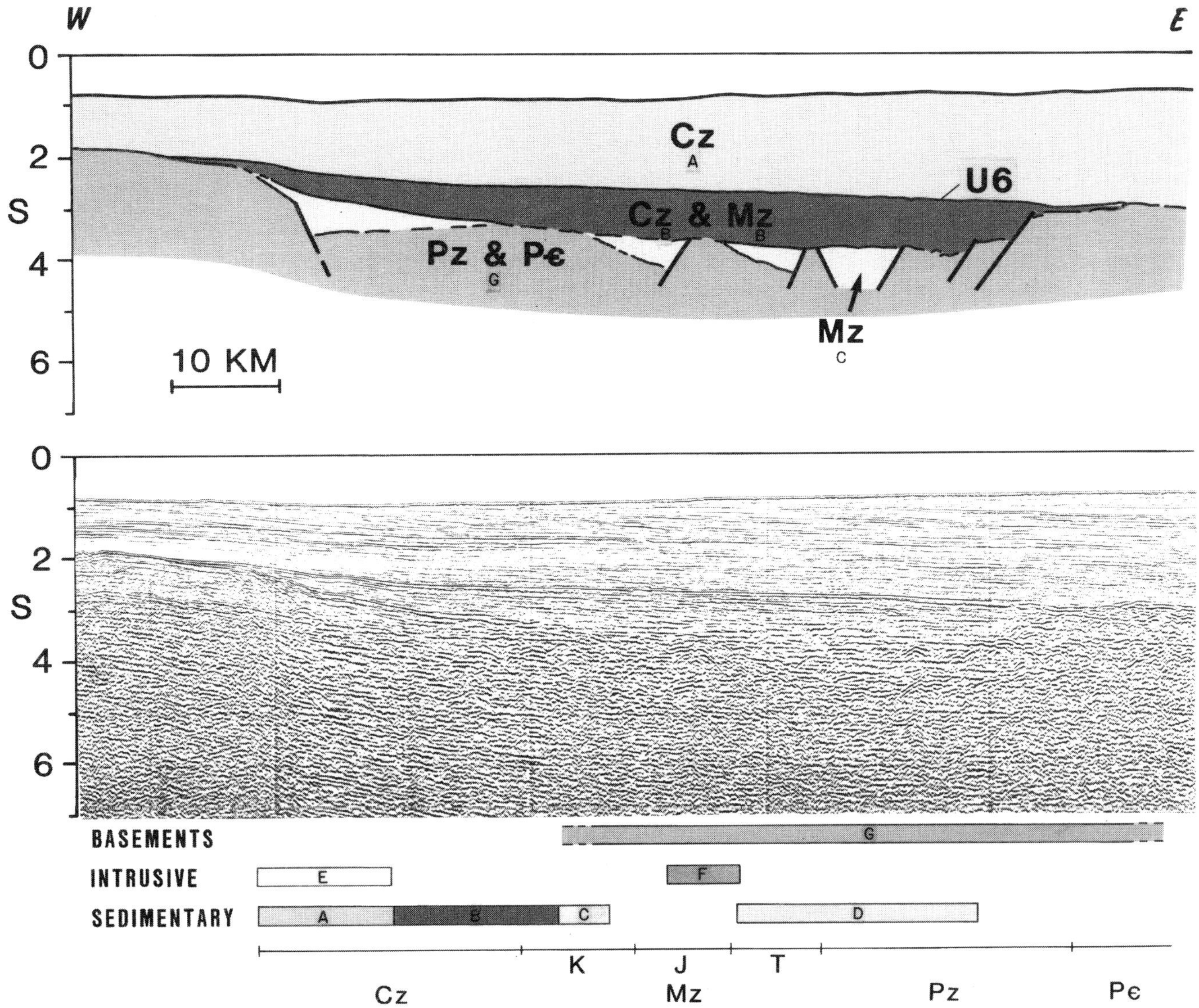

Figure 5—Multichannel seismic reflection data, with generalized interpretation, across the sediment-filled early-rift graben (e.g., below U6) and overlying prograding glacial rocks lying beneath the western edge of the the Eastern Basin. Time scale and interpreted ages are based on shallow drilling, seismic stratigraphy, and regional geology discussed in text. "Basements" refers to acoustic and igneous-metamorphic basements likely throughout the Ross Sea (Cooper et al., 1987a). See Figure 3 for location.

Victoria Land Basin (Figure 3). Sedimentary units below unconformity U6 are isolated within early-rift grabens, whereas units above U6 cover the entire region.

The age of U6 is uncertain. Rocks have not been sampled from below the unconformity, with the exception of early Paleozoic(?) basement rock at DSDP Site 270. Glacial sedimentary units only have been sampled above U6, and on basement highs where a minimum age for the unconformity would be expected. U6 is late Oligocene in age at DSDP Site 270 on the Central High (Hinz and Block, 1983), but rocks from unit V3, above unconformity U6, are of early Oligocene age at CIROS-1 on the uplifted flank of the Victoria Land Basin (Cooper et al., in press). Unconformity U6 may be time-transgressive and at least of early-to-late Oligocene (and possibly older) age.

In the Eastern Basin (Figure 5), the total sedimentary section is up to 5–6 km thick over the early-rift graben. Most (80-85%) of the Eastern Basin, however, comprises a prograding sequence of glacial-marine rocks, above U6, that dips and thickens seaward (Figure 3). The continental shelf edge is believed to have prograded 85 km seaward during the past 15 m.y. (Hinz and Block, 1983). Strata below U6 dip gently to the east and are mostly undeformed, except near the base of the sedimentary section and within small, deep half-grabens.

Seismic data from the Central Trough (Figure 6) show that the maximum sedimentary thickness is similar to that in the Eastern Basin (Figure 5), and that strata are nearly horizontal. Basement rocks beneath the center and flanks of the trough are highly eroded. Deformation of the sedimentary section is generally minor throughout the basin, and reflectors are disrupted only locally within intra-basement grabens (e.g., small grabens within acoustic basement) and over basement structures. Strata above and

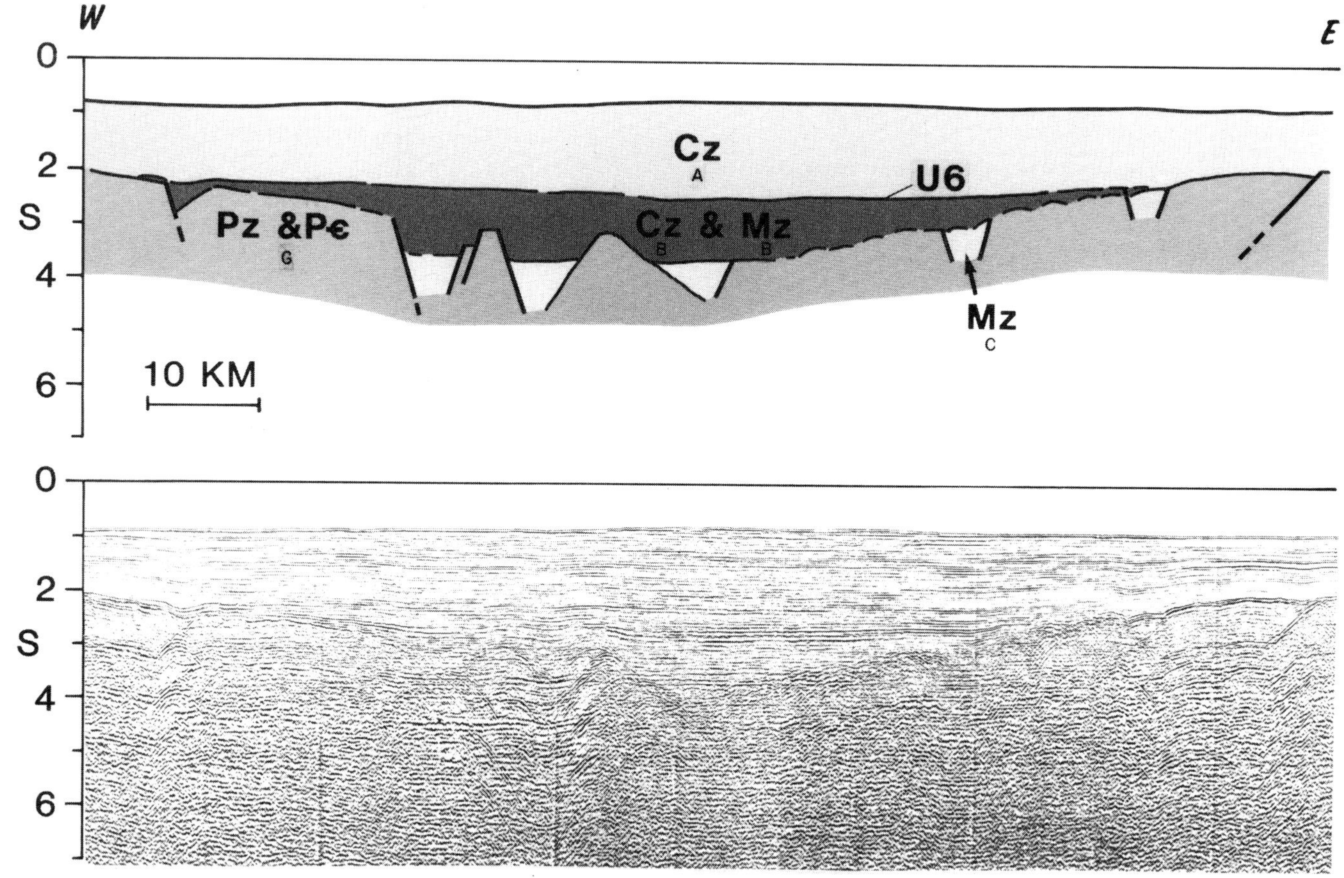

Figure 6—Multichannel seismic reflection data, with generalized interpretation, across the sediment-filled early-rift graben (e.g., below U6) and overlying part-glacial strata comprising the Central Trough. Letters beneath ages refer to inferred-time scales in Figure 5. See Figure 3 for location.

below unconformity U6 are mostly disconformable on the basin flanks, but become conformable near the basin center.

The Victoria Land Basin is characterized by a thick (up to 14 km) layered, presumed sedimentary section, by extensive basement faulting (early-rift; Figure 7), and by deformation of the sedimentary section in the Terror Rift (late-rift; Figure 8). Large-offset (1-2 km) faults along the east flank of the early-rift graben do not disrupt the nearly flat-lying sedimentary strata that fill the graben (unit V5, Figure 7) indicating that downfaulting of the basement occurred mostly prior to, rather than during, deposition of the overlying sedimentary section. Strata above and below U6 are disrupted in the Terror Rift by large normal faults that extend from basement to the sea floor (Figure 8). These faults and nearby subvolcanic intrusions are thought to be late Paleogene and Neogene (e.g., late-rift) features that developed after deposition of most of the sedimentary section. Faulting along the west side of the Victoria Land Basin has occurred, in part, during the Eocene and younger uplift of the Transantarctic Mountains, and has resulted in at least three major angular unconformities (Cooper et al., 1987a).

Tectonic History

The tectonic history of the sedimentary depocenters of the Ross Embayment region has been dominated by the Mesozoic and younger events associated with the breakup of Gondwana. Prior to breakup, the geologic record from the nearly undeformed rocks of the Beacon Supergroup of the Transantarctic Mountains indicates deposition of alluvial plain-shallow marine deposits over a broadly stable area including part of the Ross Sea (Barrett, 1981). Intrusion and extrusion of tholeiitic mafic rocks throughout the Transantarctic Mountains and contiguous Gondwana areas in Early Jurassic time (175 Ma) signaled the initial phases of continental rifting (Elliot, 1985). The continental rifting of Gondwana has resulted in several hundred kilometers of extension in the Ross Embayment (100% extension based on plate reconstructions, Lawver et al., 1987; and in press).

Two phases of rifting have been proposed for the Ross Embayment region based on regional geology, geologic histories of formerly nearby rift-basin areas of Australia/New Zealand, and relative ages derived from offshore seismic stratigraphy (Cooper et al., in press; Table 1): a) an early-rift phase from Late Jurassic or Early Cretaceous to Late Cretaceous time, and b) a late-rift phase from the Eocene to the present. The two phases are distinguished in the Ross Sea by regional differences in the vertical extent of deformation in the basement and overlying sedimentary rocks. Figure 9 illustrates a sequence of plate motions and conceptual models for evolution of Ross Sea depocenters based on the two-phase rift concept.

Crustal extension during the early-rift period is thought to have caused initial uplift, erosion, and deformation of Beacon Supergroup strata, and intrusion and extrusion of

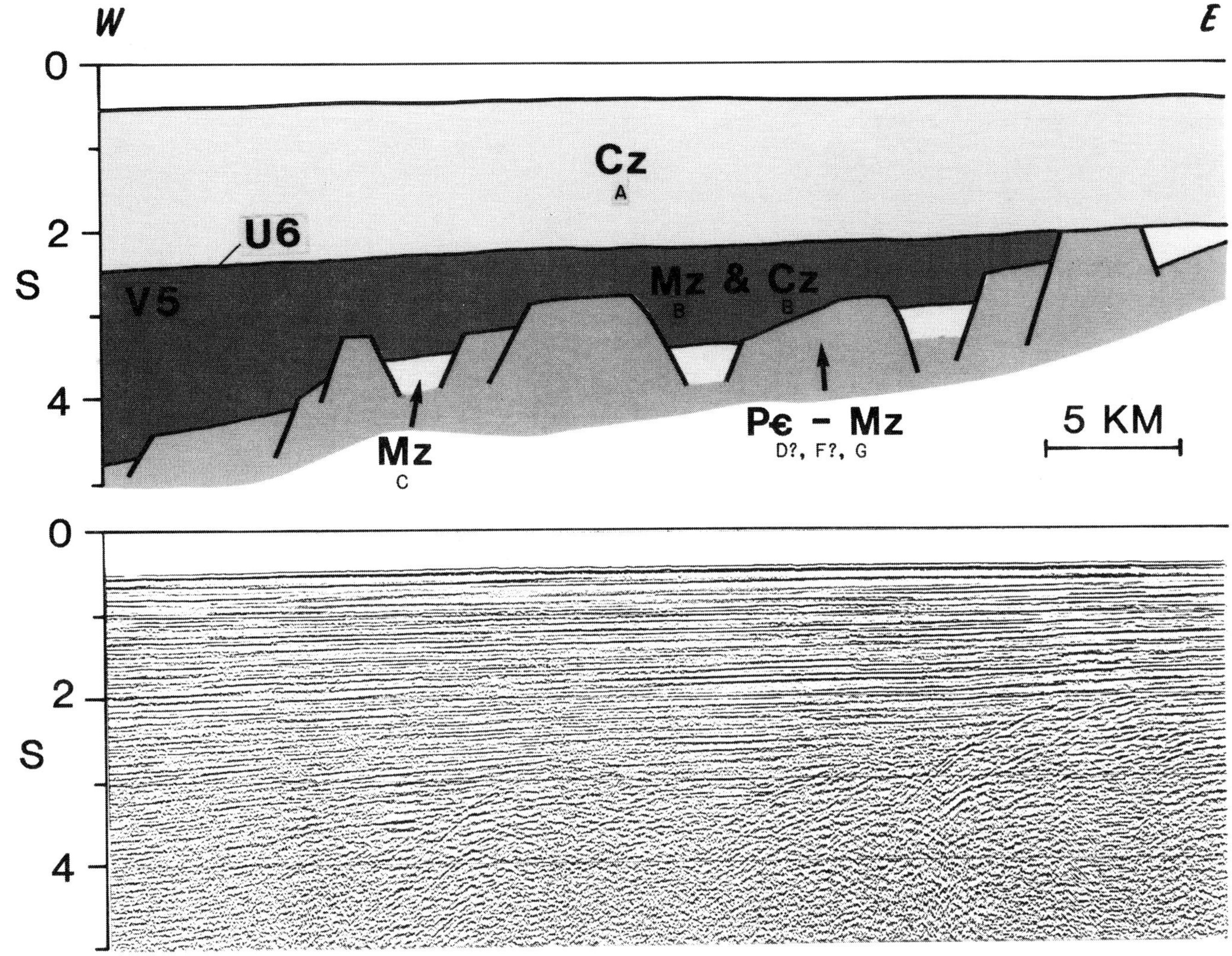

Figure 7—Multichannel seismic reflection data (Line 409A), with generalized interpretation. The profile crosses the eastern flank of the early-rift graben (e.g., below U6) and overlying part-glacial strata of the Victoria Land Basin (modified from Cooper et al., 1987a). Letters beneath ages refer to inferred-time scales in Figure 5. Line 409A lies 110 km north of generalized profile of Figure 3.

Ferrar dolerite sills and Kirkpatrick Basalt. This was followed by rapid downfaulting of the major early-rift grabens (Figure 9A). The grabens were infilled probably with nonmarine and marine sedimentary strata, although layered volcanic rocks are also possible deep within the grabens. The infilling occurred principally during Late Cretaceous and Paleogene time (Figure 9B). The late-rift period is marked by uplift of the Transantarctic Mountains, development of the Terror Rift, and volcanism in Victoria Land and Marie Byrd Land (Figure 9C).

Rifting episodes in the Ross Embayment, like those in formerly adjacent areas of Australia-Tasmania-New Zealand (Shirley, 1983; Williamson et al., 1987; Hinz et al., 1986), are a likely response to regional plate motions during early Gondwana breakup. The early-rift period corresponds with times of proposed dextral strike-slip motion parallel with the Ross Embayment (Early Jurassic, Schmidt and Rowley, 1986) and prebreakup extension between Antarctica-Australia-Tasmania (160 to 96 Ma, Veevers and Eittreim, 1988). Early rifting also includes the period of earliest seafloor spreading between Australia and Antarctica (96-80 Ma, Cande and Mutter, 1982). A major reorganization of plate motions in Eocene time is correlated with the beginning of the late-rift period (Cooper et al., in press). The reorganization is marked by a 40° change in magnetic anomaly trend south of New Zealand (50-42 Ma; Stock and Molnar, 1987) and by increased spreading rates in the Indian Ocean (43 Ma; Cande and Mutter, 1982).

The amount and orientation of regional extension in the Ross Embayment is uncertain, but plate reconstructions (Grindley and Davey, 1982; Lawver and Scotese, 1987) and regional geophysical-data interpretations (Davey et al., 1982; Cooper et al., in press) suggest that large horizontal and vertical displacements have occurred along major fault zones transverse to the Ross Embayment during the early-rift and possibly late-rift periods (Figure 9). These transverse faults have probably controlled early-rift graben development, however, the exact mechanism (e.g., failed-rift spreading [Davey et al., 1982] or simple-shear detachment faulting [Fitzgerald et al., 1987]) cannot yet be resolved.

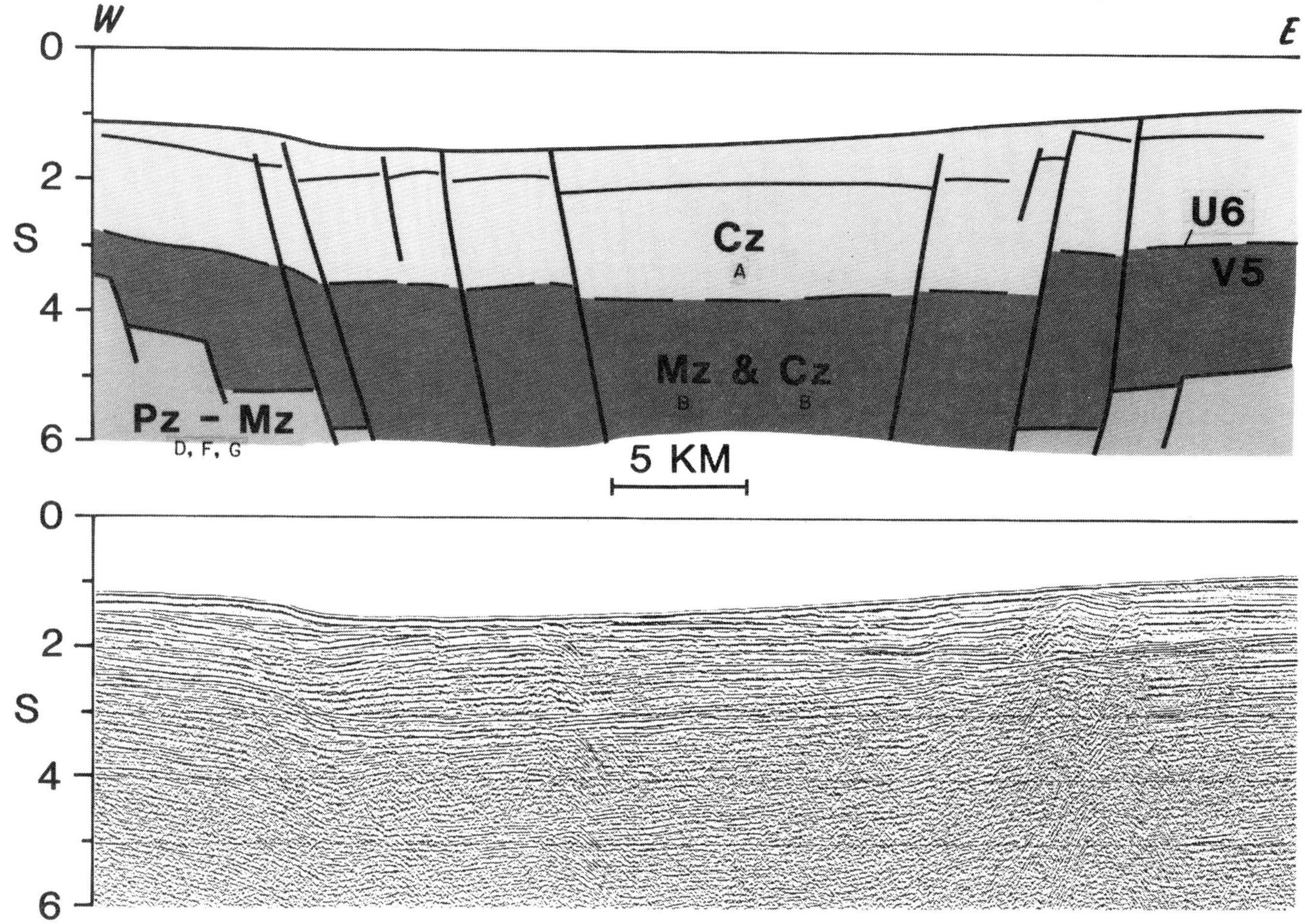

Figure 8—Multichannel seismic reflection data (Line 409B), with generalized interpretation. The profile crosses the Terror Rift within the center of the Victoria Land Basin. Here, renewed Eocene(?) and younger faulting has deformed the old sediment-filled early-rift graben (e.g., below U6). Letters beneath ages refer to inferred-age scales in Figure 5. Line 409B lies 110 km north of generalized profile of Figure 3.

Significantly, the large-displacement basement faults, along which the early-rift grabens subsided, mostly do not displace the overlying Cenozoic sedimentary sections except within the Terror Rift (Victoria Land Basin). Cenozoic (post-Eocene?) extension thus appears to have resulted in intense local deformation (and volcanism) along the edges of the Ross Embayment, in some deformation along the Central High, and in regional subsidence of all depocenters.

PETROLEUM GEOLOGY

Overview

The Ross Sea, like other segments of the Antarctic continental margin and once-contiguous margins of Gondwana, is underlain by large rift basins filled with thick sedimentary deposits that could, by analogy, have good potential for hydrocarbon generation. Research drilling and geophysical records from Antarctic margins have shown, however, that glacial sedimentary rocks comprise a significant percentage of the strata that cover the margins of the Ross Sea (Hayes and Frakes, 1975; Barrett et al., 1987, in press), Weddell Sea (Party, 1987; Barker et al., 1988), Prydz Bay (Party, 1988; Barron et al., 1989), and elsewhere. These ubiquitous glacial strata have no source-rock potential (see below), and if they persist to great depths, then hydrocarbon-generation prospects will be poor.

The hydrocarbon assessment of the Ross Sea and other Antarctic areas therefore relies heavily on the nature of preglacial sedimentary sections that are deeply buried within the rift grabens. In the Weddell Sea, good source rocks of Late Jurassic to Early Cretaceous age have been recovered (Macdonald et al., 1988; Barker et al., 1988; Kvenvolden et al., in press) from these sections. However, in the Ross Sea, the preglacial sections have not yet been sampled, in situ. Thus, all present judgments of hydrocarbon potential for the Ross Sea must be considered highly speculative.

In the following sections, we rely on geophysical data, geologic data, and two-stage rift models from the Ross Sea as well as geologic data from other Gondwana rift basins to discuss likely reservoirs, traps (structural and stratigraphic), temperature history, migration paths, and Lopatin-Waples models for hydrocarbon generation in Ross Sea depocenters. In general, conditions favorable for hydrocarbon generation and entrapment are likely throughout the Ross Sea, and especially in the Victoria Land Basin, if adequate source beds exist.

Hydrocarbon Occurrences

Hydrocarbons have been encountered in the Ross Sea region only as gases at most drill and sea-floor core sites and as an asphaltic residue in a 2-m-thick layer at the CIROS-1 drill core (Figure 1). No gas or oil seeps have been reported.

The potential economic significance of Ross Sea gases is unclear. Methane is the primary gas reported at all drill sites. At DSDP Sites 271, 272, and 273 (Figure 1) methane concentrations vary from a few parts per million (ppm) to nearly 179,000 ppm (McIver, 1975). Ethane and higher molecular-weight gases (up to 876 ppm) were also recovered at these sites. The origin of the ethane and higher-order gases has been debated. McIver (1975) notes that ethane could be attributed to thermal diagenesis and possible migration. A more recent study of DSDP cores by Claypool and Kvenvolden (1983), however, suggests that the ethane could be generated also by microbiological processes. They report high isotopic delta-13 carbon values (-78.9 to -67.5) for the methane gas, indicating to them that the Ross Sea gases are of biologic origin. An alternative explanation for the large gas concentrations is that the gases have been concentrated at shallow depths as gas hydrates (McIver, 1975; Rapp et al., 1987; Kvenvolden and Cooper, 1987). This alternative, although supported by the observed decrease in pore-water salinity with increasing depth at Site 272, is uncertain because a seismic, bottom-simulating reflector (BSR), a common indicator of gas hydrates, has not been reported from the Ross Sea (Kvenvolden and Cooper, 1987).

The asphaltic residue recovered at a subsea-floor depth of 632-634 m in the CIROS-1 well is the first direct evidence that liquid hydrocarbons have been generated in the Ross Sea (Barrett, 1989). The residue contains low (0.3%) organic carbon, is derived principally from a marine source, is thermally immature, has a complex history, and is believed to indicate the former presence of petroleum that is now dissipated (Cook and Woolhouse, 1989). The present structural setting, on the uplifted western flank of the Victoria Land Basin, and the degree of thermal maturation suggests that at least 1 kilometer of uplift and exhumation has occurred (Collen and Barrett, 1990), allowing the parent hydrocarbon to escape. The origin of the parent hydrocarbon is unknown, but it may have migrated upward from within the basin (Cook and Woolhouse, 1989).

Source Beds

Source beds do not occur in the glacial sequences sampled by Ross Sea drilling, but we consider that they may exist within preglacial sedimentary sequences. This suggestion is based on drilling data from Prydz Bay and other Gondwana rift basins, and from recycled material in the Ross Sea. Both marine and nonmarine source beds are likely in the preglacial sequences, with nonmarine more prevalent in older rift deposits (Table 1).

Early Oligocene and younger glacial rocks from the Ross Sea basins have total organic carbon (TOC) values that range from 0.05 to 0.76% and average about 0.37% (Figure 10). Sea-floor sediment samples have high, lipid-rich diatom concentrations, but low TOC values indicating that organic carbon is quickly degraded under open-ocean oxic conditions. Vitrinite reflectance values at the MSSTS-1 (R_o = 0.54; Collen and Froggatt, 1986) and CIROS-1 (R_o = 0.36; Lowery, 1989) sites are low, indicating that these rocks are immature for hydrocarbon generation (i.e., below R_o = 0.6). High vitrinite reflectance values (R_o = 1.48, 2.53, 2.68) are, however, also measured at CIROS-1, on samples believed to be recycled from Permian coals in the nearby Transantarctic Mountains (Lowery, 1989).

Adequate source beds may exist in the preglacial sedimentary sections which are likely within the deeper parts of the Ross Sea depocenters. Eocene and Upper Cretaceous coal measures are reported from areas of Tasmania and New Zealand (Campbell Plateau) that were formerly contiguous to the Ross Sea (Table 1). Also, Late Cretaceous and Eocene recycled microfossils and palynomorphs, from marine and nonmarine sources, have been sampled in the Ross Sea suggesting that source rocks of these ages may lie beneath the Ross Sea. Early Cretaceous nonmarine rocks with carbonaceous and coaly material were drilled beneath the Antarctic shelf in Prydz Bay, over 1000 km west of the Ross Sea (Figure 1) (Barron et al., 1989). Good Late Jurassic to Early Cretaceous organic-rich black shale rocks have been found in the Weddell Sea (Macdonald et al., 1988; Barker et al., 1988; Kvenvolden et al.,in press), and similar rocks are possible in the Ross Sea. Permian and Triassic coals (Beacon Supergroup), generally medium-low-volatile to anthracite, occur in the Transantarctic Mountains, and may also underlie parts of the Ross Sea (Rose and McElroy, 1987; Cooper et al., 1987a). Widespread Jurassic intrusions onshore are responsible for the high rank of the coals (Rose and McElroy, 1987), and if similar magmatic processes occurred offshore, the coals would not be likely source rocks for oil or gas.

The subsidence histories for each of the major early-rift grabens and the proximity of these grabens to marine seaways are likely to have determined whether marine or nonmarine rocks are dominant in the preglacial sedimentary sections.

Eastern Basin

The deep, early-rift grabens of the Eastern Basin open into the Southern Ocean (Hinz and Block, 1983) (Figure 1). This opening suggests that Mesozoic and Paleogene rocks, where present, may grade from nonmarine to shallow marine during basin infilling coincident with sea-floor spreading between Antarctica and the Campbell Plateau. Upper Cretaceous nonmarine and marine rocks occur on the southern edge of the formerly adjacent Campbell plateau (Table 1).

Central Trough

The axial part of the Central Trough is a narrow segmented early-rift structure that may have been isolated from marine incursions, except at the trough's northern end. Nonmarine rocks are likely during the Mesozoic to early Paleogene, with marine influence becoming more prevalent following regional subsidence. Basins along the west side of Tasmania, which may have been part of the same rift-basin system, contain shallow marine as well as non-marine coal-measure sequences during this time (Hinz et al., 1986) (Table 1).

Victoria Land Basin

The Victoria Land Basin is presently structurally isolated from the Southern Ocean by surrounding basement highs. However, prior to the Eocene uplift of the Transantarctic Mountains, marine seaways may have existed between the Victoria Land and Wilkes Basins (Cooper et al., 1987a). The recycled marine microfossils of Late Cretaceous and younger age from the western Victoria Land Basin and from sediments derived from the Wilkes Basin (Sirius Formation; Webb et al., 1984) (Table 1) suggest that marine sediments, interbedded with nonmarine sediments, may prevail in the lower part of the Victoria Land Basin. These strata could contain source beds, compositionally equivalent to those described by Fraser and Tillbury (1979) within coeval marine and nonmarine sequences in the Great Australian Bight Basin.

Reservoirs

The existence of reservoirs in the glacial and preglacial sedimentary sequences of the Ross Sea is possible, based on limited porosity measurements in glacial-marine strata and on analogy with preglacial rocks from other rift basins.

Porosity values of 1 to 46% have been measured in the Oligocene and younger glacial-marine rocks from CIROS-1 and MSSTS-1 sites, but direct porosity measurements are not reported at DSDP sites. Permeability values, however, are unknown. At MSSTS-1, large porosity variations are observed and attributed to changes in lithology and degree of calcite cementation. Upper Oligocene sandy mudstones and muddy sandstones at MSSTS-1 yield calculated porosities of 17 to 36% (Collen and Froggatt, 1986). Slightly higher porosities (up to 46%) were found in Miocene and Pliocene muddy fine sandstones. Downhole logs at CIROS-1, about 4 km from MSSTS-1, show in-situ porosities of 1 to 21% (average 13%) in early Oligocene and younger rocks (White, 1989).

The likely Mesozoic to Oligocene preglacial rocks in the deep, and unsampled, parts of the Ross Sea depocenters may have reservoir rocks in marine and nonmarine sandstones. This assumption is based on analogy with rocks drilled from coeval basins on the Campbell Plateau (Great South basin; Shirley, 1983) and western Tasmania (Hinz et al., 1986). Older Mesozoic and Paleozoic sedimentary sections may also underlie parts of the Ross Sea and contain nonmarine reservoir rocks. The quality of the reservoirs in the rift grabens could, however, be poor due to deep burial, cementation, and thermally controlled diagenesis.

Traps

Several types of structural and stratigraphic traps, whose origins are probably related to episodes of extensional deformation and glaciation, are possible in the Ross Sea. Although the relative times of trap formation can be derived from seismic reflection data, the correlation between subbasins is uncertain. Only a few acoustic units (middle Tertiary?) can be traced over basement highs and throughout the Ross Sea. The deep sedimentary units are confined within the early-rift grabens (Hinz and Block, 1983; Cooper et al., 1987a) (Figure 3).

Structural traps are most likely at great depths near the large-offset basement faults forming horsts and grabens, and at all depths within the Terror Rift, where folded and faulted strata occur. Basement faults have up to 2 km of relief (Figure 3) and most do not cut the overlying sedimentary section, with exception of the Terror Rift. These faults form the backstop against which flat-lying to gently dipping strata are deposited. The majority of early-rift downfaulting apparently occurred prior to basin infilling (Cooper et al., in press). Later sediment draping of existing basement horsts may therefore have formed deep-seated traps. Small half-grabens (1-3 km displacement) are occasionally seen on basement highs and beneath the regional unconformity (U6, Figure 3) cutting acoustic basement. Folded strata in these grabens and along the margins of the Eastern Basin may provide local traps.

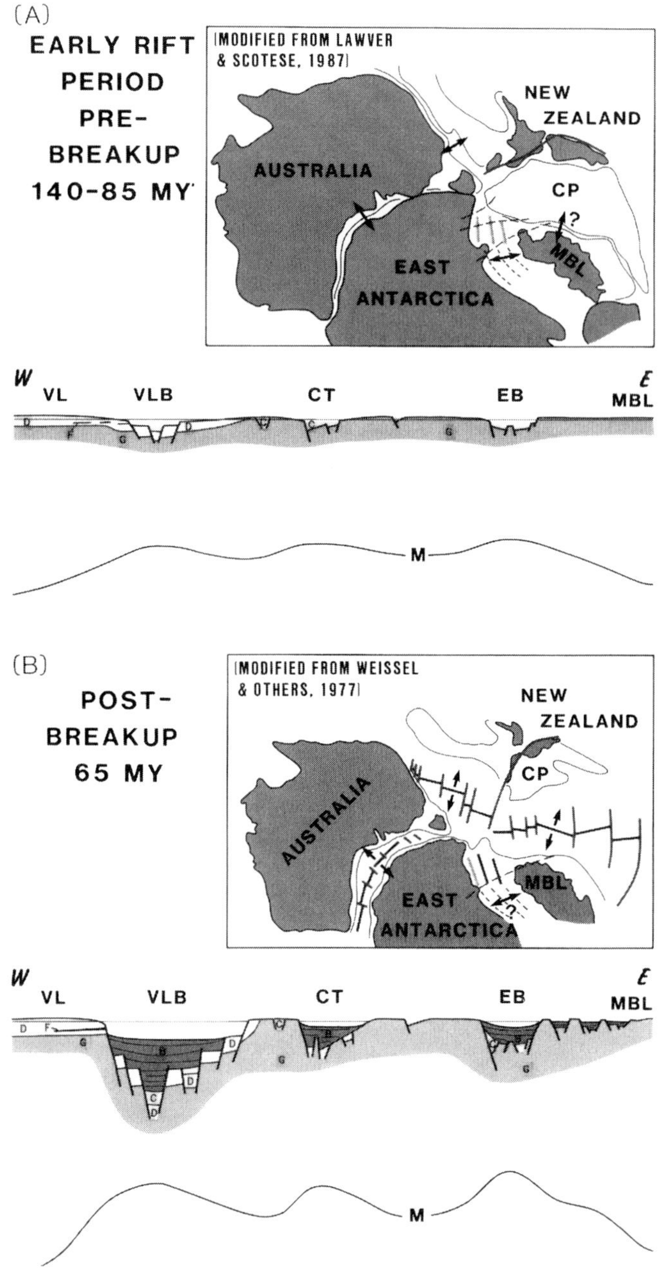

Figure 9—Conceptual models for the evolution of the Ross Sea depocenters based on seismic stratigraphy, regional Gondwana geology, and plate models: a) early-rift period—extension of continental crust and rapid downfaulting of early-rift grabens (rift stage); b) post-breakup—sea-floor spreading and major infilling of early-rift grabens (drift stage).

Within the Victoria Land Basin, traps may be associated with Eocene(?) and younger fault and subvolcanic intrusive structures of the Terror Rift that deform older (and coeval) basin strata. Deformation includes normal and listric faults, broad rotated crustal blocks with folded strata, basement uplifts, and local crustal-flexure. Traps are also likely along the west side of the Victoria Land Basin, where large basement structures and uplifted strata occur at the front of the Transantarctic Mountains. As an extreme example, base-

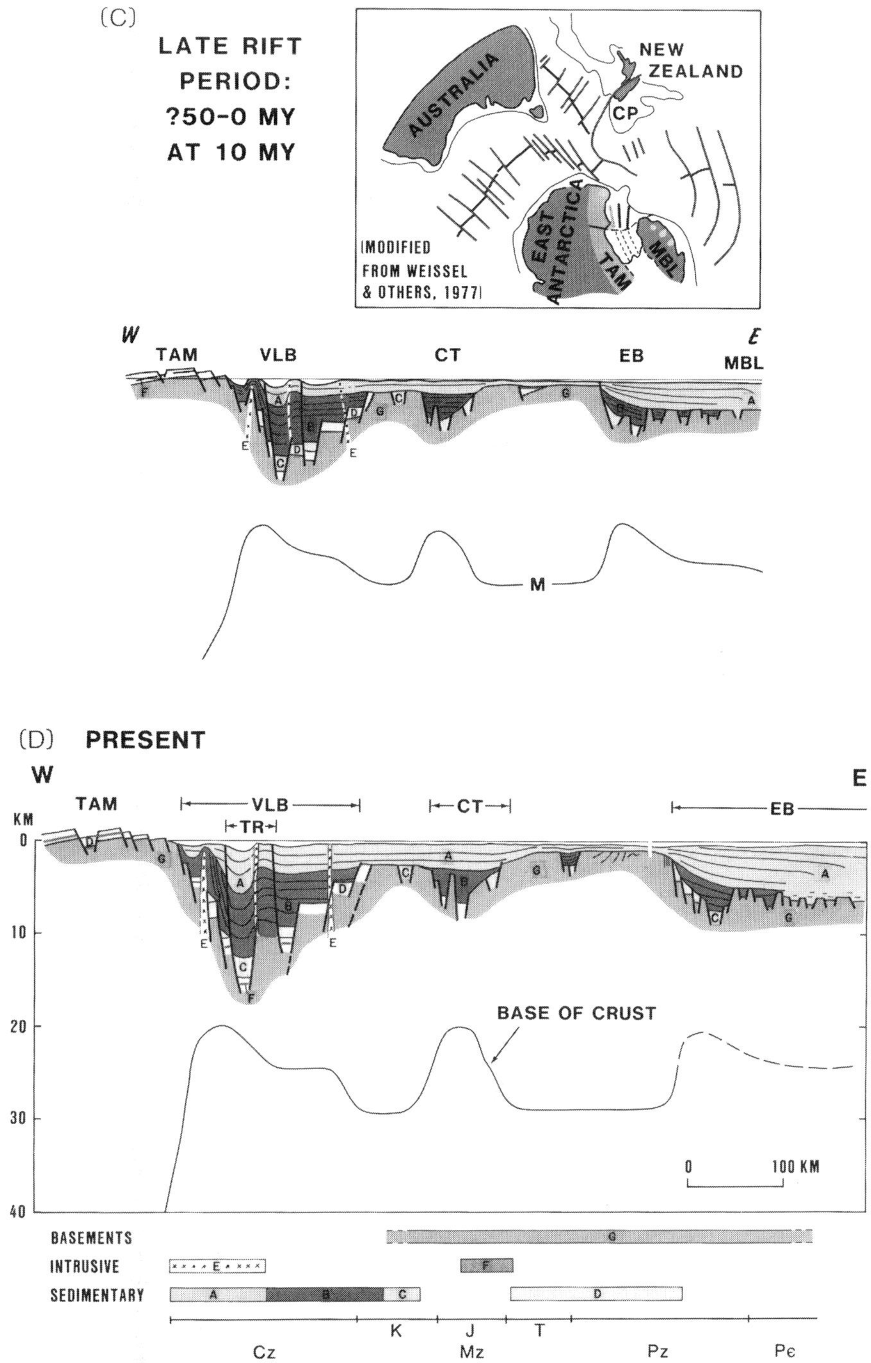

Figure 9. (Continued.) c) late-rift period—major Eocene reorganization of plate motions was followed, in the Ross Sea region, by uplift of TAM, cutting of regional unconformity (U6), development of TR, and volcanism in MBL; d) present. Abbreviations are: CP-Campbell Plateau, CT-Central Trough, EB-Eastern Basin, MBL-Marie Byrd Land, TAM-Transantarctic Mountains, TR-Terror Rift, VL-Victoria Land, VLB-Victoria Land Basin. Letters refer to inferred-time scale, which is based on shallow drilling, seismic stratigraphy, and regional geology.

ment rocks beneath the Transantarctic Mountains have been uplifted 5–10 km since early Paleogene time (50–60 Ma; Fitzgerald et al., 1987; Fitzgerald, 1989). Adjacent uplift at the MSSTS-1 and CIROS-1 drill sites is at least 1 km and possibly 2 km since Oligocene time (Collen and Froggatt, 1986).

Stratigraphic traps are likely and their location is largely controlled by relative rates of basin downfaulting/subsidence, sediment supply, and erosion. Such traps could occur at great depth along the edges of the early-rift grabens, at shallower depths within the divergent sections cut by unconformities on the edges of depocenters, and within prograding glacial-marine sections covering the Eastern Basin.

Uplifted areas with likely graben-flank, stratigraphic traps include the western Victoria Land Basin (episodic uplift since 50 Ma; Cooper et al., 1987a), the Central High (most recently subsided in late Oligocene time; Hinz and Block, 1983) and southwest Marie Byrd Land (regionally high peneplain older than 27 Ma; LeMasurier and Rex, 1982).

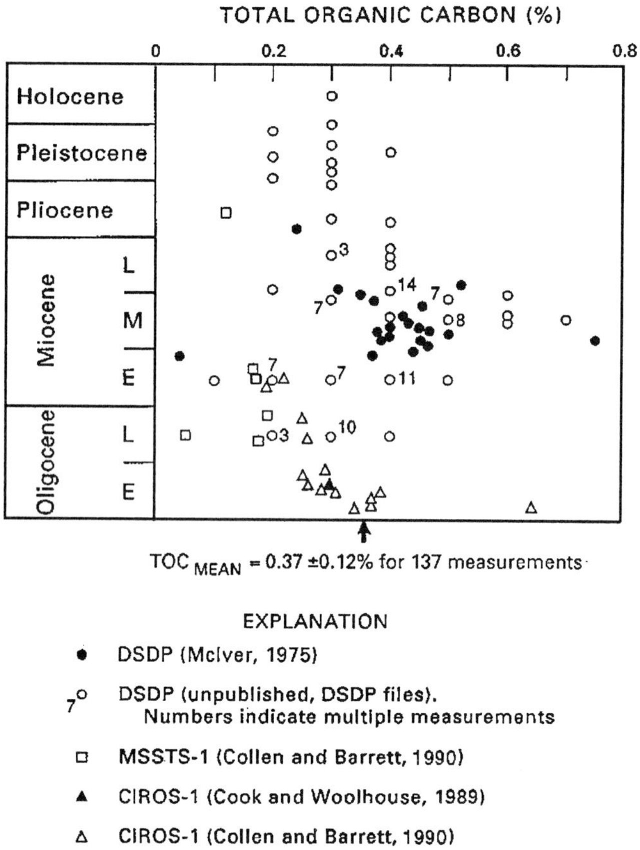

Figure 10—Total organic carbon values measured on offshore glacial-marine rocks from DSDP and CIROS drill cores.

In Ross Sea glacial sequences, stratigraphic traps may exist within the diamictite (impermeable) and sand (permeable) sequences caused by intermittent ice-load-overcompaction and open-marine deposition at the front of former ice sheets. Traps may also occur at buried glacial-erosion unconformities. The Eastern Basin and the western Ross Sea shelf edge are underlain by thick (up to 4 km) prograding glacial deposits with numerous unconformities (Hinz and Block, 1983; Zayatz et al., 1990; Cooper et al., 1990a) (Figure 3). Downhole logging of similar glacially eroded and prograding glacial deposits beneath the outer shelf of Prydz Bay, Antarctica, showed highly variable porosities and lithologies, within 500 m of the sea floor (Barron et al., 1989; Cooper et al., 1990b) (Figure 1). Extensive prograding glacial strata are not found within inner-shelf depocenters (Central Trough and Victoria Land Basin), where these strata are mostly flat lying to gently dipping and glacially eroded.

Gas hydrates are possible in the Ross Sea and could form stratigraphic traps within the uppermost part of the sedimentary section. However, the evidence for hydrates is equivocal—drilling at DSDP Site 272 supports their presence, but a bottom-simulating reflector (BSR) has not been observed in the Ross Sea (Kvenvolden and Cooper, 1987).

Temperature History

The temperature history of the Ross Sea depocenters is a critical factor affecting hydrocarbon generation, yet this history is poorly known. Present heat flow measurements and evolutionary models for the Ross Sea indicate that sub-surface temperatures are now, and probably have been, above normal from the Mesozoic to the present due to extensive crustal rifting.

Subsurface temperatures for the Ross Sea basins are known only from sea-floor, heat-flow measurements and downhole temperature readings in lower Oligocene to Holocene glacial deposits at a few sites in the Ross Embayment (Figure 11) (summarized by Blackman et al., 1987). Heat-flow measurements are generally high (50 to 164 mW/m^2; 1.2 to 3.9 HFU), and most values lie above the mean for continental areas (57 mW/m^2; 1.4 HFU). Near-sea-floor temperature gradients are also high (35 to over 110°C/km). The temperature gradients are higher in sea-floor cores (in excess of 60°C/km) than at drill sites (35-50°C/km), in large part because of low thermal conductivity of the near-sea-floor glacial sediment. Blackman et al. (1987) note that the near-sea-floor gradients in the Victoria Land Basin have not been significantly affected by probable Neogene ice grounding and erosion. White (1989) reports a temperature gradient of 40°C/km in the CIROS-1 hole.

The generally high heat-flow measurements throughout the Ross Embayment, although limited, suggest that the entire region is tectonically active. However, seismicity and seismic reflection studies suggest that shallow (0-10 km) crustal deformation is presently concentrated principally along the western (VLB) and eastern (MBL) sides of the embayment. Deeper flexural or ductile deformation may be occurring along shear zones at the western edge (Stern and ten Brink, 1989) or beneath the embayment (Fitzgerald et al., 1987), which is underlain by thinned continental crust. The hypothesis of deep shear zones is, however, without seismic confirmation.

Existing seismic reflection data support a three-stage thermal-subsidence history of the Ross Sea region. Stage 1 is marked by regional crustal stretching and downfaulting of the narrow and deep early-rift grabens (i.e., early-rift stage). During stage 2, crustal subsidence by cooling caused broad crustal sags above the grabens (i.e., drift stage). All Ross Sea depocenters were affected by stages 1 and 2. Stage 3, which appears, from seismic data, to be confined principally to the eastern and western edges of the Ross Sea is, like stage 1, marked by renewed crustal stretching with downfaulting and intrusions (e.g., late-rift stage). The Terror Rift in the Victoria Land Basin formed during stage 3.

Hydrocarbon-Generation Models

Lopatin-Waples models indicate that hydrocarbons could presently be generated in all Ross Sea depocenters at sub-sea-floor depths of 2.5 to 4.0 km, if adequate source rocks exist. Preglacial sedimentary rocks that fill early-rift grabens are likely at these depths, except over the central Victoria Land Basin and most of the Eastern Basin where glacial strata prevail. The depth to the Ross Sea "oil window" is highly dependent on the postulated regional thermal history, which is poorly known. Our depths give a likely shallow limit for the position of the oil window. Deeper strata within the early-rift grabens have passed through this window.

The Lopatin-Waples method, which relates hydrocarbon generation solely to temperature and time (Waples, 1980), has been used previously in the Ross Sea (Hinz and Block, 1983; Cook and Davey, 1984) because detailed information on source rocks does not exist. Our models (Figure 12) differ from previous ones because we assume a longer Mesozoic rifting history for the Ross Embayment, a higher regional heat-flow value, and an older Paleogene age for unconformity U6 based on new and reinterpreted data.

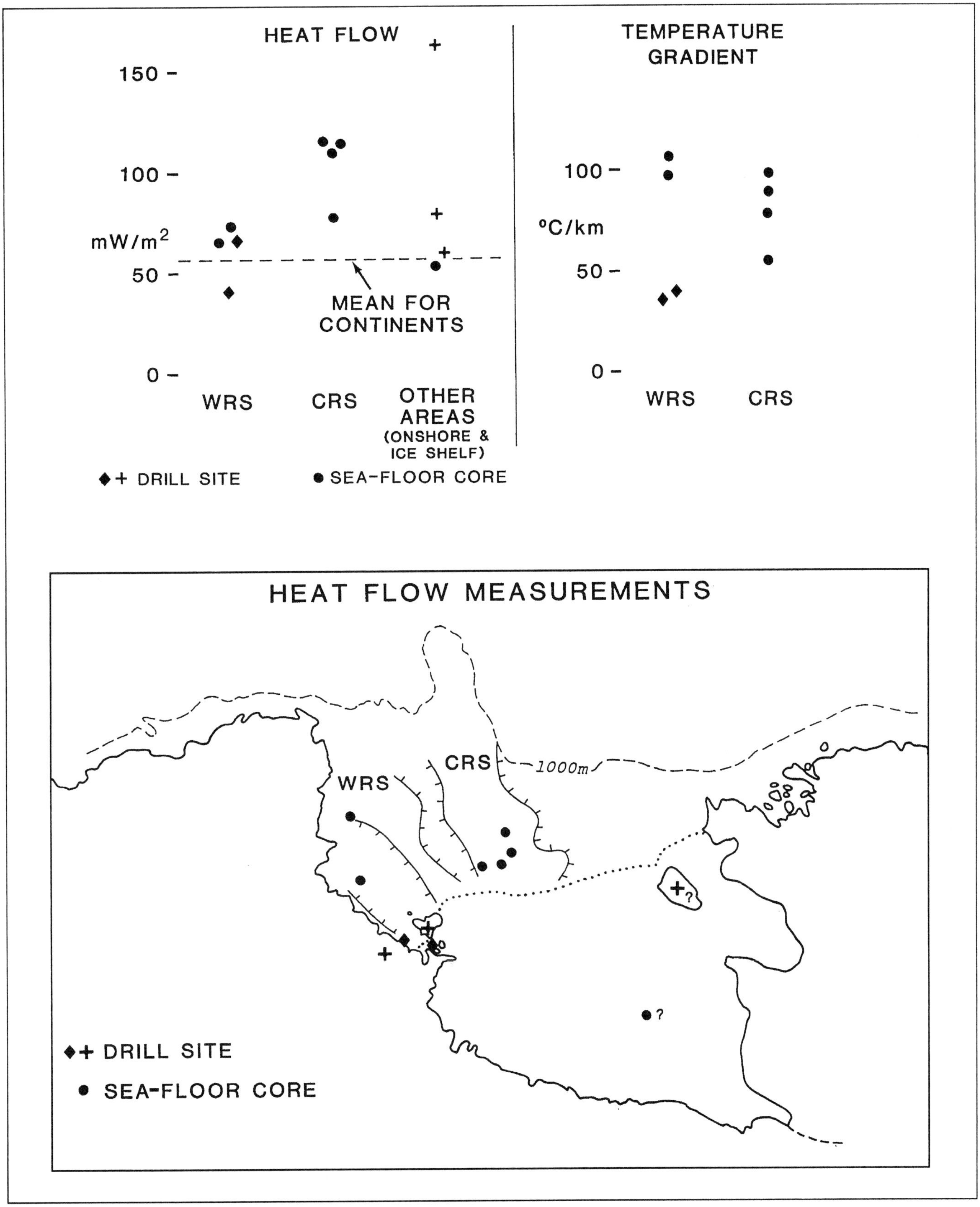

Figure 11—Heat flow and temperature gradient data from the Ross Embayment region (modified from Blackman et al., 1987).

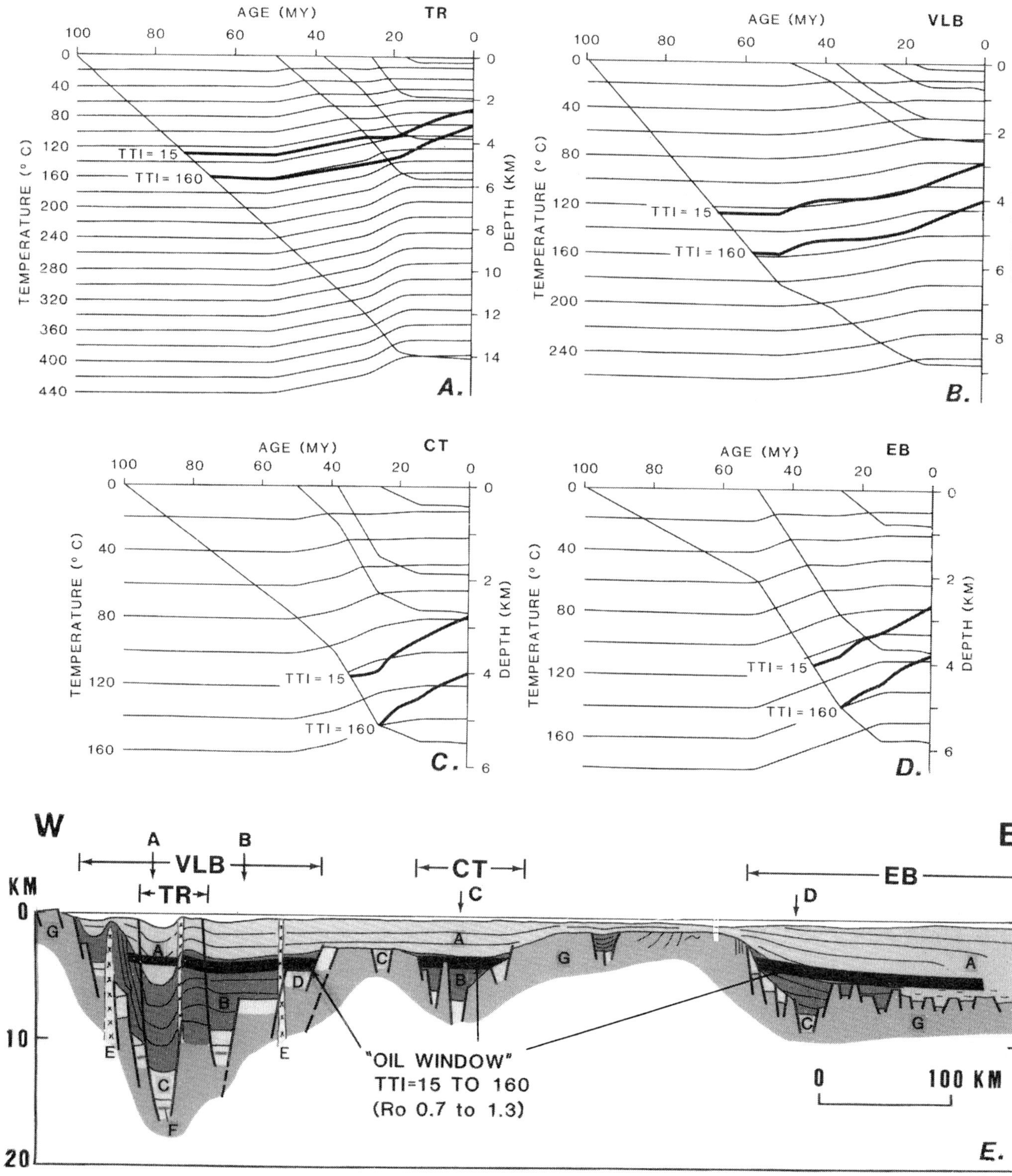

Figure 12—Hydrocarbon-generation models for the Ross Sea and cross section showing location of likely oil window. The models are based on the temperature-time (TTI) method of Lopatin-Waples (Waples, 1980) and parameters given in Table 2 and the text. The likely oil window lies between TTI values of 15 and 160. A) Model for Terror Rift in the Victoria Land Basin; B) Model for the east flank of the Victoria Land Basin; C) Model for the Central Trough; D) Model for the Eastern Basin. E) Cross section showing computed location of oil window (black band) beneath Ross Sea depocenters. Letters (A-D) above cross section indicate location of models; letters on section refer to inferred-time scales from Figure 9.

Table 2: Parameters Used for Lopatin-Waples (TTI) Hydrocarbon-Maturation Models (Figure 12).

I. TIME PARAMETERS

AGE (MY)	HORIZON/ EVENT[1]	VICTORIA LAND BASIN[2]	CENTRAL TROUGH[3]	EASTERN BASIN[3]
0	Sea floor	x	x	x
4-14	Ross Sea unconformity (40 m below sea floor)	x	x	x
18	Unconformity/ Increase in glaciation	V1/V2	x	-
26	Unconformity/ Drake passage opening	V2/V3	x	x
38	Unconformity/ Early E.Antarctic glaciation	V3/V4	x	-
50	Unconformity U6/ Begin late-rift period	V4/V5	U6	U6
100	Breakup unconformity(?) (Top of acoustic basement)	V5/V7	x	x

1 = Events are based on Hinz and Kristoffersen (1987) and references in text
2 = Acoustic units V1-V7 from Cooper et al. (1987a)
3 = Ages modified from Hinz and Kristoffersen (1987) and Sato et al. (1984) assuming time-transgressive unconformities that are older in basin center than on basement highs.
x = Age horizon is shown in model
- = Age horizon is not shown in model

II. TEMPERATURE PARAMETERS

ROCK TYPE	CONDUCTIVITY (K) W/m/°C	TEMPERATURE GRADIENT (Tg)[4] °C/km
Marine glacial	1.7	41
Marine/shallow marine	2.0	35
Terrestrial/fluvial	2.5	28

[4] The temperature gradient is calculated from: $T_g = Q/K$. where Q, the heat flow, is assumed to have been constant (70 mW/m2) for the period 100 My to 0 My.

The parameters we use for the Lopatin-Waples models are given in Table 2. The ages of horizons are based on drilling results, basin-evolution models, and acoustic stratigraphy from multichannel seismic reflection data (Hinz and Block, 1983; Cooper et al., 1987a) (Table 2). Present-day horizon depths, uncorrected for sediment compaction, are used to reconstruct an approximate subsidence-history curve. Paleo-temperatures within the depocenters are based on measured heat flows and on assumed thermal conductivities (e.g., marine and nonmarine rocks) and paleo heat flows. Calculated time-temperature values (TTI) of 15 and 160 are used in the subsidence models for the onset and end of oil generation, respectively (Waples, 1980).

The hydrocarbon-generation zone (TTI = 15 to 160; oil window) presently lies at depths of 2.5 to 4.0 km, and varies by less than 600 m vertically between depocenters (Figure 12). In general, the oil window depths are more sensitive to changes in the basin's temperature history than to shifts in the relative time of sedimentation and subsidence. For example, from the TTI model of Hinz and Block (1983) (Figure 8) for the Eastern Basin, the effect of increasing the temperature gradient from 25 to 30°C/km is to push the oil window up by about 900 m. Whereas, increasing the age of unconformity U6 from 29 Ma (Hinz and Block, 1983) to 50 Ma (Figure 12) pushes the oil window deeper by only 200 m. The present oil window location near the edges of the depocenters is uncertain, because the amount and timing of vertical motion there is poorly defined.

The oil window depths of Figure 12 are up to 1600 m (55%) shallower than those of Cook and Davey (1984) and Hinz and Block (1983), principally because our heat-flow and temperature-gradient values are larger (up to 40%) and have been acting over a longer period (100 m.y instead of 55-80 m.y). Our shallower depths place the present oil window mostly below unconformity U6 with exception of the Eastern Basin and central part of the Victoria Land Basin where the window lies principally above U6. If unconformity U6 is the base of Cenozoic glacial deposits, then the position of the oil window above or below U6 is important to possible present-day hydrocarbon generation because source beds are not likely in glacial deposits. The position of the oil window in Figure 12 is a minimum likely depth within the Ross Sea. Hydrocarbons have probably been generated in Cenozoic and Mesozoic(?) times within the great thickness of rift-related rocks that lie below the present oil window and have previously reached peak hydrocarbon generation.

Migration

Any hydrocarbons that may have been generated in the Ross Sea are likely to have migrated laterally along dipping strata that are common at the edges of depocenters. The asphaltic residue at CIROS-1 is the only confirmed evidence of this migration. Direct vertical-migration pathways to the deeply buried early-rift graben strata below unconformity U6 are also likely, but only in parts of the Victoria Land Basin.

The downfaulted flanks of the early-rift grabens and the upfaulted front of the (late-rift) Transantarctic Mountains provide numerous likely pathways, within fault zones and dipping strata, for migrating hydrocarbons. The CIROS-1 drill site penetrated the basinward-dipping strata on the west flank of the Victoria Land Basin, and encountered the 2-m-thick asphaltic residue in Oligocene glacial strata that is believed evidence for hydrocarbon migration (Barrett, 1989; Cook and Woolhouse, 1989). Here, liquid hydrocarbons may have migrated beneath the drill site from deeper parts of the basin, but have now escaped. Similar basinward-dipping strata (pathways) with numerous unconformities (possible traps) are found along the entire western edge of the Victoria Land Basin.

A few areas, such as within the Terror Rift and along the east edge of the Victoria Land Basin, have late-rift fault zones or intrusive structures that could provide direct pathways for hydrocarbons to migrate upward from deep within the early-rift grabens to shallow levels. Elsewhere in the Ross Sea, however, seismic reflection data do not show late-rift faults in the sedimentary section. Gentle stratal dips below unconformity U6 would also permit hydrocarbons to migrate into graben-edge traps near faulted basement rocks or local disconformities.

The thick glacial deposits in the Eastern Basin dip seaward and could provide lateral pathways for hydrocarbon migration, however, evidence for lateral continuity of beds is equivocal. The large concentration of gases recovered at DSDP Sites 271 and 272 has been cited as evidence of possible migration of these gases from within the basin (McIver, 1975); however, the likelihood that the gases are of biogenic origin (Claypool and Kvenvolden, 1983) suggests that significant migration is not required. Diamictites recovered at Antarctic drill sites where grounded ice sheets have existed intermittently have variable composition, porosity and compaction (Barrett, 1986, 1989; Barron et al., 1989). Yet, acoustic reflectors within the glacial sequences are, in places, continuous over tens of kilometers. The likelihood of extensive lateral hydrocarbon migration within the prograding glacial sequences is thus uncertain.

SUMMARY AND CONCLUSIONS

The Ross Sea and other areas of the Ross Embayment are underlain by early-rift grabens that contain up to 8 km of sedimentary and, perhaps, volcanic(?) rocks, which are buried beneath up to 6 km of mostly glacial-marine strata. These early-rift grabens may have evolved, prior to Eocene time, much like the offshore rift basins of New Zealand, Tasmania, and southeast Australia, which formerly were part of Gondwana and lay adjacent to the Ross Sea.

Our preferred geologic model for the Ross Sea has an early-rift period (Early to Late Cretaceous) of crustal extension, rapid graben-downfaulting, and minor nonmarine and shallow-marine sedimentation (i.e., rift stage). A Late Cretaceous and early Paleogene period followed, with broad regional subsidence and infilling of the early-rift grabens by likely marine and nonmarine strata (i.e., drift stage); major seaways may have existed during this period. During the final stages of continental separation between Antarctica and Australia, renewed rifting initiated in the Ross Sea, and has extended from Eocene time into the present (i.e., late-rift period); uplift and volcanism in the Transantarctic Mountains, development of the Terror Rift, and volcanism in Marie Byrd Land occurred during this period. Glaciation in post-Eocene time has resulted in glacial-marine sediments that may be up to 6 km thick across the Ross Sea. Major rift periods are believed to be linked with major changes in plate motions.

Our assessment of possible Ross Sea hydrocarbons is based on offshore seismic reflection data, on limited in-situ geologic samples of lower Oligocene and younger glacial-marine rocks, on a few recycled Late Cretaceous and Paleogene glacial erratics, and on analogy with formerly nearby Gondwana basins. These data suggest that source rocks are not likely in the thick glacial-marine sections covering the Ross Sea, but source beds probably exist within preglacial (Eocene and older) rocks deep within the rift grabens. Glacial-marine sections may have adequate porosity for local reservoirs, but permeability is unknown. Reservoirs probably exist within preglacial strata, based on analogy with other Gondwana rift basins. Structural and stratigraphic traps are likely throughout the Ross Sea, in particular along basin flanks, near horst blocks, within the Terror Rift, and at numerous unconformities. The temperature-depth histories of the Ross Sea depocenters, which now have above-average heat flow, are suitable for generation of hydrocarbons at sub-sea-floor depths of about 2.5 to 4.0 (TTI = 15 to 160; Waples, 1980), if adequate source beds are present. The principal pathways for migration of hydrocarbons that may have been generated within the deep early-rift grabens, below unconformity U6, are within steeply dipping or faulted sections along the western and eastern sides of the Victoria Land Basin and the Terror Rift. Extensive pathways for lateral migration may not exist in the prograding glacial sequences of the Eastern Basin.

Although the major Ross Sea early-rift grabens may have had pre-Eocene structural and sedimentologic histories similar to those of other Gondwana rift basins, the middle Eocene(?) and younger histories of Ross Sea depocenters have been dominated by glacial sedimentation not recorded elsewhere. The Ross Sea early-rift grabens each have different thermal-subsidence and glacial-sedimentation histories that are likely to affect their hydrocarbon potential. The Eastern Basin is the broadest of the Ross Sea depocenters because of thick prograding glacial strata, however, the hydrocarbon potential may be lower than expected (for the size of the basin) because only a small area (15-20%) is underlain by a major early-rift graben (Figure 3). In contrast, the Victoria Land Basin is in area smaller than the Eastern Basin but it is underlain entirely by a major early-rift graben that has additionally experienced late-rift deformation.

Existing geologic data are insufficient to accurately document the existence and location of liquid hydrocarbons in the Ross Sea. However, geologic models of the Ross Sea and comparisons of the Ross Sea with formerly contiguous, rift-basin areas of Gondwana (Figure 13) indicate to us that liquid hydrocarbons are probably present beneath the Ross Sea. The areas with greatest hydrocarbon potential include the deep preglacial sedimentary sections of the early-rift grabens and the major late-rift structures of the Victoria Land Basin. A more definitive assessment of Ross Sea hydrocarbons requires detailed geologic information about the deeply buried preglacial rocks lying within the early-rift grabens.

ACKNOWLEDGMENTS

We appreciate the cooperation of colleagues from several countries (Australia, France, Italy, Japan, New Zealand, Norway, United Kingdom, United States, USSR, West Ger-

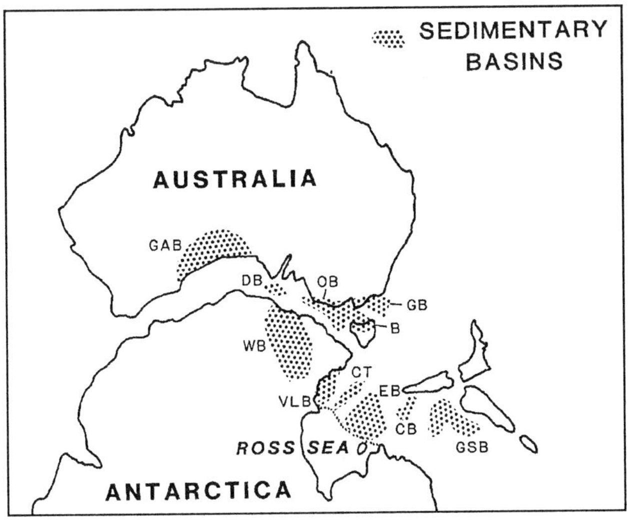

Figure 13—Plate reconstruction of Gondwana showing the location of major late Mesozoic sedimentary rift basins near the Ross Sea (modified from Truswell, 1983). BB-Bass Basin, CB-Campbell Basin, CT-Central Trough, DB-Duntroon Basin, EB-Eastern Basin, GB-Gippsland Basin, GAB-Great Australian Bight Basin, GSB-Great South Basin, OB-Otway Basin, VLB-Victoria Land Basin, WB-Wilkes Basin.

many) who have openly discussed and provided copies of unpublished geologic and geophysical data relevant to this chapter. Helpful reviews of the manuscript were made by John Behrendt and Steve Eittreim, Dave Macdonald, Peter Barrett, and John Kingston.

REFERENCES CITED

Anderton, P.W., N.H. Holloway, N.H. Engstrom and J.C. Ahmad, 1982, Evaluation of geology and hydrocarbon potential of the Great South and Campbell basins, New Zealand: New Zealand Geological Survey open-file, Petroleum Report no. 828, 25 p.

Barker, P. F. et al., 1988, Proceedings of the Ocean Drilling Program, Initial Reports, Volume 113: College Station, TX, Ocean Drilling Program, 785 p.

Barrett, P.J., 1981, History of the Ross Sea region during the deposition of the Beacon Supergroup 400-180 million years ago: Journal of the Royal Society of New Zealand, v. 11(4), p. 447-458.

Barrett, P.J., ed., 1986, Antarctic Cenozoic history from the MSSTS-1 drillhole, western McMurdo Sound: New Zealand Department of Scientific and Industrial Research Miscellaneous Bulletin No. 237, 174 p.

Barrett, P. J., ed., 1989, Antarctic Cenozoic history from CIROS-1 drillhole: New Zealand Department of Scientific and Industrial Research Bulletin 245, 255 p.

Barrett, P.J., D.P. Elston, D.M. Harwood, B.C. McKelvey and P.N. Webb, 1987, Mid-Cenozoic record of glaciation and sea-level change on the margin of the Victoria Land basin, Antarctica: Geology, v. 15, p. 634-637.

Barrett, P.J., M.J. Hambrey and P.R. Robinson, in press, Cenozoic glacial and tectonic history from CIROS-1, McMurdo Sound, *in* M.R.A. Thomson, J.A. Crame and J.W. Thomson, eds., Geological Evolution of Antarctica: Cambridge, Cambridge University Press.

Barron, J., et al., 1989, Proceedings of the Ocean Drilling Program, Initial Reports, Volume 119: College Station, TX, Ocean Drilling Program, 942 p.

Behrendt, J.C., 1983, Are there petroleum resources in Antarctica?, *in* J.C. Behrendt, ed., Petroleum and mineral resources of Antarctica: USGS Circular 909, p. 3-24.

Behrendt, J. C., 1989, Geophysical and geological research in Antarctica related to the assessment of petroleum and mineral resources and potential environmental hazards, *in* R. Wolfrum, ed., Antarctic Challenge III: Berlin, Duncker & Humblot, p. 165-178.

Bentley, C.R., 1983, Crustal structure of Antarctica from geophysical evidence—a review, *in* R.L. Oliver, P.R. James, and J.B. Jago, eds., Antarctic Earth Science: Canberra, Australian Academy of Science, p. 491-497.

Blackman, D.K., R.P.Von Herzen and L.A. Lawver, 1987, Heat flow and tectonics in the western Ross Sea, Antarctica, *in* A.K. Cooper and F.J. Davey, eds., The Antarctic Continental Margin: Geology and Geophysics of the Western Ross Sea: Circum-Pacific Council Energy and Mineral Resources, Earth Science Series; v. 5B, Houston, p. 179-190.

Bradshaw, J.D. and M.G. Laird, 1983, The Pre-Beacon geology of Northern Victoria Land: a review, *in* R.L. Oliver, P.R. James and J.B. Jago, eds., Antarctic Earth Science: Canberra, Australian Academy of Science, p. 98-101.

Brancolini, G., 1990, Preliminary description of Italian geophysical surveys in the Ross Sea and Antarctic Peninsula areas: 1988 through 1990: *in* A.K. Cooper and P.N. Webb, conveners, International Workshop on Antarctic Offshore Acoustic Stratigraphy (ANTOSTRAT): Overview and Extended Abstracts: USGS open-file report 90-309, p. 90–97.

Cameron, P. J., 1981, The petroleum potential of Antarctica and its Continental Margin: Australian Petroleum Exploration Association Journal, v. 21, p. 99-111.

Cande, S. C. and J.C. Mutter, 1982, A revised identification of the oldest sea-floor spreading anomalies between Australia and Antarctica: Earth Planetary Science Letters, v. 58, p. 151-160.

Claypool, G.E. and K.A. Kvenvolden, 1983, Methane and other hydrocarbon gases in marine sediment: Annual Reviews Earth Planetary Science, v. 11, p. 299-327.

Collen, J.D. and P.C. Froggatt, 1986, Depth of burial and hydrocarbon source potential, *in* P.J. Barrett, ed., Antarctic Cenozoic history from the MSSTS-1 drillhole, McMurdo Sound: New Zealand Department of Scientific and Industrial Research Bulletin 237, p. 161-164.

Collen, J.D. and P.J. Barrett, 1990, Petroleum geology from the CIROS-1 drill hole, McMurdo sound: Implications for the potential of the Victoria Land Basin, Antarctica, this volume.

Cook, R.A. and F.J. Davey, 1984, The hydrocarbon exploration of the basins of the Ross Sea, Antarctica, from modeling of the geophysical data: Journal of Petroleum Geology, v. 7(2), p. 213-226.

Cook, R.A. and A.D. Woolhouse, 1989, Analysis of the oil residue, CIROS-1 core hole, McMurdo Sound, Antarctica, *in* P.J. Barrett, ed., Antarctic Cenozoic history from CIROS-1 drillhole, McMurdo Sound: New Zealand Department of Scientific and Industrial Research Bulletin 245.

Cooper, A.K., D.M. Mann, J.R. Childs and F.J. Davey, 1986, Marine geophysical data from the western Ross Sea, Antarctica: USGS open-file report 86-634, 6 p.

Cooper, A.K., F.J. Davey and J.C. Behrendt, 1987a, Seismic stratigraphy and structure of the Victoria Land basin western Ross Sea, Antarctica, *in* A.K. Cooper and F.J. Davey, eds., The Antarctic Continental Margin: Geology and Geophysics of the Western Ross Sea: Circum-Pacific Council Energy and Mineral Resources, Earth Science Series; v. 5B, Houston, p. 27-76.

Cooper, A.K., F.J. Davey and G.R. Cochrane, 1987b, Structure of extensionally rifted crust beneath the western Ross Sea and Iselin Bank, Antarctica, from sonobuoy seismic data, *in* A.K. Cooper and F.J. Davey, eds., The Antarctic Continental Margin: Geology and Geophysics of the Western Ross Sea: Circum-Pacific Council Energy and Mineral Resources, Earth Science Series; v. 5B, Houston, p. 93-118.

Cooper, A.K., F.J. Davey and K. Hinz, in press, Crustal extension and origin of sedimentary basins beneath the Ross Sea and Ross Ice shelf, Antarctica: *in* M.R.A. Thomson, J.A. Crame and J.W. Thomson, eds., Geological Evolution of Antarctica: Cambridge, Cambridge University Press.

Cooper, A.K., H. Stagg, G. Leitchenkov and J.R. Childs, 1990, Cenozoic seismic stratigraphy of Prydz Bay from seismic-reflection and downhole log data: *in* A.K. Cooper and P.N. Webb, conveners, International Workshop on Antarctic Offshore Acoustic Stratigraphy (ANTOSTRAT): Overview and Extended Abstracts: USGS open-file report 90-309, p. 119–127.

Cooper, A.K., P.J. Barrett, V.V. Traube, K. Hinz and H. Stagg, 1990, Prograding Cenozoic sedimentary sequences of the Antarctic continental margin: *in* A.K. Cooper and P.N. Webb, conveners, International Workshop on Antarctic Offshore Acoustic Stratigraphy (ANTOSTRAT): Overview and Extended Abstracts: USGS open-file report 90-309, p. 109–118.

Craddock, C., 1982, Geologic map of Antarctica, *in* C. Craddock, ed., Antarctic Geoscience: Madison, University of Wisconsin Press, 1 map.

Dalziel, I. W. D., 1982, West Antarctica: Problem child of Gondwanaland: Tectonics, v. 1(1), p. 3-19.

Davey, F.J., 1985, The Antarctic margin and its possible hydrocarbon potential: Tectonophysics, v. 114, p. 443-470.

Davey, F.J., 1987, Geology and structure of the Ross Sea region, *in* A.K. Cooper and F.J. Davey, eds., The Antarctic Continental Margin: Geology and Geophysics of the Western Ross Sea: Circum-Pacific Council Energy and Mineral Resources, Earth Science Series, v. 5B, Houston, p. 1-16.

Davey, F.J. and A.K. Cooper, 1987, Gravity studies of the Victoria Land basin and Iselin Bank, *in* A.K. Cooper and F.J.Davey, eds., The Antarctic Continental Margin: Geology and Geophysics of the Western Ross Sea: Circum-Pacific Council Energy and Mineral Resources, Earth Science Series, v. 5B, Houston, p. 119-138.

Davey, F.J., D.J. Bennett and R.E. Houtz, 1982, Sedimentary basins of the Ross Sea, Antarctica: New Zealand Journal of Geology and Geophysics, v. 25, p. 245-255.

Domack, E.W., W.W. Fairchild and J.B. Anderson, 1980, Lower Cretaceous sediment from the East Antarctic continental shelf: Nature, v. 287, no. 5783, p. 625-626.

Drewry, D. J., 1983, Antarctica: Glaciological and geophysical folio: Cambridge, Scott Polar Research Institute, University of Cambridge, 8 maps.

Elliot, D.H., 1985, Physical geography—geological evolution, *in* D.W.H. Walton and W.N. Bonner, eds., Key Environments—Antarctica: New York, Pergamon Press, p. 39-61.

Elliot, D., 1988, Antarctica—Is there any oil and natural gas: Oceanus, v. 31, no. 2, p. 32-38.

Etheridge, M. A., J.C. Branson and P.G. Stuart-Smith, 1985, Extensional basin-forming structures in Bass Strait and their importance for hydrocarbon exploration: Australian Petroleum Exploration Association Journal, v. 25, p. 344-361.

Fitzgerald, P.G., 1989, Uplift and formation of Transantarctic Mountains: Application of apatite fission track analysis to tectonic problems, abs., Abstracts of 28th International Geological Congress, Washington, D.C., June 9-19, 1989, p. 1–419.

Fitzgerald, P.G., T. Sandiford, P.J. Barrett and A.J.W. Gleadow, 1987, Asymmetric extension associated with uplift and subsidence in the Transantarctic Mountains and Ross Embayment: Earth Planetary Science Letters, v. 81, p. 67-78.

Fraser, A.R. and L.A. Tillbury, 1979, Structure and stratigraphy of the Ceduna Terrace region, Great Australian Bight, Australia: Australian Petroleum Exploration Association Journal, v. 19, p. 53-65.

Grindley, G.W., 1981, Precambrian rocks of the Ross Sea region: Journal of the Royal Society of New Zealand, v. 11(4), p. 411-423.

Grindley, G.W. and F.J. Davey, 1982, The reconstruction of New Zealand, Australia, and Antarctica (review paper), *in* C. Craddock, ed., Antarctic Geoscience: Madison, University of Wisconsin Press, p. 15-30.

Hayes, D. E. and F.J. Davey, 1975, A geophysical study of the Ross Sea, Antarctica, *in* Initial Reports of the Deep Sea Drilling Project: v. 28, Washington, (U.S. Government Printing Office), p. 887-907.

Hayes, D. E. and L.A. Frakes, 1975, General Synthesis, Deep Sea Drilling Project Leg 28, *in* Initial Reports of the Deep Sea Drilling Project: v. 28, Washington, (U.S. Government Printing Office), p. 919-941.

Hinz, K. and M. Block, 1983, Results of geophysical investigations in the Weddell Sea and in the Ross Sea, Antarctica: Proceedings of 11th World Petroleum Congress (London), PD 2, no. 1, p. 1-13.

Hinz, K., J.B. Willcox, M. Whiticar, H.R. Kudrass, N.F. Exon and D.A. Feary, 1986, The west Tasmanian margin: an underrated petroleum province?, *in* R.C. Glenie, ed., Second southeastern Australia oil exploration symposium: Melbourne, Petroleum Exploration Society of Australia, p. 395-410.

Hinz, K. and Y. Kristoffersen, 1987, Antarctica—Recent advances in the understanding of the continental shelf: Geologisches Jahrbuch, Reihe E, Heft 37, p. 1-54.

Houtz, R. and F.J. Davey, 1973, Seismic Profiler and sonobuoy measurements in Ross Sea, Antarctica: Journal of Geophysical Research, v. 78, no. 17, p. 3448-3468.

Jankowski, E.J. and D.J. Drewry, 1981, The structure of West Antarctica from geophysical studies: Nature, v. 291, p. 17-21.

Jankowski, E.J., D.J. Drewry and J.C. Behrendt, 1983, Magnetic studies of upper crustal structure in West Antarctica and the boundary with East Antarctica, *in* R.L. Oliver, P.R. James and J.B. Jago, eds., Antarctic Earth Science: Canberra, Australian Academy of Science, p. 197-203.

Karl, H.A., E. Reimnitz and B.D. Edwards, 1987, Extent and nature of Ross Sea unconformity in the western Ross Sea, Antarctica, *in* A.K. Cooper and F.J. Davey, eds., The Antarctic Continental Margin: Geology and Geophysics of the Western Ross Sea: Circum-Pacific Council Energy and Mineral Resources, Earth Science Series, v. 5B, Houston, p. 77-92.

Kennett, J.P. and R.E. Houtz, 1975, *in* Initial reports of the Deep Sea Drilling Project, Leg 29: Washington, (U.S. Government Printing Office), 1197 p.

Kim, Y., L.D. McGinnis and R.H. Bower, 1986, The Victoria Land basin: part of an extended crustal complex between East and West Antarctica: Washington, American Geophysical Union, Geodynamics Series, v. 14, p. 323-330.

Kvenvolden, K.A. and A.K. Cooper, 1987, Natural gas hydrates of the offshore circum-Pacific margin—a future energy resource?, *in* Proceedings of Circum-Pacific Council Energy and Mineral Resources Symposium, August 1986: Houston, Circum-Pacific Council for Energy and Mineral Resources.

Kvenvolden, K.A., F.D. Hostettler and T.J. Frank, in press, Hydrocarbons in sediment of the Weddell Sea, Antarctica, *in* P.F. Barker and J.P. Kennett, eds., Proceedings of the Ocean Drilling Program, Scientific Results, Volume 113: College Station, TX, Ocean Drilling Program.

Lawver, L.A. and C.R. Scotese, 1987, A revised reconstruction of Gondwanaland, *in* G.D. McKenzie, ed., Gondwana Six: Structure, tectonics and geophysics: Washington, American Geophysical Union Geophysical Monograph 40, p. 17-24.

Lawver, L. A., J. Royer, D.T. Sandwell and C.R. Scotese, in press, Evolution of the Antarctic continental margins *in* M.A. Thompson, J.A. Crame and J. W. Thompson, eds., Geological Evolution of Antarctica: Cambridge, Cambridge University Press.

LeMasurier, W.E. and D.C. Rex, 1982, Volcanic record of Cenozoic glacial history in Marie Byrd Land and western Ellsworth Land: Revised chronology and evaluation of tectonic factors, *in* C. Craddock, ed., Antarctic Geoscience: Madison, University Wisconsin Press, p. 725-734.

Lowery, J. H., 1989, CIROS-1 organic geochemistry—coal particles and laminae *in* P.J. Barrett, ed., Antarctic Cenozoic History from CIROS-1 Drillhole, McMurdo Sound: New Zealand Department of Scientific and Industrial Research Bulletin 245, p. 219-222.

Macdonald, D.I.M., P.F. Barker, S.W. Garrett, J.R. Ineson, D. Pirrie, B.C. Storey, A.G. Whitham, R.R.F. Kinghorn and J.E.A. Marshall, 1988, A preliminary assessment of the hydrocarbon potential of the Larsen Basin, Antarctica: Marine and Petroleum Geology, v. 1(5), p. 34-53.

McGinnis, L.D., ed., 1981, Dry Valley drilling project: Washington, American Geophysical Union Antarctic Research Series volume 33, 470 p.

McGinnis, L. D., R.H. Bowen, J.M. Erickson, B.J. Allred and J.L. Kreamer, 1985, East-West antarctic boundary in McMurdo Sound: Tectonophysics, v. 114, p. 341-356.

McIver, R.D., 1975, Hydrocarbon gases in canned core samples from Leg 28 Sites 271, 272, and 273, Ross Sea, *in* D.E. Hayes and L.A. Frakes et al., Initial Report of the Deep Sea Drilling Project, v. 28, Washington, (U.S. Government Printing Office), p. 815-817.

Party, Leg 113 Shipboard Scientific, 1987, Glacial history of Antarctica: Nature, v. 328, p. 115-116.

Party, Leg 119 Shipboard Scientific, 1988, Early glaciation of Antarctica: Nature, v. 333, p. 303-304.

Rapp, J.B., K.A. Kvenvolden and M. Golan-Bac, 1987, Hydrocarbon Geochemistry of sediments offshore from Antarctica, *in* A.K. Cooper and F.J. Davey, eds., The Antarctic Continental Margin: Geology and Geophysics of the Western Ross Sea: Circum-Pacific Council Energy and Mineral Resources, Earth Science Series, v. 5B, Houston, p. 217-224.

Rose, G. and C.T. McElroy, 1987, Coal potential of Antarctica, Bureau of Mineral Resources, Geology and Geophysics, Resource Report 2, 19 p.

Sandford, R.M., 1980, Exploration results off South New Zealand: Oil and Gas Journal, v. 78(5), p. 83-90.

Sato, S., N. Asakura, T. Saki, N. Oikawa and Y. Kaneda, 1984, Preliminary results of geological and geophysical surveys in the Ross Sea and in the Dumont D'urville Sea, off Antarctica: Memoirs of National Institute of Polar Research Special Issue No. 33, p.66-92.

Savage, M.L. and P.F. Ciesielski, 1983, A revised history of glacial sedimentation in the Ross Sea region, *in* R.L. Oliver, P.R. James and J.B. Jago, eds., Antarctic Earth Science: Canberra, Australian Academy of Science, p. 555-559.

Schmidt, D.L. and P.D. Rowley, 1986, Continental rifting and transform faulting along the Jurassic Transantarctic rift, Antarctica: Tectonics, v. 5, p. 279-291.

Shirley, J.E., 1983, Oil exploration, prospects along northwest flank of Great South Basin, New Zealand: Oil and Gas Journal, March 21, p. 199-206.

Stern, T. A. and U.S. ten Brink, 1989, Flexural uplift of the Transantarctic Mountains: Journal of Geophysical Research, v. 94, p. 10315-10330.

St. John, B., 1986, Antarctica—geology and hydrocarbon potential, *in* M. Halbouty, ed., Future Petroleum Provinces of the World: Houston, AAPG Memoir 40, p. 55-100.

Stock, J. and P. Molnar, 1987, Revised history of early Tertiary plate motion in the south-west Pacific: Nature, v. 325, p. 495-499.

Tessensohn, F., in press, The Cenozoic Ross Sea rift system—definition and comparison with the East African Rift, *in* M.R.A. Thomson, J.A. Crame and J.W. Thomson, eds., Geological Evolution of Antarctica: Cambridge, Cambridge University Press.

Truswell, E.M., 1983, Geological implications of recycled palynomorphs in Continental Shelf sediments around Antarctica, *in* R.L. Oliver, P.R. James and J.B. Jago, eds., Antarctic Earth Science: Canberra, Australian Academy of Science, p. 394-399.

Veevers, J.J. and S.L. Eittreim, 1988, Reconstruction of Antarctica and Australia at breakup (95 ± 5 Ma) and before rifting (160 Ma): Australian Journal of Earth Science, v.35, p.355-362.

Waples, D.W., 1980, Time and temperature in petroleum formation: application of Lopatin's method to petroleum exploration: AAPG Bulletin, v.64, p. 916-926.

Webb, P.N., 1981, Late Mesozoic-Cenozoic geology of the Ross Sector, Antarctica: Journal of the Royal Society of New Zealand, v. 11(4), p. 439-446.

Webb, P. N., D.M. Harwood, B.C. McKelvey, J.H. Mercer and L.D. Stott, 1984, Cenozoic marine sedimentation and ice-volume variation on the east Antarctic craton: Geology, v.12, p. 287-291.

Weissel, J.K., D.E. Hayes and E.M. Herron, 1977, Plate tectonics synthesis: The displacements between Australia, New Zealand, and Antarctica since the Late Cretaceous: Marine Geology, v. 25, p. 231-277.

White, P., 1989, Downhole geophysical logs of CIROS-1, *in* P.J. Barrett, ed., Antarctic Cenozoic history from CIROS-1 drillhole, McMurdo Sound: New Zealand Department of Scientific and Industrial Research Bulletin 245.

Williamson, P.E., C.J. Pigram, J.B. Colwell, A.S. Scherl, K.L. Lockwood and J.C. Branson, 1987, Review of stratigraphy, structure, and hydrocarbon potential of Bass Basin, Australia: AAPG Bulletin, v. 71, p. 253-280.

Zayatz, I., M. Kavun and V. Traube, 1990, The Soviet geophysical research in the Ross Sea: *in* A.K. Cooper and P.N. Webb, conveners, International Workshop on Antarctic Offshore Acoustic Stratigraphy (ANTOSTRAT): Overview and Extended Abstracts: USGS open-file report 90-309, p. 283–290.

Chapter 5

Multichannel Seismic Reflection Surveys Over the Antarctic Continental Margin Relevant to Petroleum Resource Studies

John C. Behrendt
U.S. Geological Survey
Denver, Colorado, U.S.A.

ABSTRACT

More than 100,000 km of marine multichannel seismic profiles have been acquired over the continental margin of Antarctica since 1976 by scientific research programs of Australia, Brazil, France, Italy, Japan, Norway, Poland, United Kingdom, United States, U.S.S.R. and West Germany. Although scientific results are reported for most of these data, they also are relevant to petroleum resource assessment. Because of the one or two orders of magnitude greater cost of standard land survey techniques in Antarctica compared with marine techniques in areas of open water, there will likely be no great amount of coverage on the interior of the Antarctic ice sheet. Despite this, several countries are experimenting in a research mode using land systems, and deep crustal reflection surveys at carefully selected interior sites will probably be made soon.

INTRODUCTION

The recently adopted (June 2, 1988) Convention on the Regulation of Antarctic Mineral Resources Activities (CRAMRA) has focused attention on scientific research carried out mostly over the last decade using modern marine geophysical survey technology. Although conventional wisdom has suggested that future development of petroleum resources in Antarctica would be over the continental margin (e.g., Behrendt, 1983; Behrendt, in press), and the multichannel seismic reflection surveys over Antarctica, shown in Figure 1, are exclusively over the continental margin areas, development of petroleum beneath basins underlying the grounded ice sheet is probably feasible as well. Nonetheless, because land multichannel seismic reflection surveys over the ice sheet are one or two orders of magnitude more expensive than marine surveys, it appears likely that there will be no extensive land surveys within the next 10 years or so. A research program using deep crustal reflection profiles in carefully selected areas will probably be started soon despite high cost.

The reported shows of gas in Deep Sea Drilling Project (DSDP) holes in the Ross Sea (Hayes and Frakes, 1975) stimulated political and scientific interest which led to the negotiation of CRAMRA. Recent coring by the Ocean Drilling Program (ODP) on the Antarctic margin, discussed elsewhere in this volume, has also directed attention to petroleum resource possibilities over the Antarctic continental margin (Behrendt, in press).

This short chapter itemizes the multichannel seismic surveys over the continental margin shown in Figure 1. Examples of areal coverage by several countries (Japan, West Germany, USSR and United States) are shown in Figure 2. The Ross Sea and Antarctic Peninsula area profiles are indicated at a larger scale in Figures 3 and 4. More extensive discussion is given by Behrendt (in press) and Hinz and Kristoffersen, 1987.

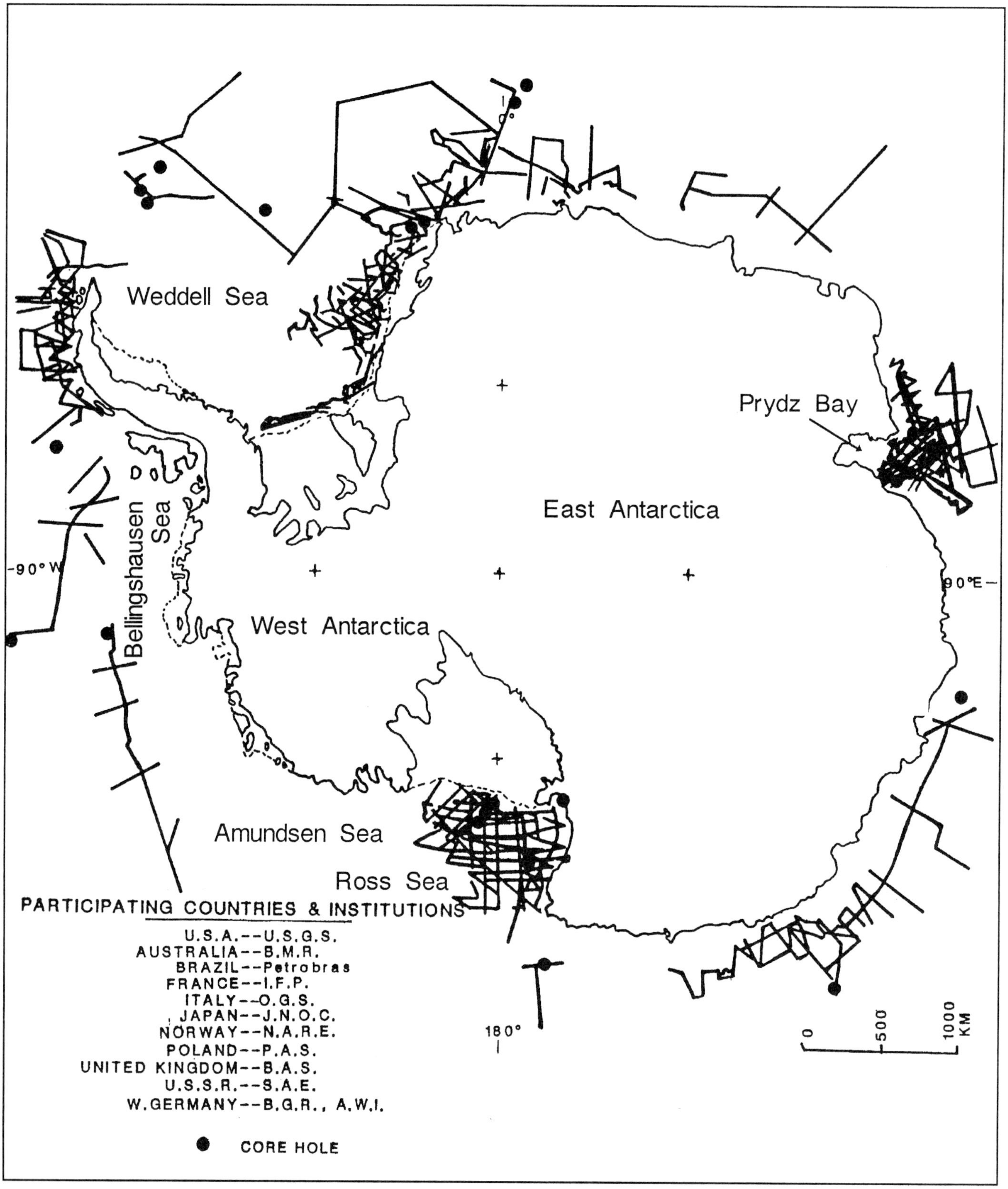

Figure 1—Map of Antarctica showing locations of marine multichannel seismic reflection profiles collected between 1976 and 1988. Locations of all known research core holes, e.g., DSDP and ODP, are indicated; however, locations in Ross Sea and Prydz Bay are obscured by seismic lines.

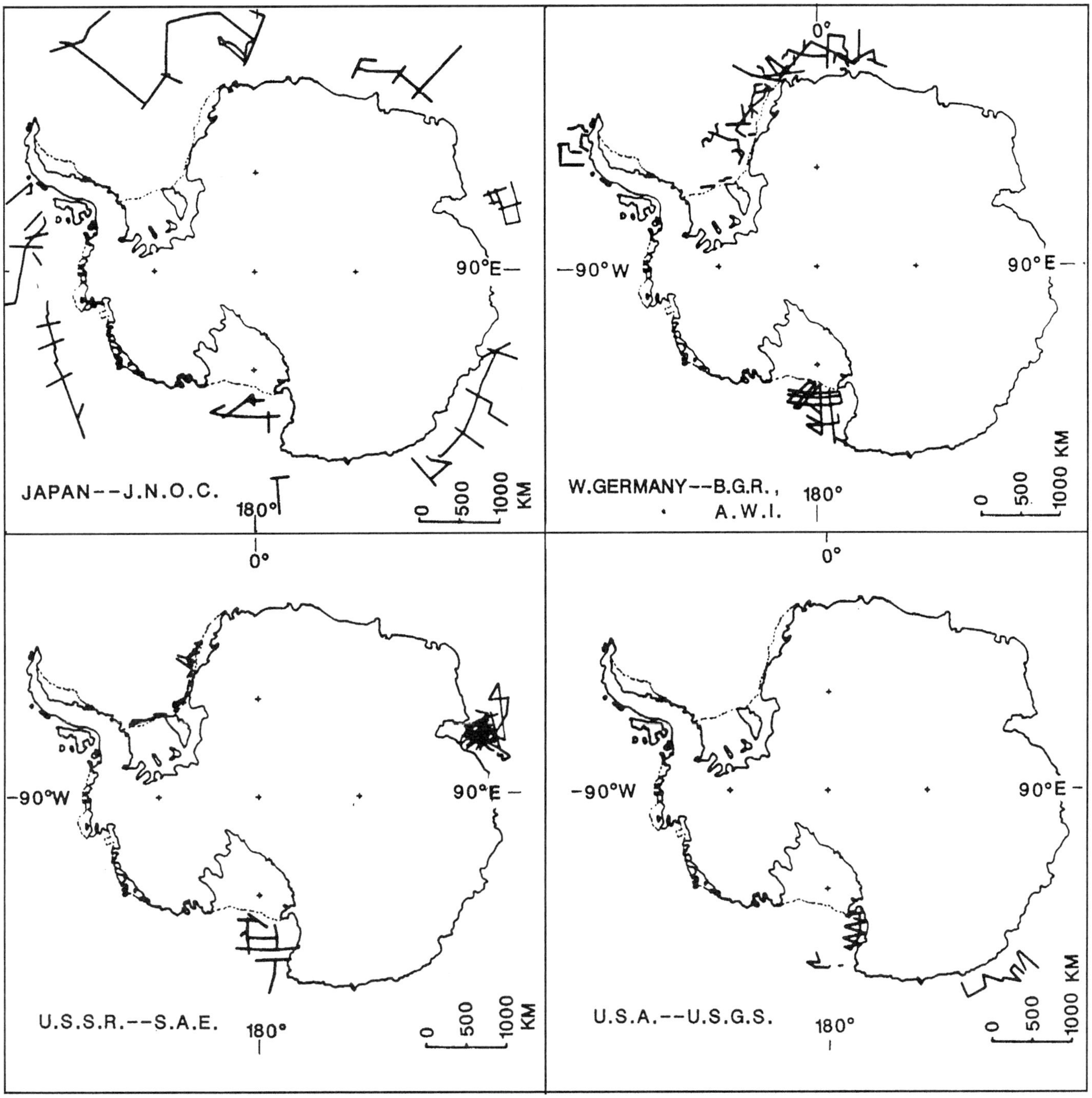

Figure 2—Maps showing examples of multichannel data coverage from Figure 1 by Japan, West Germany, USSR and United States.

Ross Sea

In 1980, the West German Federal Institute for Geosciences and Natural Resources (BGR) acquired 6100 km of 48-channel data using the *Explora* over the Ross Sea continental shelf (Hinz and Block, 1983). In 1981-1982, the Institut Francais du Petrol (IFP), using the same ship *Explora*, collected about 1500 km of 48-channel data in the Ross Sea area (J. Wannesson, 1982, personal communication). No results from this work are available. In 1983 the Japanese National Oil Company (JNOC) ship *Hakurei-Maru* also collected multichannel reflection profiles (6-fold) in the Ross Sea discussed by Sato et al. (1984). In February 1984, the U.S. Geological Survey (USGS) collected 2350 km of 24-channel reflection profiles (using the vessel *S. P. Lee*) in the Ross Sea area (Cooper et al., 1987). Kim et al. (1986) reported on a short 48 km long (not shown in Figures 1, 2) long multichannel reflection (CDP) profile in McMurdo Sound recorded using land techniques on sea ice. The Soviet Antarctic Expedition (SAE) collected 4300 km of 24-fold seismic reflection profiles over the Ross Sea margin in 1987 using the R/V *Geolog Dmitry Nalivkin* (Grikurov, 1987).

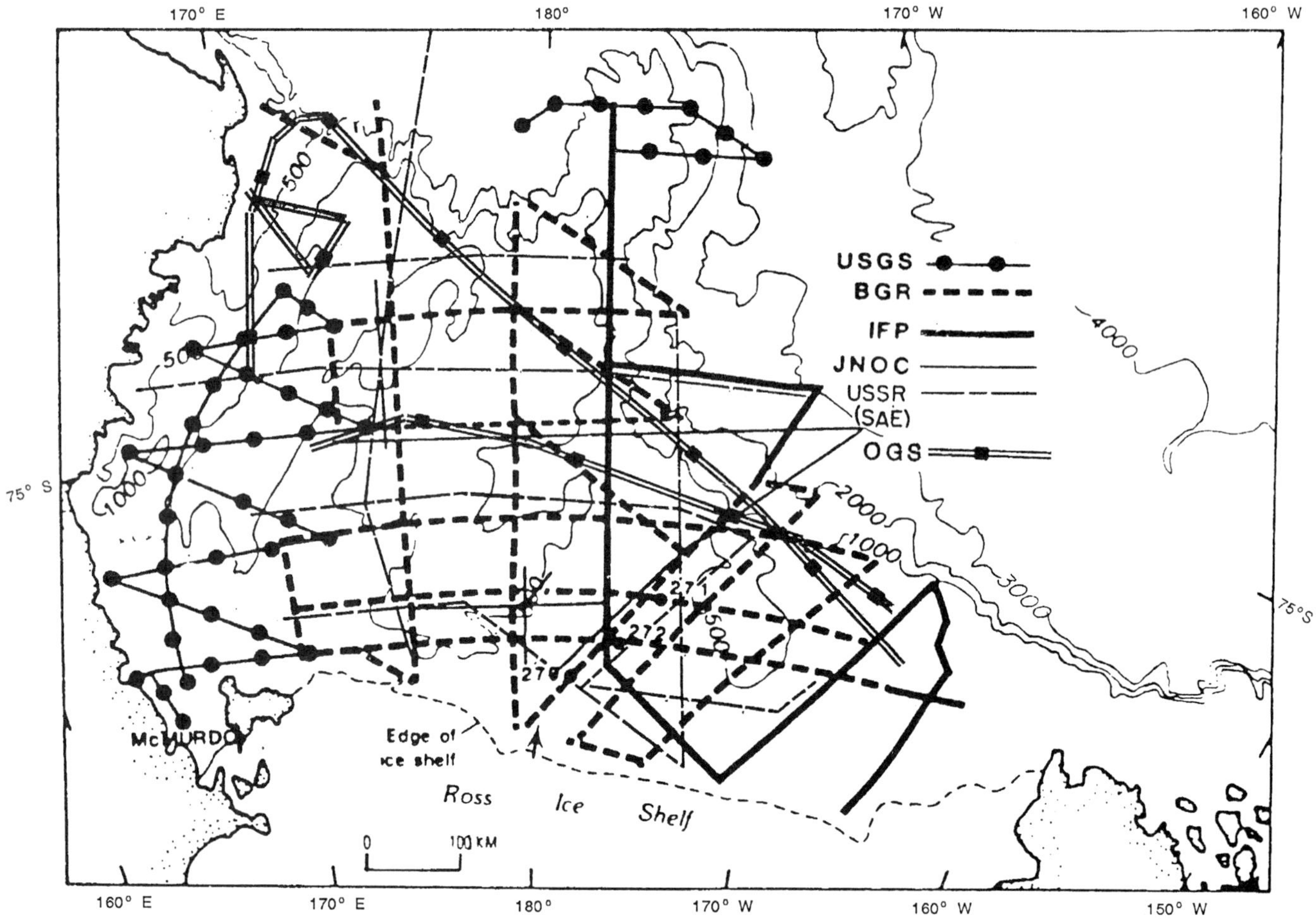

Figure 3—Map showing location of multichannel seismic reflection profiles from Figure 1 in Ross Sea area. Bathymetry in meters. DSDP holes 270, 271, 272 are indicated.

Results from 1987-1988 and 1988-1989 cruises in the Ross Sea are not available, but reports presented to the Scientific Committee on Antarctic Research (SCAR) Working group on Solid Earth Geophysics indicated *Explora* collected 2500 km of 48-channel data over the Ross Sea shelf in early 1988 and an additional few thousand kilometers in early 1989 for the Italian Observatoria Geofisico Sperimentale (A. Cooper, 1988, personal communication). The SAE also collected data in the Ross Sea in January-February 1989 (G. Grikurov, 1988, personal communication). As of 1989, there is an approximate total of 30,000 km of profiles.

Antarctic Peninsula-Bellingshausen Sea

The Polish research vessel *Kopernik* made 1100 km multichannel seismic reflection measurements in the Bransfield Strait in 1979-1980 where the profiles in Figure 1 are identified (Guterch et al., 1985). In 1987, Petrobras of Brazil, using the research vessel *Almirante Camera,* collected about 5000 km of 72-channel reflection profiles from the Bransfield Strait area to Adelaide Island along the Antarctic Peninsula (Gamboa et al., 1987). The Japanese ship (JNOC) *Hakurei-Maru* collected more than 3280 km of 12-channel (3-fold) reflection data in the Bellingshausen Sea area in 1981 (Kimura, 1982). In 1987, *Hakurei-Maru* acquired about 2800 km of reflection profiles in the Amundsen Sea (JNOC, 1987, personal communication). The British Antarctic Survey (BAS) and University of Birmingham cooperatively made 3700 km of 24-fold multichannel reflection profiles in the Bransfield Strait-Scotia Arc area near the northeast end of the Antarctic Peninsula (P. Barker, personal communication, 1985; and University of Birmingham, Antarctic Marine Group, personal communication, 1985). Additional surveys in this area in 1987-1988 were reported to SCAR. BAS (UK) collected an unknown amount of data, using *Discovery* (British Antarctic Survey, 1988). Petrobras (Brazil) acquired 560 km of 72-channel data (Brazil, 1988). The FRG *Polarstern* (Alfred Wegner Institute, AWI, with University of Kiel) collected about 1350 km of reflection profiles in the Bransfield Strait area in late 1987 (R. Meissner, 1988, personal communication).

Weddell Sea

In 1976-1977 the Norwegian Antarctic Research Expedition (NARE) acquired 16-channel data along lines shown in Figure 1 (Haugland, 1982; Haugland et al., 1985). The SAE collected 12-channel seismic reflection data (G.E. Grikurov, 1982, personal communication) partly over this area along the tracks shown as far west as 60°W from 1980-1982. In

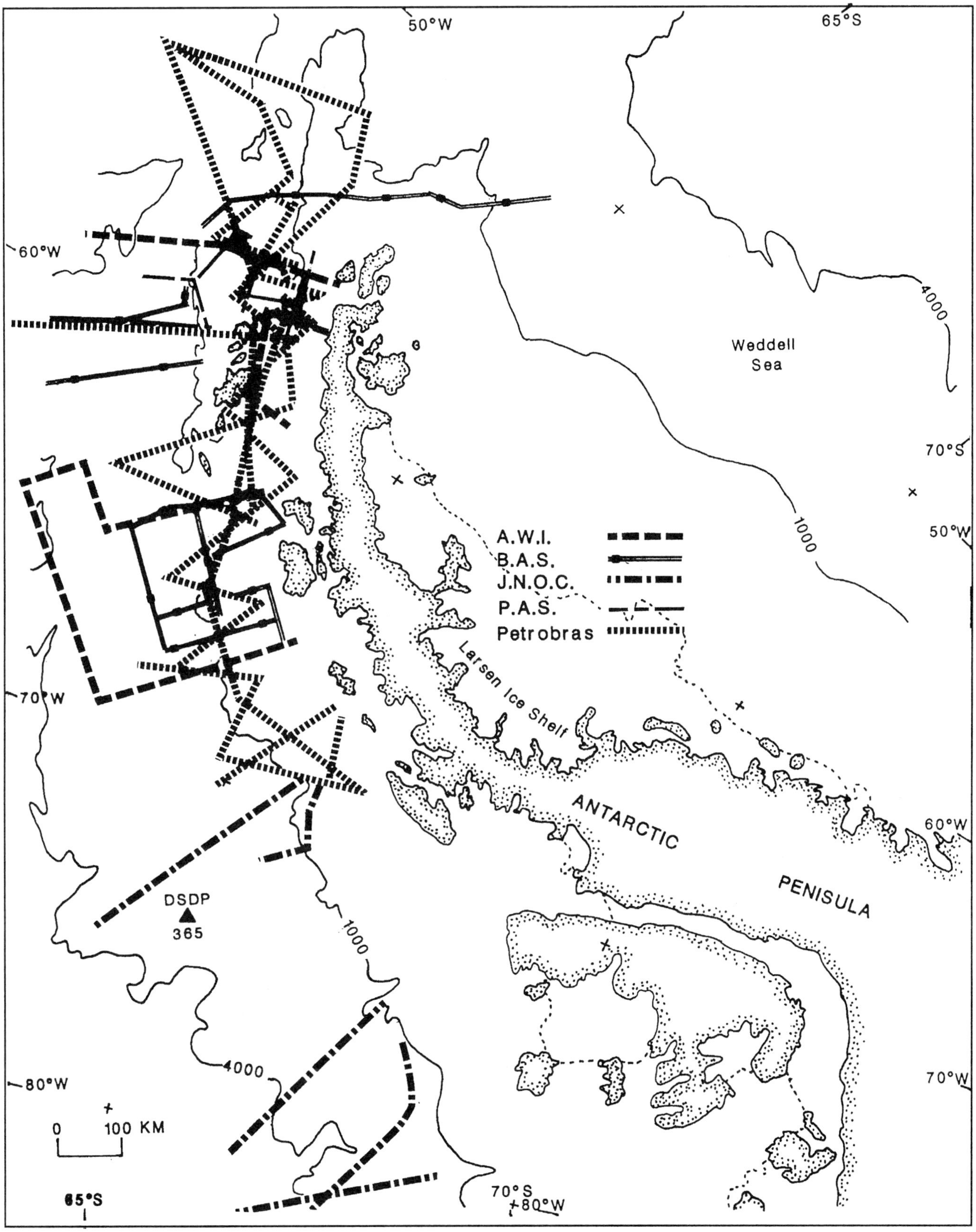

Figure 4—Map showing location of multichannel seismic reflection profiles from Figure 1 in Antarctic Peninsula area.

1978, the BGR collected 5854 km of 48-channel data over the continental shelf between 25°W and 20°W (Figure 1) (Hinz and Krause, 1982). The FRG ship *Polarstern* made 6263 km of multichannel seismic profiles in the Weddell Sea area (Hinz, 1986). In 1982-1983, the R/V *Hakurei-Maru* from Japan National Oil Corporation (JNOC) collected 24-channel reflection profiles in the southern Weddell Sea to the north of the BGR-NARE-SAE surveys over oceanic crust along the track lines shown from Okuda et al. (1983).

Prydz Bay—East Antarctica

The Australian Bureau of Mineral Resources (BMR) collected about 5000 km of 6-channel reflection data on closely spaced lines over the continental margin in the area offshore of the Amery Ice Shelf, between 55°E and 80°E during 1981-1982 (Stagg et al., 1983; Stagg, 1985). In 1985-1986, the SAE surveyed about 1000 km of multichannel lines in the Prydz Bay area (Grikurov, 1986). The SAE acquired 3100 km of 24-channel data in the Prydz Bay area in 1986-1987 and an additional 3000 km there in 1987-1988 (Grikurov, 1988). In 1984-1985, the *Hakurei Maru* collected 24-channel, 6-fold seismic reflection data in the bay area north of the Australian BMR survey (Mizukoshi et al., 1986). All of this activity in recent seasons in the Prydz Bay area, although certainly indicating strong scientific interest, attests to petroleum resources assessment interest as well.

In the 1981-1982 season, *Explora* collected 48-channel reflection data in Wilkes Land for IFP along 3000 km of lines shown in Figure 1, between 135°E and 155°E, over the East Antarctica continental margin (Wannesson et al., 1985). In 1984, the *Hakurei-Maru* collected 24-channel lines off Wilkes Land near the 140°E meridian (Tanahashi et al., 1987) (Figure 1). The USGS ship *S. P. Lee* also obtained 1800 km of 24-channel data over the Wilkes Land margin in January 1984 (Eittreim and Smith, 1987).

CONCLUSION

I have indicated briefly the multichannel surveys conducted over the Antarctic continental margin since 1976. Several countries (e.g., US, UK, FRG, NZ) are experimenting with the use of land techniques on the grounded Antarctic ice sheet to study the poorly known subglacial basins. The U.S. Geological Survey 1984 cruise to the Ross Sea cost about $300/km for acquisition (with mobilization from New Zealand). I estimate, in comparison, that 24- to 96-channel data collected in the interior of Antarctica would cost somewhere between $3000-30,000/km using standard land techniques including multigeophones per channel and closely spaced shot holes. Use of a land streamer such as the one being experimentally developed in FRG (F. Thyssen, personal communication, 1989) could reduce these costs significantly. However, marine multichannel reflection surveys over the continental margin using substantially less expensive techniques will probably collect orders of magnitude more data in the coming decade than any likely land surveys. At present about 100,000 km of marine multichannel data exist over the continental margin collected by the various countries' research programs described here.

REFERENCES CITED

Behrendt, J.C., 1983, Are there petroleum resources in Antarctica? *in* J. C. Behrendt, ed., Petroleum and Mineral Resources of Antarctica: USGS Circular 909, p. 3-24.

Behrendt, J.C., in press, Scientific studies relevant to the question of Antarctica's petroleum resource potential, *in* R.J. Tingey, ed., The Geology of Antarctica: Oxford, Oxford University Press.

Brazil, 1988, Report to SCAR Working Group, Unpublished.

British Antarctic Survey, 1988, Report to SCAR Working Group, Unpublished.

Cooper, A.K., F.J. Davey and J.C. Behrendt, 1987, Seismic stratigraphy and structure of the Victoria Land Basin, western Ross Sea, *in* A.K. Cooper and F.J. Davey, eds., Geology and Geophysics of the Western Ross Sea: Circum-Pacific Council Energy and Resources Earth Science Series, 5B, Houston, p. 37-76.

Eittreim, S.L. and G.L. Smith, 1987, Seismic sequences and their distribution on the Wilkes Land margin, *in* S.L. Eittreim and M.A. Hampton, eds., The Antarctic continental margin, Geology and Geophysics of Offshore Wilkes Land: Circum-Pacific Council Energy and Mineral Resources Earth Science Series, 5A, p. 15-44.

Gambôa, L.A., J.J. Rodrigues, V.H. Viviane and P.R. Holdando, 1987: Preliminary results of the first Brazilian Geophysical Expedition to Antarctica, Petrobras, 3 p.

Grikurov, G.E., 1986, Report to XIX SCAR on activity of the Soviet representative in the Report to Geology SCAR Working Group during 1982-1986, 7 p. Unpublished.

Grikurov, G.E., 1987, A brief report on the Soviet Research activities in the Ross Sea in 1986-1987: Report to SCAR Geology Working Group. Unpublished.

Grikurov, G.E., 1988, USSR Report to SCAR Working Group. Unpublished.

Guterch, A., 1986, Polish National Report 1983-86, to Solid Earth Geophysics Working Group—Subcommittee on Antarctic Margin Studies, SCAR Working Group. Unpublished.

Guterch, A., M. Grad, T. Janik, E. Perchuc and J. Pajchel, 1985, Seismic studies of the crustal structure in West Antarctica, 1979/1980-Preliminary results: Tectonophysics 114, p. 411-429.

Hayes, D.E. and L.A. Frakes, 1975, General synthesis Deep Sea Drilling Project Leg 28, *in* D. E. Hayes and L.A. Frakes, eds., Initial Reports of Deep Sea Drilling Project, v. 28, p. 919-942.

Haugland, K., 1982, Seismic reconnaissance survey in the Weddell Sea, *in* C. Craddock, ed., Antarctic Geoscience: Madison, University of Wisconsin Press, p. 405-413.

Haugland, K., I. Kristoffersen and A. Velde, 1985, Seismic investigations of the Weddell Sea embayment: Tectonophysics 114, p. 293-312.

Hinz, K., 1986, Report on the cruise "Antarctica-IV/3" of F.S. Polarstern Part 1: Results of BGR's Geophysical Investigations in the Weddell Sea, 6th December 1985-13th March 1986, unpublished.

Hinz, K. and M. Block, 1983, Results of geophysical investigations in the Weddell Sea and in the Ross Sea, Antarctica; Explorations in New Regions, Eleventh Petroleum Congress of London, p. PD 2 (1) 1-13.

Hinz, K. and W. Krause, 1982, The continental margin of Queen Maud Land Antarctica: Seismic sequences, structural elements and geological development; Geol. Jahrbuch, Reihe E., E. Schweizerbartliche Verlagsbuchhandlung, no. E23, p. 17-41.

Hinz, K. and I. Kristoffersen, 1987, Antarctica, recent advances in understanding of the continental shelf: Geologisches Jahrbuch Reihe E., Heft 37, 54 p.

Kim, Y., L.D. McGinnis and R.H. Bower, 1986, The Victoria Land Basin: part of an extended crustal complex between East and West Antarctica, *in* M. Barazangi and L. Brown,

eds., Reflection Seismology, The Continental Crust: American Geophysical Union Geodynamic Series 14, p. 323-330.
Kimura, K., 1982, Geological and Geophysical Survey in the Bellingshausen Basin, off Antarctica: Antarctic Research 75, p. 12-24.
Mizukoshi, I., H. Sunouchi, T. Saki, S. Sato and M. Tarahushi, 1986, Preliminary report of geological and geophysical survey off Amery Ice Shelf, East Antarctica, Unpublished Report to SCAR, 13 p.
Okuda, Y., T. Ymazaki, S. Sato, T. Saki and N. Oikawa, 1983, Framework of the Weddell Sea, Basin inferred from the new geophysical and geological data: Memoirs of Polar Research Special Issue No. 28, Proceedings of the Third Symposium on Antarctic Geosciences, 1982, p. 93-114.
Sato, S., N. Asakura, T. Saki, N. Oikawa and Y. Kaneda, 1984, Preliminary results of geological and geophysical surveys in the Ross Sea and in Dumont D'Urville Sea off Antarctica: Memoirs of National Institute of Polar Research, Special Issue 33, p. 66–92.
Stagg, H.M.J., 1985, The structure and origin of Prydz Bay and MacRobertson Shelf, East Antarctica: Tectonophysics 114, p. 315-340.
Stagg, H.M.J., D.C. Ramsay and R. Whitworth, 1983, Preliminary report of a marine geophysical survey between Davis and Mawson stations, 1983, *in* R.L. Oliver, P.R. James and J.B. Jago, eds., Antarctic Earth Sciences: Australian Academy of Science, p. 527-532.
Tanahashi, M., T. Saki, N. Oikawa and S. Sato, 1987, An interpretation of the multichannel seismic reflection profiles across the continental margin of the Dumont d'Urville Sea off Wilkes Land, East Antarctica, *in* S.L. Eittreim and M.A. Hampton, eds., The Antarctic continental margins: Geology and geophysics of offshore Wilkes Land: Circum-Pacific Council Energy and Mineral Resources Earth Science Series, 5A, p. 1-14.
University of Birmingham, 1985, Antarctic Marine Group: Cruise Report RRS Discovery Cruise, B 154, 44 p.
Wannesson, J., M. Pelras, B. Petitperrin, M. Perret and J.A. Segoufin, 1985, Geophysical transect of the Adelie margin, East Antarctica: Marine and Petroleum Geology 2, p. 192-201.

Chapter 6

Geology and Petroleum Potential of the Adelie Coast Margin, East Antarctica

J. Wannesson
Institut Français du Pétrole
Rueil–Malmaison, France

ABSTRACT

On the Adelie Coast continental margin, a multichannel seismic survey has revealed the presence of a thick sedimentary basin, beneath the outer continental shelf and upper slope, that may exceed 6000 m. This basin results from the creation and evolution of a continental margin, initiated about 100 million years ago from the separation of Australia and Antarctica. Beneath the outer shelf, which is 400-500 m deep, the sedimentary series consists of four units separated by three major unconformities:

– A prerift unit including a Precambrian basement, possible Paleozoic and early Mesozoic sediments and a Mesozoic synrift sequence;

– An early postrift unit, ranging in age from Cenomanian to Eocene, assumed to consist mainly of fluviatile to deltaic clastics;

– An Upper Eocene to Oligocene unit in a shallow marine environment;

– A Neogene glacial prograding unit.

The early postrift unit is considered to be a promising petroleum target based on comparison to other passive margins.

INTRODUCTION

The purpose of this chapter is to take advantage of the seismic lines recently recorded in East Antarctica: (1) to describe accurately the sedimentary history of the area and (2) to make a qualitative estimate of its petroleum potential.

The geology of Antarctica is poorly known because approximately 98% of this 14 million km^2 continent is covered with a thick ice sheet. Outcrops comprising 2% or about 300,000 km^2—nearly equivalent to the British Isles—are also unevenly distributed. They are mainly located in the Transantarctic Mountains and in the Antarctic Peninsula. Hardly better known are the Antarctic continental margins where about 50,000 km of multichannel seismic profiles have been recorded in the past 10 years, in limited zones, mainly the Weddell Sea, the Ross Sea and the margin of Wilkes Land/Adelie Land (Hinz, 1978; Haugland, 1982; Hinz and Krause, 1982; Kimura, 1982; Okuda et al., 1983; Stagg et al., 1983; Hinz and Block, 1984; Eittreim et al., 1984; Sato et al., 1984; Wannesson et al., 1985; Haugland et al., 1985; Tsumuraya et al., 1985; Hinz et al., 1986).

Some scientific drilling was also done on the Antarctic margins in the early 1970s (Hayes and Frakes, 1975; Craddock and Hollister, 1976) and in 1987-1988 (ODP Legs 113 and 119).With this package of data in hand, it seems rather difficult to assess the overall petroleum potential of Antarctica. The oil crises in the middle and late 1970s, however, provided the incentive for some assessments (Wright and

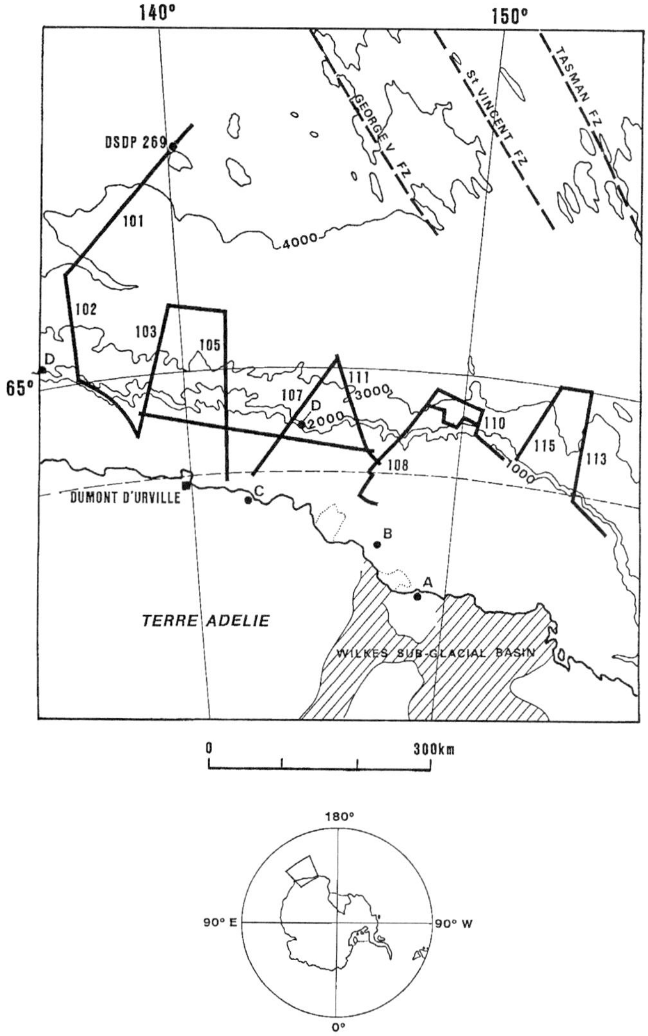

Figure 1—Location map of the IFP-CEPM Antarctique-1982 cruise. Bathymetry from the Gebco map 5.18 (1980). Sedimentary rock indications and subglacial basin extension from Davey (1985). Fracture zone identification from Hayes (1972).

Williams, 1974; Piper, 1976; Splettstoesser, 1977; Rivera, 1977; Dugger, 1978; Zumberge, 1979; Holdgate and Tinker, 1979; Ivanhoe, 1980; Cameron, 1981; Quigg, 1983; Apostolescu and Wannesson, 1982; Behrendt, 1983; Cook and Davey, 1984; Hinz and Block, 1984; Tessensohn, 1984; Davey, 1985; St. John, 1986) mainly by comparison with areas supposed to be close to Antarctica prior to the Gondwana breakup in Mesozoic times. Figures for potential reserves have been computed, which may be completely unrealistic.

In recent years, better estimates of hydrocarbon generation have been made possible in some Antarctic areas, essentially offshore, with the availability of good multichannel seismic lines coupled with improved structural and geochemical modeling techniques. It must be pointed out, however, that the lack of drilling data on the upper margin concerning the ages and facies of the seismic units remains a key problem.

THE ADELIE COAST MARGIN SURVEY

During the austral summer of 1981-1982, a French group operated by the Institut Français du Pétrole conducted a seismic survey on the Antarctic margin from 136°E to 156°E (Wannesson et al., 1985) (Figure 1). The survey consisted of 18 lines with a total of 3190 km. The survey vessel *Explora* (Prakla-Seismos) used a 2000 in^3 airgun array and a 2400 m, 48-trace streamer. The shotpoint interval was 50 m, giving a 24-fold coverage. Gravity and magnetic field were also recorded. The seismic profiles afforded the first complete picture of the continental margin of eastern Antarctica opposite Australia. Line 102, west of the survey zone, reveals the main features of the margin (Figure 2).

From the morphological standpoint, the continental shelf is over 100 km wide, with the southern end of the line being more than 30 km from the coast. This shelf is 400 to 750 m deep, and the shallowest part corresponds to the outer shelf. This reverse arrangement, like the erosion of the superficial sedimentary series, is characteristic of the Antarctic environment and is due to the ice cap load, and its advances and retreats. The upper continental slope is steep. The rise is characterized by many canyons and sedimentary structures associated with turbidity currents and deep underwater currents. The sedimentary series is thick, especially below the outer shelf, where it exceeds 4 s TWT, or more than 6000 m. It has four units separated by three major regional unconformities (Figure 2):

– Unit A comprises the deepest sediments observed on the section; they appear beneath the major unconformity U1 and are laid out in many faulted blocks rising in steps towards the south. Its thickness cannot be estimated since the basement is difficult to delineate.

– Unit B fills the relief on unconformity U1 and pinches out on it toward the south; its upper boundary is unconformity U2. It reaches a maximum thickness of 2500 m beneath the upper continental slope.

– Unit C, displayed between unconformities U2 and U3, consists of subparallel strata over an average thickness of 2500 m. It overlies unit B and onlaps unconformity U1.

– Unit D, the uppermost unit, which reaches a maximum development of 2500 m below the outer continental shelf, consists of two prograding sequences.

This series cannot be dated directly due to the absence of boreholes on the upper margin. Onshore, the outcrops consist essentially of Precambrian eruptive and metamorphic rocks. There are only four indications of sedimentary rocks in the survey area, which range in age from late Paleozoic to early Tertiary:

– Beacon Supergroup (Devonian to Jurassic) sedimentary and volcanic outcrop at Horn Bluff, onshore southeast of line ATC 108 (A on Figure 1);

– nonmarine Lower Cretaceous brecciated siltstone cored on the shelf, in the 1400-m-deep Mertz trough, a submarine fjord lying south of line ATC 108 (Domack et al., 1980) (B on Figure 1);

– sedimentary erratics of probable post-Beacon age collected at Cape Denison, onshore south of line ATC 107 (Mawson, 1940) (C on Figure 1);

– recycled Cretaceous and early Tertiary palynomorphs in glacial muds cored west of line ATC 102 and along line ATC 107, which suggest reworking of nearby Mesozoic sediments (Truswell, 1982) (D on Figure 1).

Based on these indications, it seems reasonable to assess the presence of these sediments on the major part of the shelf and upper slope and hence in some of the units observed on the seismic lines. On the lower continental rise, a limited stratigraphic check was also obtained on line ATC

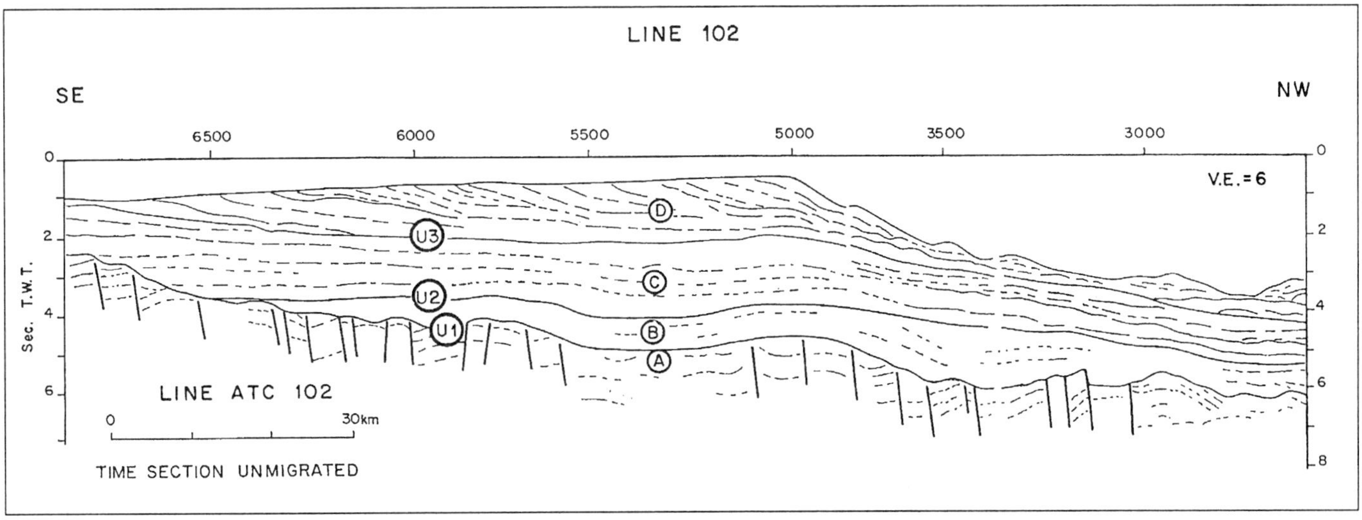

Figure 2—Simplified geoseismic section of line ATC 102 on the shelf and slope off Adelie Coast. Proposed age of unconformities is: U1 = 95-100 Ma; U2 = 42 Ma; U3 = 25 Ma. Unit A is pre-Cenomanian; Unit B is Late Cretaceous to early Eocene; Unit C is middle Eocene to Oligocene; Unit D is Oligocene to Quaternary. Numbers along top of section are shot points.

101 (SP 805) from DSDP Site 269 (Figure 3). This well was drilled through 958 m of distal turbidites of Neogene age down to a series tentatively defined as upper Oligocene to lower Miocene, corresponding to the top of seismic Unit C. We then consider Units A and B as pre-late Oligocene.

An indirect approach to the age of the oldest sedimentary series observed on the Adelie margin lines was then made by interpreting the magnetic field anomalies, especially along profiles connecting the mid-Indian ridge to the continental margin (Schlich and Patriat, 1967; Wannesson et al., 1985) (Figure 4). Interpretation reveals the presence of anomaly 34, dated 83 Ma and the magnetic quiet Upper Cretaceous zone before 83 Ma (Figure 5). This interpretation is in accordance with the revised magnetic anomaly identification by Cande and Mutter (1982) and the model of Veevers (1986) in the southeastern Indian Ocean. The extent of the magnetic quiet zone toward the continent depends on the location of the continent-ocean boundary (COB). Controversies exist among authors about the location of the COB. We place it at the northern edge of the magnetic trough (Talwani and Eldholm, 1973; Rabinowitz, 1976) (Figure 6) situated at the base of the continental slope (ATC 102, SP 2700), some 100 km south of anomaly 34. If we extrapolate the computed expansion rates for the anomaly 19-34 period to that point, the oceanic breakup between Australia and Antarctica occurred at about 100 Ma, at the end of the Albian. The earliest oceanic sediments of the Antarctic margin would thus have been deposited at that time. Our magnetic interpretation also indicates a major change in spreading rates at the time of anomaly 19 in the middle Eocene around 42 Ma. This change is probably reflected on the continental margin by a major unconformity.

The comparison with the Australian margin (Griffiths, 1971; Boeuf and Doust, 1975; Deighton et al., 1976; Konig, 1980; Falvey and Mutter, 1981; Davey 1985) (Figure 7), especially the age of the oldest series (nonmarine Early Cretaceous) and the existence of unconformities in the Albian and the Eocene, provides arguments in favor of the following interpretation (Figure 2).

Unit A, topped by the breakup unconformity U1 around 100 Ma, is a complex unit of prerift and synrift formations. In addition to the eruptive and metamorphic Precambrian basement, the prerift could contain Paleozoic and Mesozoic sedimentary series of the Beacon Supergroup. Reflectors in some faulted blocks of Unit A suggest the existence of such sedimentary series. The Albo Aptian synrift is probably also preserved in places at the top of this unit, as attested by the nonmarine Early Cretaceous siltstone core collected by Domack (1980) on the inner continental shelf.

Overlying Unit B corresponds to the period of thermal subsidence associated with the narrow ocean stage with a low spreading rate which prevailed between the late Albian and the middle Eocene. The sedimentary deposits associated with this stage of the history of the margin are probably predominantly detrital (deltaic and fluviatile). At the top of this unit, unconformity U2 can be traced on the slope and on the rise beyond the COB down to DSDP hole 269, where it lies at about 0.2 sec TWT below the deepest sediments drilled (probable late Oligocene to early Miocene).

More open ocean conditions prevailed starting in the middle Eocene. The sediments deposited on the Antarctic margin during this period form Unit C, predominantly carbonates and fine detrital sediments. The Antarctic glaciation onset, which convergent geological observations place in the Oligocene around 30-25 Ma, is reflected by both a regression and a radical change in sedimentation. The configuration of the upper Unit D clearly reflects this change with large sedimentary influxes in low sea level periods, and by the erosion of the inner shelf in the recent past.

This interpretation agrees relatively well with that by Tanahashi et al. (1987). On the other hand, it differs from the interpretation proposed by Eittreim and Smith (1987) who consider that most of what I interpret as Late Cretaceous oceanic crust is attenuated continental crust. They define U1 as rift onset unconformity, U2 as breakup unconformity, and U3 as Eocene unconformity respectively. I believe that U2 is younger than the breakup unconformity since this horizon does not downlap on the oceanic basement near the COB but can be traced farther north on the lower margin on top of a sedimentary unit several hundred meters thick, resting

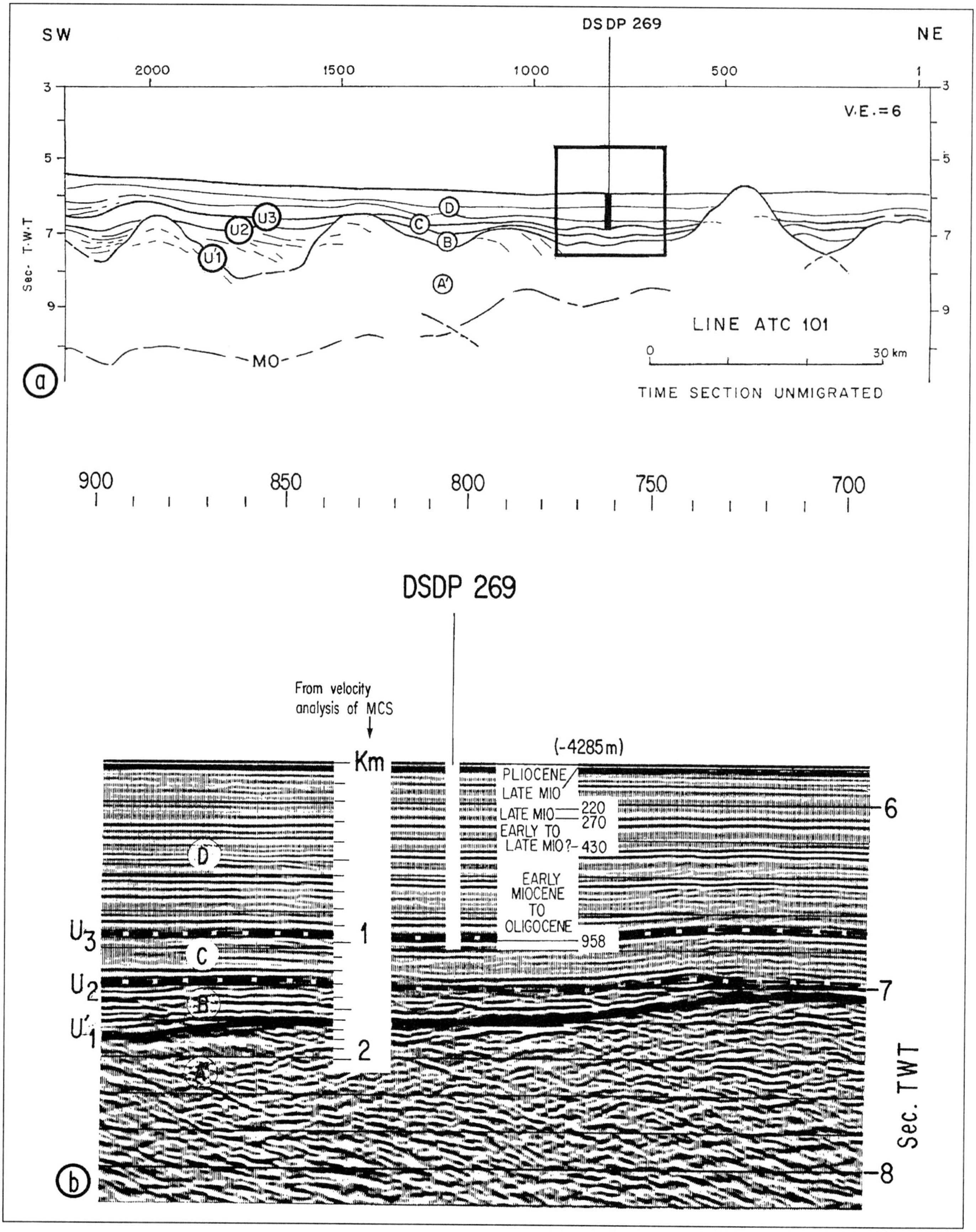

Figure 3—A. Geoseismic section of line ATC 101 on the lower continental rise off Adelie Coast, in the area of DSDP hole 269. B. Interpretation of line ATC 101 and correlation with DSDP hole 269. Same legend as on Figure 1. MO = Moho.

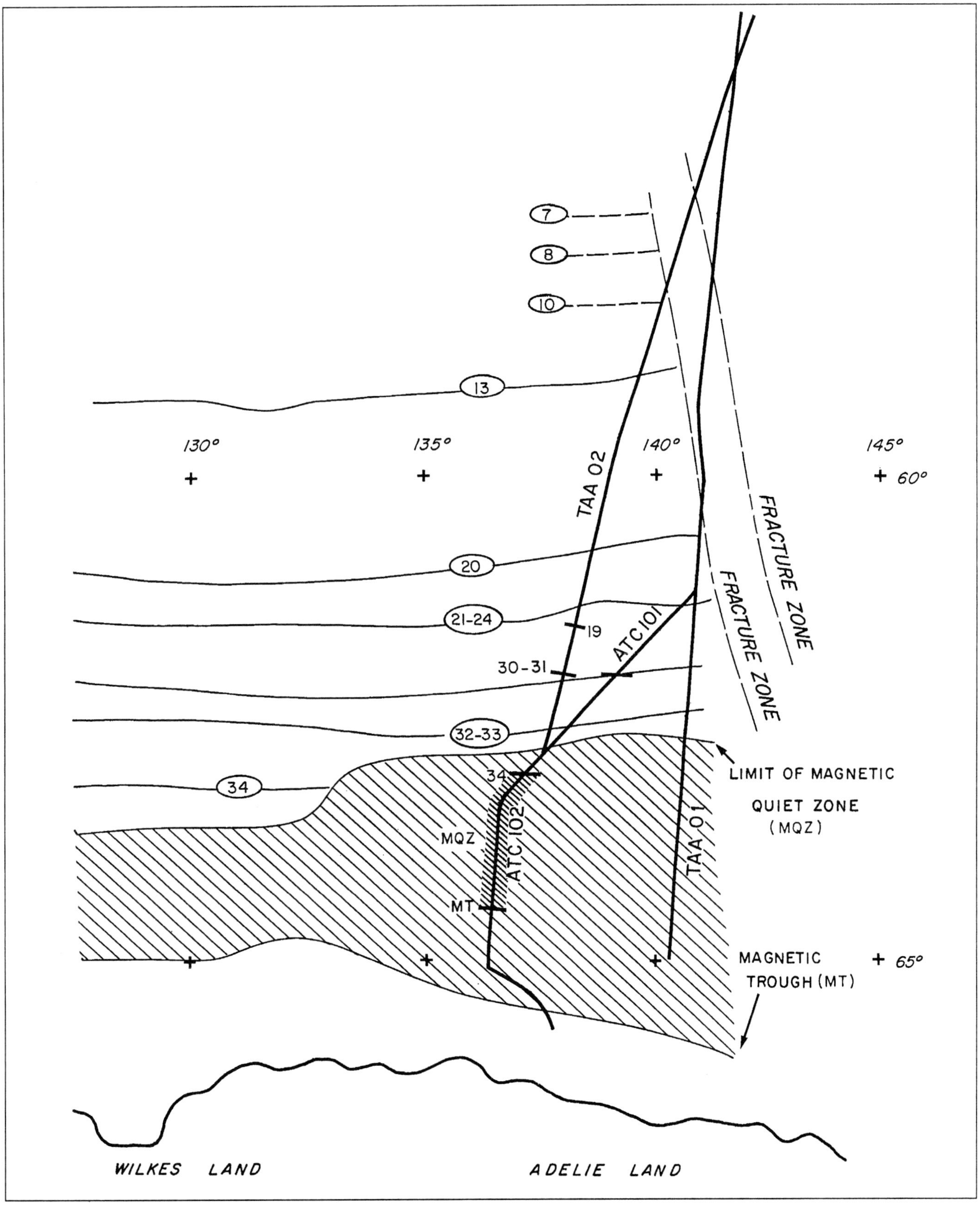

Figure 4—Location of magnetic profiles discussed in text: ATC 101/ATC 102 (Explora 1982); TAA 01/TAA 02 (Thala Dan 1965). Magnetic anomaly, magnetic quiet zone, and magnetic trough locations are from Konig (1980). Circled magnetic anomaly identification is from Cande and Mutter (1982). Magnetic anomaly identification along lines TAA 02 and ATC 101-ATC 102 are from Wannesson et al.(1985).

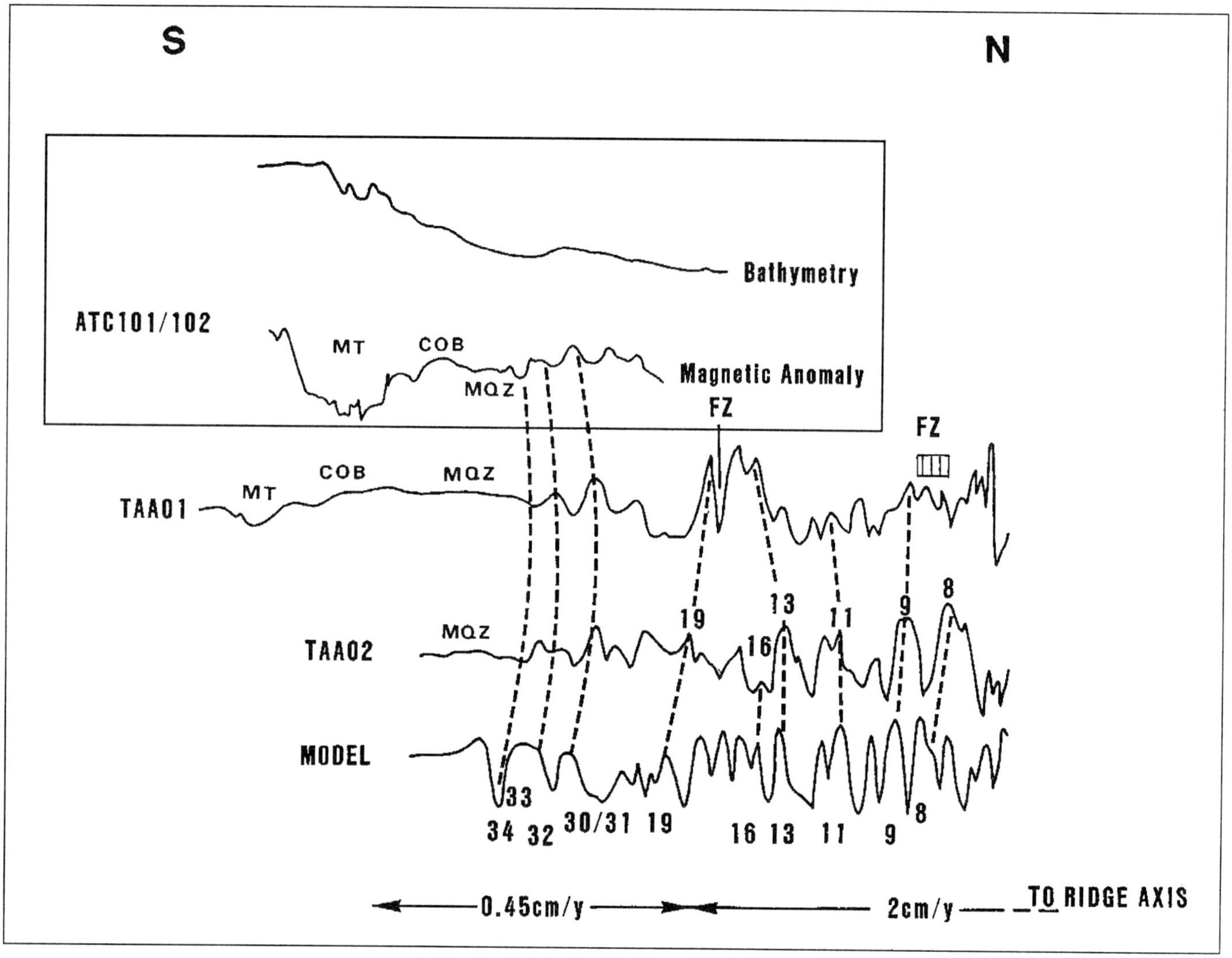

Figure 5—Interpretation of magnetic profiles off Adelie Coast from a two-spreading-rates model of 2 cm yr back to anomaly 19 and 0.45 cm yr from anomaly 19 to anomaly 34. MQZ = magnetic quiet zone; MT = magnetic trough; COB = continent-ocean boundary.

on well-identified oceanic basement. Furthermore, major faulting on the upper margin does not extend upwards into Unit B, which seems therefore to postdate the breakup.

The other profiles recorded on this margin show a comparable structure and sedimentary succession (Figure 8). We note on some lines a broad bulge of the unconformity U1 below the continental shelf and the draping of that unconformity by the early postrift Unit B. The similarity of all dip sections extends to line ATC 113 to the east, where the first of a series of major fracture zones (George V FZ or St Vincent FZ; Hayes, 1972) (Figure 1) marks the eastern boundary of the purely divergent East Antarctic margin.

WHAT PETROLEUM POTENTIAL?

What is the petroleum potential of the sedimentary series visible on this continental shelf? On this type of passive margin, we have learned from our experience, especially along the African Atlantic margins, that most of the petroleum objectives are located in the synrift and in the early postrift series. Given the proposed hypotheses concerning the structural and sedimentary history of this margin, it is possible to simulate the evolution of its petroleum potential by geological and geochemical modeling (Ungerer et al., 1984; Ungerer and Pelet, 1987). For different simulations, we assumed the possible existence of source rocks in these synrift and early postrift series as follows:

– continental organic matter (type 3 organic matter, Tissot et al., 1974) in the synrift unit below U1;

– marine organic matter (type 2) in the early postrift unit between U1 and U2.

We also assumed the possible existence of lacustrine organic matter (type 3) in the prerift Beacon Supergroup below U1 and of marine organic matter (type 2) in the Eocene/Oligocene unit between U2 and U3. The corresponding geological and geochemical parameters were introduced in backstripping and maturation computer programs developed at IFP. The result (Figure 9) showed that all the prerift- and synrift-supposed source rocks reached the gas window before the Oligocene. By contrast, the Late Cretaceous early postrift series which drapes the reliefs of unconformity U1 lies in the oil window beneath most of the continental shelf. Unit C has not reached the oil generation stage.

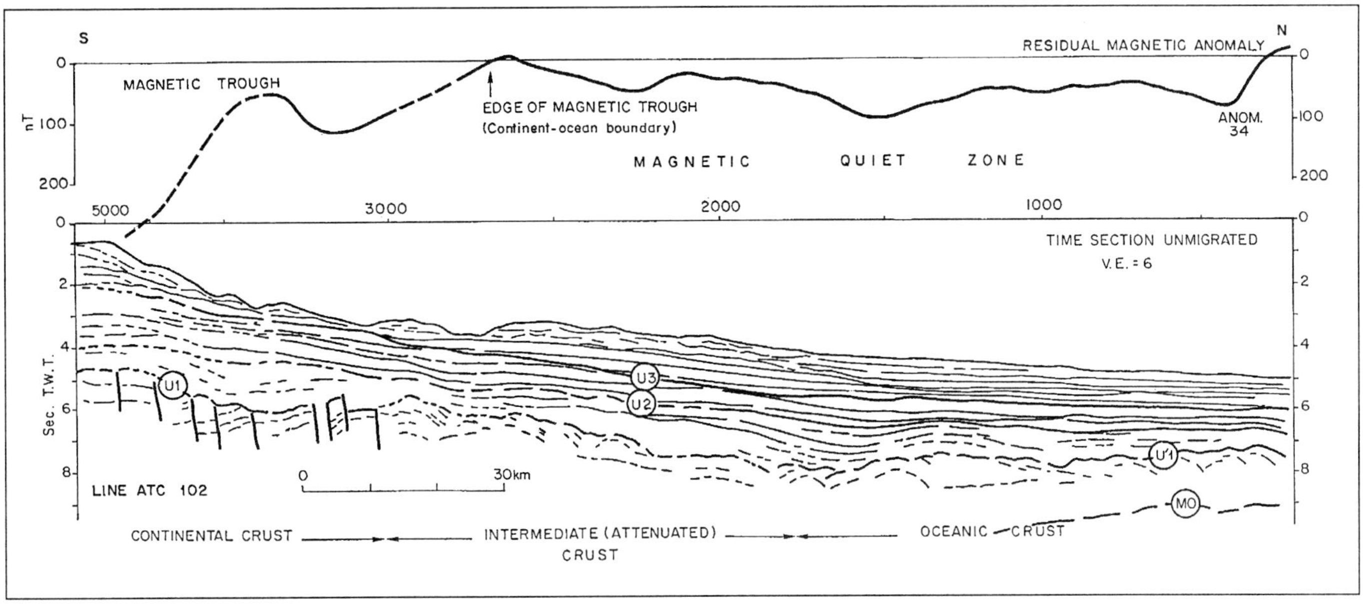

Figure 6—Simplified geoseismic section of line ATC 102 on the continental slope and rise off Adelie Coast in the area of the assumed continent-ocean boundary. Correlation with the residual magnetic anomaly interpretation. Unconformity identification is shown on Figures 2 and 3.

It seems possible to extrapolate this geological and petroleum scheme to the rest of the East Antarctic margin which experienced a comparable history; in other words, the margin associated with the opening of the southeastern Indian Ocean between Australia and Antarctica. As a first approximation, the area involved extends from 90°E to 155°E, namely between the South Kerguelen Plateau and the Transantarctic Mountains, including a shelf area of about 300,000 km^2. Hydrocarbon generation on this Antarctic margin appears to be likely with a certain number of reasonable geological hypotheses. Any attempt, however, to go further in any quantitative evaluation will be unrealistic until additional scientific drilling on the upper continental margin provides solid information on ages and composition of strata.

ACKNOWLEDGMENTS

The author wishes to thank the Comité d'Etudes Pétrolières Marines for giving permission to publish the seismic lines from the Antarctique 1982 cruise. The geological and geophysical interpretation was made by a group consisting of M. Pelras (SNEAP), B. Petitperrin (SNEAP), M. Perret (TOTAL-CFP), J. Segoufin (IPGP) and the author. The geochemical modeling was done by R. Vially (IFP). Interpretation of the magnetism was done by J. Segoufin (IPGP).

REFERENCES

Apostolescu, V. and J. Wannesson, 1982, Y a-t-il du pétrole dans l'Antarctique: La Recherche, v. 13, n. 138, p. 1340-1342.

Behrendt, J. C., 1983, Are there petroleum resources in Antarctica?: USGS, Circular 909, p. 3-24.

Boeuf, M. G. and H. Doust, 1975, Structure and development of the southern margin of Australia: Australian Petroleum Exploration Association Journal, v. 15, p. 33-43.

Cameron, P. J., 1981, The petroleum potential of Antarctica and its continental margin: Australian Petroleum Exploration Association Journal, v. 21, part 1, p. 99-111.

Cande, S. C. and J. C. Mutter, 1982, A revised identification of the oldest seafloor spreading anomalies between Australia and Antarctica: Earth Planetary Science Letters, v. 58, p. 151-160.

Cook, R. A. and F. J. Davey, 1984, The hydrocarbon exploration of the basins of the Ross Sea, Antarctica, from modeling of the geophysical data: Journal of Petroleum Geology, v. 7, n. 2, p. 213-226.

Craddock, C. and C. D. Hollister, 1976, Geologic evolution of the Southeast Pacific Basin, *in* C. D. Hollister et al., eds., Initial reports of the Deep Sea Drilling Project, v. 35: Washington, (U.S. Government Printing Office), p. 723-743.

Davey, F. J., 1985, The Antarctic margin and its possible petroleum potential: Tectonophysics, v. 114, p. 443-470.

Deighton, J., D. A. Falvey and D. J. Taylor, 1976, Depositional environments and geotectonic framework: southern Australia continental margin: Australian Petroleum Exploration Association Journal, v. 16, p. 25-36.

Domack, E. W., W. W. Fairchild and J. B. Anderson, 1980, Lower Cretaceous sediment from the East Antarctic continental shelf: Nature, v. 287, p. 625-626.

Dugger, J. A., 1978, Exploiting Antarctic mineral resources-technology, economics, and the environment: University of Miami Law Review, v. 33, n. 2, p. 315-339.

Eittreim, S. L., A. K. Cooper and Scientific Staff, 1984, Marine geological and geophysical investigations of the Antarctic continental margin: USGS Circular 935.

Eittreim, S. L. and G. L. Smith, 1987, Seismic sequences and their distribution on the Wilkes Land margin, *in* S. L. Eittreim and M. A. Hampton, eds., The Antarctic Continental Margin: Geology and Geophysics of Offshore Wilkes Land, Circum-Pacific Council Energy and Mineral Resources, Earth Science Series, v. 5A: Houston, p. 15–43.

Falvey, D. A and J. C. Mutter, 1981, Regional plate tectonics and the evolution of Australia's passive continental mar-

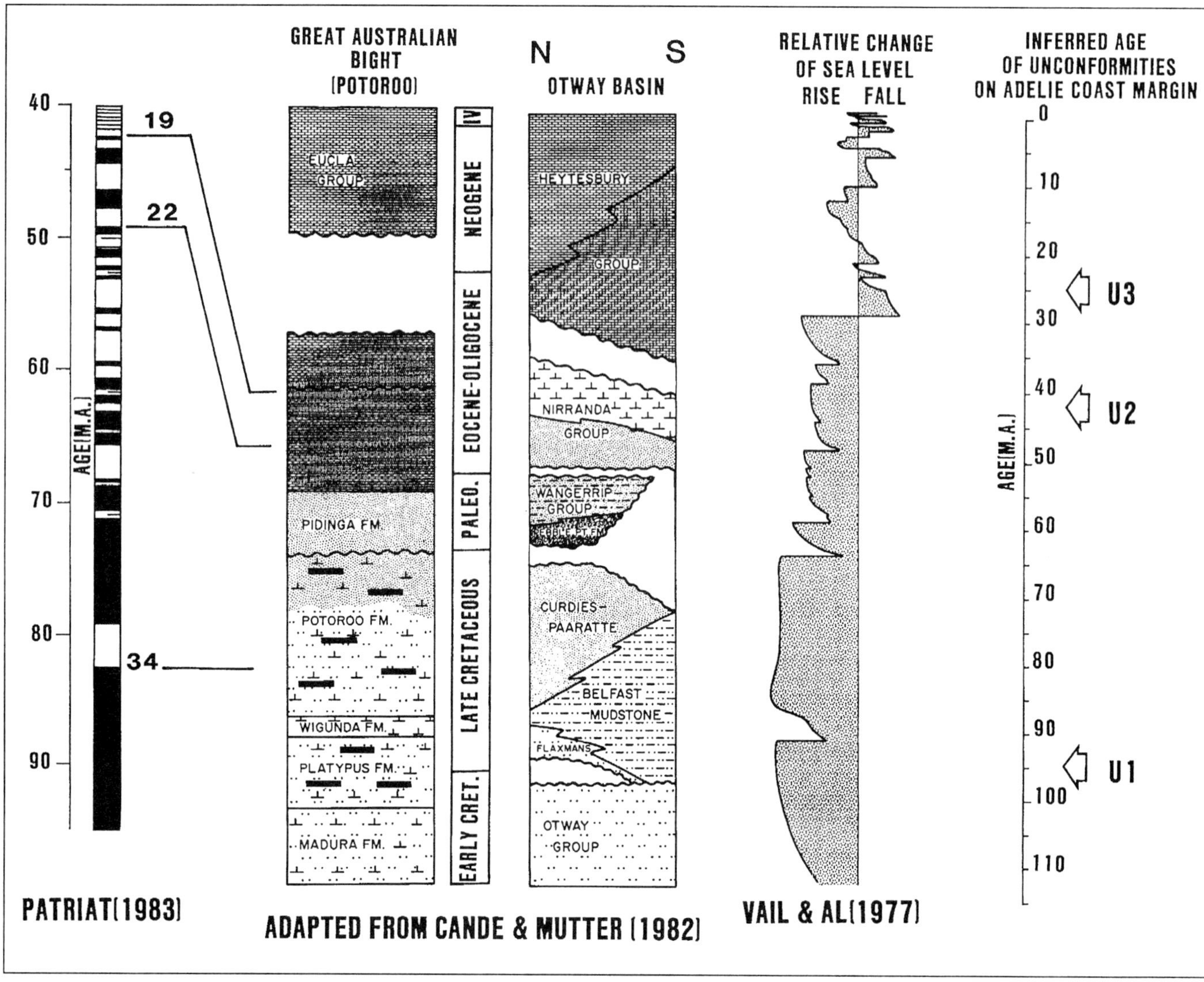

Figure 7—Stratigraphy of the Otway Basin and Great Australian Bight from Cande and Mutter (1982). Comparison with magnetic reversals (Patriat, 1983), relative sea level curve (Vail et al., 1977) and inferred age of unconformities on the Adelie Coast margin.

gins: Bureau of Mineral Resources Journal of Australian Geology and Geophysics, v. 6, p. 1-29.

Griffiths, J. R., 1971, Continental margin tectonics and evolution of south-east Australia: Australian Petroleum Exploration Association Journal, v. 11, p. 75-79.

Haugland, K., 1982, Seismic reconnaissance in the Weddell Sea, *in* C. Craddock, ed., Antarctic Geoscience: Madison, University of Wisconsin Press, p. 405-413.

Haugland, K., Y. Kristoffersen and A. Velde, 1985, Seismic investigations in the Weddell Sea Embayment: Tectonophysics, v. 114, p. 293-313.

Hayes, D. E., 1972, Introduction: Marine geophysics of the Southeast Indian Ocean, *in* D. E. Hayes, ed., The Australian-New Zealand sector: American Geophysical Union, Antarctic Research Series, v. 19, p. 119-124.

Hayes, D. E. and L. A. Frakes, 1975, General synthesis, Deep Sea Drilling Project Leg 28, *in* C. D. Hayes et al., eds., Initial Reports of the Deep Sea Drilling Project, v. 35: Washington, (U.S. Government Printing Office), p. 919-942.

Hinz, K., 1978, Geophysikalische Untersuchungen in Antarktischen Gewassern mit M.S. Explora: Meerestechnik v. 9, n. 3, p. 83-87.

Hinz, K. and W. Krause, 1982, The continental margin of Queen Maud Land/Antarctica: Seismic sequences, structural elements and geological development: Geologisches Jahrbuch, v. E 23, p. 17-41.

Hinz, K. and M. Block, 1984, Results of geophysical investigations in the Weddell Sea and in the Ross Sea, Antarctica: Proceedings 11th World Petroleum Congress, London, 1983: New York, Wiley, p. 279-291.

Hinz, K., et al., 1986, Report on the cruise "Antarctica IV/3" of F.S. Polar Stern: Bundesanstalt für Geowissenschaften und Rohstoffe Report 99194, 156 p.

Holdgate, M. W. and J. Tinker, 1979, Oil and other minerals in the Antarctic: London, House of Print, 51 p.

Ivanhoe, L. F., 1980, Antarctica: Operating conditions and petroleum prospects: Oil and Gas Journal, v. 78, n. 52, p. 212-220.

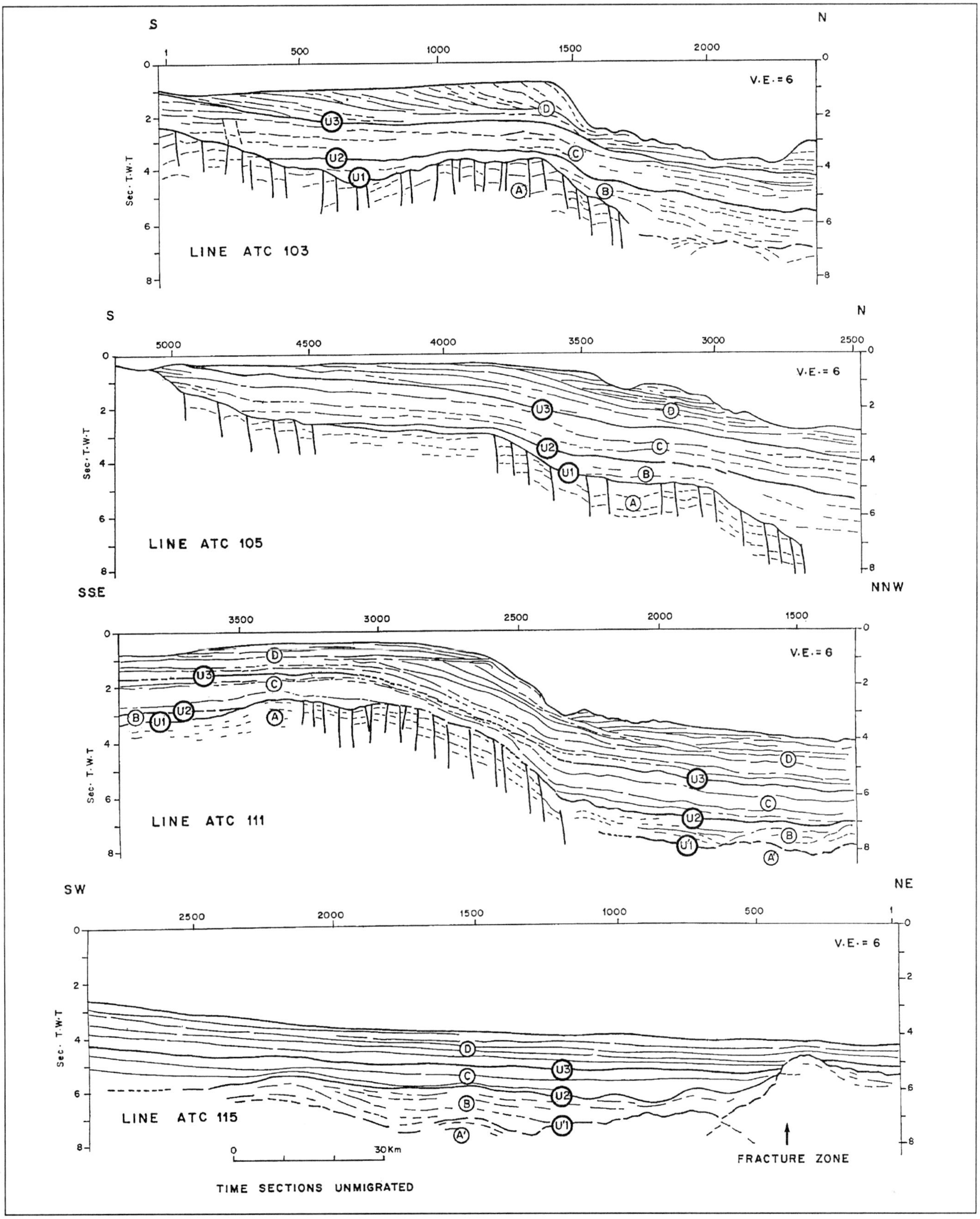

Figure 8—Simplified geoseismic section of lines ATC 103, ATC 105, ATC 111, and ATC 115 on the Adelie Coast/Wilkes Land margin. See location on Figure 1.

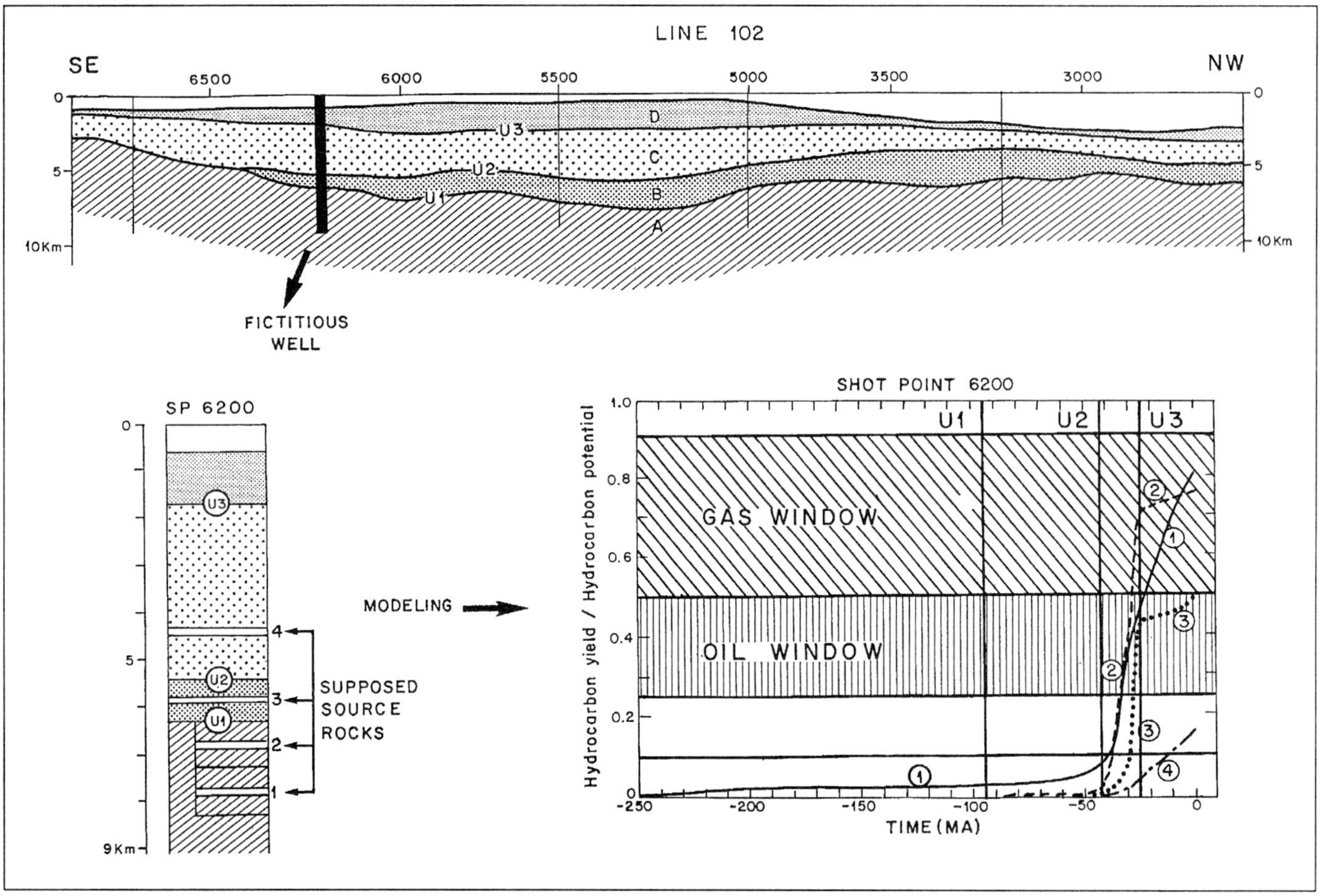

Figure 9—Modeling of source rock maturation on the Adelie Coast margin along seismic line ATC 102.

Kimura, K., 1982, Geological and geophysical survey in the Bellingshausen Basin, off Antarctica: Antarctic Record, n. 75, p. 12-24.

Konig, M., 1980, Geophysical investigations in the southern continental margin of Australia and the conjugate sector of east Antarctica: Ph.D. dissertation, Columbia University, New York, 337 p.

Mawson, D., 1940, Sedimentary rocks. Australasian Antarct. Exp. 911-1914: Scientific Report Series A, n. 4 (2), p. 347-367.

Okuda, Y., T. Yamazaki, S. Sato, T. Saki and N. Oikawa, 1983, Framework of the Weddell Basin inferred from the new geophysical and geological data: Reprint of the National Institute of Polar Research, Tokyo, October 1983.

Patriat, P., 1983, Evolution du systeme de dorsales de l'Ocean Indien: Thèse d'Etat, Universite Pierre et Marie Curie, Paris VI, 350 p.

Piper, A., 1976, Frozen mineral waters: Nature, v. 260, p. 182-183.

Quigg, P. W., 1983, A pole apart, the emerging issue of Antarctica: New York, McGraw-Hill, 299 p.

Rabinowitz, P.D., 1976, Geophysical study of the continental margin of southern Africa: GSA Bulletin, v. 87, p. 1643-1653.

Rivera, R. C., 1977, Perspectivas de desarollo de recursos de hidrocarburos en la Antartica, *in* F. O. Vicuna and A. S. Araya, eds., Desarollo de la Antartica: Santiago, Editorial Universitaria, Instituto de Estudias Internacionales de la Universitad de Chile, p. 265-271.

Sato, S., N. Asakura, T. Saaki, N. Oikawa and Y. Kaneda, 1984, Preliminary results of geological and geophysical surveys in the Ross Sea and in the Dumont d'Urville Sea off Antarctica: National Institute Polar Research, Tokyo, Special Issue 33, p. 66-92.

Schlich, R. and P. Patriat, 1967, Profils magnétiques sur la dorsale médio océanique "indo-pacifique": Annales de Géophysique, v. 23, p. 63-78.

St. John, B., 1986, Antarctica-Geology and hydrocarbon potential, *in* M. T. Halbouty, ed., Future Petroleum Provinces of the World: AAPG Memoir 40, Tulsa, p. 55-100.

Splettstoesser, J. F., 1977, Offshore development for oil and gas in Antarctica: 4th International Conference Port and Ocean Engineering Under Arctic Conditions, University of Newfoundland Memoir, p. 1-10.

Stagg, H. M. J., D. C. Ramsay and R. Whitworth, 1983, Preliminary report on a marine geophysical survey between Davis and Mawson stations, 1982, *in* R. L. Oliver, P. R. James and J. B. Jago, eds., Antarctic Earth Science, Canberra, Australian Academy Science, p. 527-532.

Talwani, M. and O. Eldholm, 1973, The boundary between continental and oceanic basement at the margin of rifted continents: Nature, v. 241, n. 5388, p. 325-330.

Tanahashi, M., T. Saki, N. Oikawa and S. Sato, 1987, An interpretation of the multichannel seismic reflection profiles across the Continental margin of the Dumont d'Urville Sea, off Wilkes Land, East Antarctica, *in* S. L. Eittreim and M. A. Hampton, eds., The Antarctic Continental Margin: Geology and Geophysics of Offshore Wilkes Land, Circum-Pacific Council Energy and Mineral Resources, Earth Science Series, v. 5 A, Houston, p. 1-13.

Tessensohn, F., 1984, Present knowledge of non-living resources in the Antarctic, possibilities for their exploitation and scientific perspectives, *in* R. Wolfrum and K. Bockslaff, eds., Antarctic Challenge, Berlin, Duncker and Hublot, p. 189-210.

Tissot, B., B. Durand, J. Espitalié and A. Combaz, 1974, Influence of nature and diagenesis of organic matter in formation of petroleum: AAPG Bulletin, v. 58, n. 3, p. 499-506.

Truswell, E., 1982, Palynology of seafloor samples collected by the 1911-1914 Australasian-Antarctic expedition: implications for the geology of coastal East Antarctica: Journal Geological Society of Australia, v. 29, p. 343-356.

Tsumuraya, Y. M., T. Saki, T. Machihara and N. Asakura, 1985, Preliminary report on the geological and geophysical surveys off Wilkes Land, Antarctica in 1983-1984: Memoirs–National Institute of Polar Research, Special Issue, n. 37, p. 48-62.

Ungerer, P. and R. Pelet, 1987, Extrapolation of oil and gas formation kinetics from laboratory experiments to sedimentary basins: Nature, v. 327, p. 52-54.

Ungerer, P. et al., 1984, Geological and geochemical models in oil exploration: principles and practical examples, *in* G. Demaison and R. J. Murris, eds., Petroleum Geochemistry and Basin Evaluation, AAPG Memoir 35, Tulsa, p. 53-78.

Vail, P. R., R. M. Mitchum and S. Thomson III, 1977, Seismic stratigraphy and global changes of sea level, part 3: Relative changes of sea level from coastal onlap, *in* C. E. Payton, ed., Seismic stratigraphy: Application to hydrocarbon exploration, AAPG Memoir 26, Tulsa, p. 63-97.

Veevers, J. J., 1986, Breakup of Australia and Antarctica estimated as mid-Cretaceous (95 ± 5 Ma) from magnetic and seismic data at the continental margin: Earth Planetary Science Letters, v. 77, p. 91-99.

Wannesson, J., M. Pelras, B. Petitperrin, M. Perret and J. Segoufin, 1985, A geophysical transect of the Adelie Margin, East Antarctica: Marine and Petroleum Geology, v. 2, p. 192-201.

Weissel, J. K. and D. E. Hayes, 1972, Magnetic anomalies in the southeast Indian Ocean, *in* D. E. Hayes, ed., Antarctic Oceanology II: The Australian New Zealand sector, American Geophysical Union, Antarctic Research Series, v. 19, p. 165-196.

Wright, N. A. and P. L. Williams, 1974, Mineral resources of Antarctica: USGS Circular n. 705, 29 p.

Zumberge, J. H., 1979, Mineral resources and geopolitics in Antarctica: American Scientist, v. 67, p. 68-77.

Chapter 7

A Tectonic Chart for the Southern Ocean Derived from Geosat Altimetry Data

Jean-Yves Royer[1]
Lisa M. Gahagan
Lawrence A. Lawver
Cathy L. Mayes[2]
Dirk Nürnberg[3]
David T. Sandwell[4]
Christopher R. Scotese[5]

Institute for Geophysics
The University of Texas at Austin
Austin, Texas, U.S.A.

[1]Now at Laboratoire de Géodynamique sous-marine, CEROV, Villefranche sur mer, France
[2]Now at Sierra Geophysics Inc., Houston, Texas, U.S.A.
[3]Now at GEOMAR, Christian Albrechts University, Kiel, Federal Republic of Germany
[4]Now at Scripps Institution of Oceanography, La Jolla, California, U.S.A.
[5]Now at Shell Oil Company, Bellaire Research Center, Houston, Texas, U.S.A.

ABSTRACT

We present a new tectonic fabric map of the Southern Ocean south of 45°S derived from Geosat altimeter profiles and published bathymetric charts and magnetic anomaly picks. The interpretation of the Geosat data is based on an analysis of the first derivative of the geoid profiles (i.e., vertical deflection profiles). To improve the accuracy and resolution of the vertical deflection profiles, 22 repeat cycles from the first year of the Geosat/Exact Repeat Mission (Geosat/ERM) were averaged. At wavelengths less than about 200 km, the vertical deflection is highly correlated with sea-floor topography and thus reveals major features in areas that were previously unsurveyed. The density of the Geosat data is greatest in the high latitudes where lineated bathymetric features such as fracture zones, spreading ridges, trenches, and rifted margins stand out. To construct the tectonic fabric chart, the Geosat data are analyzed in combination with available shipboard bathymetric data and magnetic anomaly identifications.

INTRODUCTION

Over the past decade, satellite altimetry has become an important technique for studying the tectonics of remote ocean areas (Haxby et al., 1983; Sandwell, 1984). Spacecraft such as Seasat and Geosat use pulse limited radars, along with very accurate orbits, to measure the topography of the ocean surface (see, Johns Hopkins Applied Physics Laboratory Technical Digest, 1987). Since the sea surface is nearly an equipotential surface of the Earth's gravity field (marine geoid), variations in sea surface topography reveal variations in gravitational potential. At short wavelengths (< 200 km), the topography of the sea surface mimics the sea-floor topography. The accuracy and resolution of these satellite-derived gravity measurements are now comparable to state-of-the-art shipboard gravity measurements. Thus satellite altimeters provide important reconnaissance information over vast areas of uncharted sea floor such as the Southern Ocean and Antarctic continental margins.

The close correlation between the short wavelengths of the geoid (20-200 km) and the sea-floor topography allows altimetry data to be used to chart tectonic features of the ocean floor such as fracture zones, seamounts, and spreading ridges. This information is critical in the Southern Ocean where there are large regions containing few if any shipboard data. Here we present a new tectonic map of the oceans south of 45°S derived from a combination of Geosat altimeter profiles and published bathymetric charts, and magnetic anomaly picks. The map provides new detailed information on the dispersal of the Gondwana fragments from Antarctica and on the tectonic evolution of the South Pacific Ocean basin.

Recent interpretation of vertical deflection profiles (i.e., the along-track derivative of the geoid profiles) (Gahagan et al., 1988) from Seasat data have permitted the construction of a preliminary tectonic chart of the world's ocean floor. The next step in the analysis of the altimeter data is to compare the lineations, identified in the vertical deflection data, with the available bathymetric data and marine magnetic anomalies to derive a tectonic fabric map of the ocean floor. Such an approach has already been applied in the South Pacific (Mayes et al., 1990) and in the Indian Ocean (Royer et al., 1989). In this chapter, we complement the latter analyses by including the tectonic fabric of the far South Atlantic. The combined maps form a new general tectonic chart for the Southern Ocean. This article is a companion paper to that of Sandwell and McAdoo (1988), which presented the Geosat altimeter data collected around Antarctica and to the paper by Lawver et al. (in press) which describes the evolution of the Antarctic continental margins.

Virtually every type of plate boundary, continental margin, oceanic island, and deep-sea basin are found in the Southern Ocean south of 45°S (Figure 1, oversize enclosure). The Antarctic plate covers about 75% of this area; the remaining 25% is shared by the South American, African, Australian, Pacific, and Scotia plates. The present-day plate boundaries include the following sea-floor spreading ridges: the South Atlantic Ridge, the South American-Antarctic Ridge, the Southwest Indian Ridge, the Southeast Indian Ridge, the Pacific-Antarctic Ridge, the Chile Rise, and the South Sandwich backarc spreading center. They include the following transform boundaries: the Shackleton Fracture Zone and the north and south Scotia Ridges which are the transform fault system that bounds the Scotia Sea; and the following trenches: the Hjort Trench south of New Zealand, the Chile Trench, and the South Sandwich Trench. The deep-sea basins in the Southern Ocean are the Weddell Abyssal Plain, the African-Antarctic Basin, the Enderby Basin, the Australian-Antarctic Basin, the Amundsen and Bellingshausen Abyssal Plains off West Antarctica, the Mornington Abyssal Plain west of South America, and the Scotia Sea. In addition, the Southern Ocean is the site of many submarine platforms of various origin among which the largest are the Falkland Plateau, the Conrad Rise, the Kerguelen Plateau, the Chatham Rise, and the Campbell Plateau.

Summary of the Tectonic History of the Southern Ocean

Except for the Southeast Pacific, the opening of the Southern Ocean resulted in the dispersal of the Gondwana fragments. This presentation of the major events in the formation of the Southern Ocean goes clockwise from the Weddell Sea as the ocean floor surrounding the Antarctic margins becomes younger.

Many models have been proposed for the early opening of the Weddell Sea and the motion of the Antarctic Peninsula relative to South America and East Antarctica (see discussion in Lawver et al., in press), and the early history is still controversial. LaBrecque and Barker (1981) and LaBrecque and Cande (1986) have identified Late Jurassic magnetic anomalies in the Weddell Sea (M25 to M29; 157 to 160 Ma). In order to reconcile the direction of motion inferred by these magnetic lineations with the clockwise motion of the Antarctic Peninsula suggested by paleomagnetic results (Grunow et al., 1987), the age of these magnetic lineations would have to be younger, i.e., Early Cretaceous. A younger age and the orientation of these lineations would be compatible with regard to the development of a triple junction with initiation of sea-floor spreading between South America and Africa. Since most of the conjugate half of the Weddell Sea Basin anomalies have been subducted beneath the Scotia Sea, the age discrepancy is difficult to resolve. In contrast, the evolution of the South American-Antarctic Ridge since the Late Cretaceous (Chron 34, 84 Ma) is better understood (Barker and Jahn, 1980; Bergh and Barrett, 1980; LaBrecque and Barker, 1981; LaBrecque and Keller, 1982; Lawver and Dick, 1983; Barker and Lawver, 1988). Two major changes in spreading direction have been recognized: one gradual change starting at Chron 31 (68 Ma, Paleocene) and one abrupt change at Chron 6 (20 Ma, early Miocene).

The South Atlantic began to open in the Early Cretaceous and shows symmetric Mesozoic magnetic anomalies (M4 to M0; 124 to 118 Ma) in the Argentina and Cape Basins, and in the Georgia Basin and the Natal Valley (Rabinowitz and LaBrecque, 1979; Goodlad et al., 1982; Martin et al., 1982). Except for several ridge jumps south of the Falkland-Agulhas Fracture Zone which have progressively eliminated most of the initial 1400 km offset across this fracture zone, the geometry of the South Atlantic spreading has remained very stable with only a few small changes in the direction of spreading (Cande et al., 1988). The largest ridge jump occurred at Chrons 31/28 (68/64 Ma) when the ridge jumped from the Agulhas Basin westward to Meteor Rise (Barker, 1979; LaBrecque and Hayes, 1979). This reorganization at Chrons 31/28 occurred contemporaneously with a slowing down of the spreading rate in the South Atlantic and with drastic changes in rates or directions of spreading in the Indian Ocean.

The early separation of Africa and Antarctica is well documented by symmetric Mesozoic magnetic anomalies (M16 to M0, 142 to 118 Ma) identified off Dronning Maud Land (Antarctica) (Bergh, 1977, 1987) and in the Mozambique Channel (Ségoufin, 1978; Simpson et al., 1979). Since then, sea-floor spreading along the Southwest Indian Ridge has been continuous with medium to slow spreading along a general north-south direction (Bergh and Norton, 1976; LaBrecque and Hayes, 1979; Patriat, 1979, 1987; Bergh and Barrett, 1980; Sclater et al., 1981; Fisher and Sclater, 1983; Bergh, 1986; LaBrecque and Cande, 1986). Two major changes in the direction of motion occurred between Chrons 32 (74 Ma) and 24 (56 Ma) (Patriat et al., 1985; Royer et al., 1988).

The age of the rifting of India and Sri Lanka from Antarctica is not well known but probably began between Chrons M10 and M0 (130 and 118 Ma). No magnetic anomalies have yet been identified off eastern India or in the Enderby Basin off Antarctica. The magnetic anomaly pattern off western Australia dates the separation of Greater India as Chron M9/M10 (129/130 Ma) (Markl, 1974, 1978; Veevers et al., 1985). The cessation of sea-floor spreading in the Somali Basin between Africa and the Madagascar-India block just subsequent to Chron M0 (118 Ma) (Ségoufin and Patriat, 1980) gives a lower limit for the initiation of motion between India and Antarctica. Paleogeographic reconstructions of the Mesozoic basins in the Indian Ocean (Lawver et al., 1985) provide further indication that India cannot have been separated from Antarctica earlier than Chron M10 and that their separation probably occurred before Chron M0 (Lawver et al., 1985; and in press). The next event in the development of the southern Indian Ocean corresponds with a major reorganization of the spreading centers during the mid-Cretaceous (~96 Ma). At that time, India separated from Madagascar and began its northward drift toward Asia that is later documented by the Paleogene sequence of magnetic anomalies in the Mascarene, Madagascar, and Crozet Basins (McKenzie and Sclater, 1971; Schlich, 1975, 1982; Norton and Sclater, 1979; Patriat, 1987). The breakup between Australia and Antarctica is also dated at 95 Ma (Cande and Mutter, 1982; Veevers, 1986). Sea-floor spreading in the Australian-Antarctic Basin is characterized by very slow spreading rates from Chron 34 (84 Ma) to Chron 20 (45 Ma) (Cande and Mutter, 1982). The most recent major tectonic event to affect the Indian Ocean basins occurred in the middle Eocene when India collided with Eurasia. Directions of motion changed dramatically in the central Indian Ocean (Schlich, 1975, 1982; Patriat, 1987). Sea-floor spreading initiated between the Kerguelen Plateau and Broken Ridge (Mutter and Cande, 1983; Royer and Sandwell, 1989) as the spreading rates increased between Australia and Antarctica (Weissel and Hayes, 1972; Cande and Mutter, 1982; Vogt et al., 1983). This is also the time when sea-floor spreading began south of the Tasman Sea (Weissel et al., 1977).

Prior to 100 Ma, a subduction zone extended from north of New Zealand, which was attached to Marie Byrd Land, to the tip of the Antarctic Peninsula (Barker, 1982). In the southwest Pacific, the breakup between South New Zealand and Marie Byrd Land occurred in two steps during the Late Cretaceous. Chatham Rise split from the Campbell Plateau and Marie Byrd Land prior to Chron 34 (84 Ma), opening up the Bounty Trough (Davey, 1977). Chatham Rise and the Campbell Plateau then rifted away from Marie Byrd Land at approximately Chron 34 (84 Ma) (Mayes et al., 1990). The sea-floor magnetic anomaly pattern in the South Pacific records two major plate boundary reorganizations that occurred after the Late Cretaceous (Christoffel and Falconer, 1972; Herron, 1972; Molnar et al., 1975; Herron and Tucholke, 1976; Weissel et al., 1977; Barker, 1982; Cande et al., 1982; Cande and Leslie, 1986; Stock and Molnar, 1987). From Chron 32 (74 Ma) to Chron 25 (59 Ma), at least four different spreading centers were active (Figure 2A): one between the Pacific plate and the Marie Byrd Land part of West Antarctica, one between the Pacific and Bellingshausen-West Antarctic plates, one between the Pacific and Farallon plates and one between the Bellingshausen and Aluk plates. A fifth plate boundary between the Bellingshausen plate and Marie Byrd Land has not been documented. Between Chron 25 and Chron 21 (50 Ma), the Bellingshausen plate and a part of the Pacific plate transferred onto the Marie Byrd Land or Antarctic plate (Figure 2B), leading to a Pacific-Antarctic-Farallon-Aluk plate system. The next major reorganization took place in the late Oligocene between Chron 7 (26 Ma) and Chron 6 (21 Ma). The Farallon plate split into the Nazca and Cocos plate while the southern tip of the Chile Rise (Antarctic/Nazca) started subducting under South America. The collision of the Antarctic/Aluk spreading center with the Antarctic Peninsula resulted in the stabilization of the Antarctic continental margin. As a result of these collisions, changes in the spreading direction are observed simultaneously along the East Pacific Rise (Pacific/Nazca), the Chile Rise, and the Pacific-Antarctic Rise.

The development of the Scotia Sea and dispersal of the continental fragments of the Scotia Arc are extremely complex (e.g., review by Barker and Dalziel, 1983). At least four different spreading systems have been identified in the Scotia Sea (Barker, 1972; Barker and Burrell, 1977; Hill and Barker, 1980; British Antarctic Survey, 1985). The oldest magnetic anomaly identified in the Scotia Sea is Chron 10 (30 Ma) (LaBrecque and Rabinowitz, 1977). The most recent reorganization of the spreading system occurred at Chron 4a (8 Ma) when the spreading ridge west of the South Sandwich Arc became active.

Tectonic Fabric of the Southern Ocean

Figure 3 (oversize enclosure) presents the lineated structures of the Southern Ocean interpreted from the vertical deflection profiles (Geosat). The procedure to produce such a map is fully described in Mayes et al. (1990) and Royer et al. (1989). The vertical deflection profiles are presented in Sandwell and McAdoo (1988). Since the Geosat coverage in the southern polar region, where seasonal sea ice can corrupt the radar altimeter data, is more complete than Seasat coverage (Geosat has operated for two austral summers while Seasat only operated during one austral winter), our interpretations are based solely on Geosat data. In order to reduce the noise level of the vertical deflection profiles from Geosat, 22 exact repeat profiles (1 year) have been averaged; this also improves coverage in areas of temporary sea ice. After averaging, a digital, band-pass filter was used to eliminate wavelengths greater than 4000 km and smaller than 20 km. The 164 km spacing of the Geosat profiles at the Equator decreases toward the high latitudes, being 82 km at 60°S. The uniform coverage and high density of the Geosat data are particularly suitable for tectonic fabric mapping.

Our interpretation, shown in Figure 3, is based on averaged and filtered vertical deflection profiles. The peaks (blue symbols) and troughs (red symbols) of the vertical deflection data correspond to maxima and minima in the gradient of the marine geoid, respectively. Peaks and troughs in vertical deflection are well correlated with the gradient in the bathymetry (Figure 4). After identifying the maxima and minima along the profiles, we interpret and connect peaks and troughs from profile to profile (Figure 5). These interpreted lineations (fabric) reflect lineated features of the sea floor such as trenches, fracture zones, and spreading ridges. Continental shelf breaks, which are often poorly

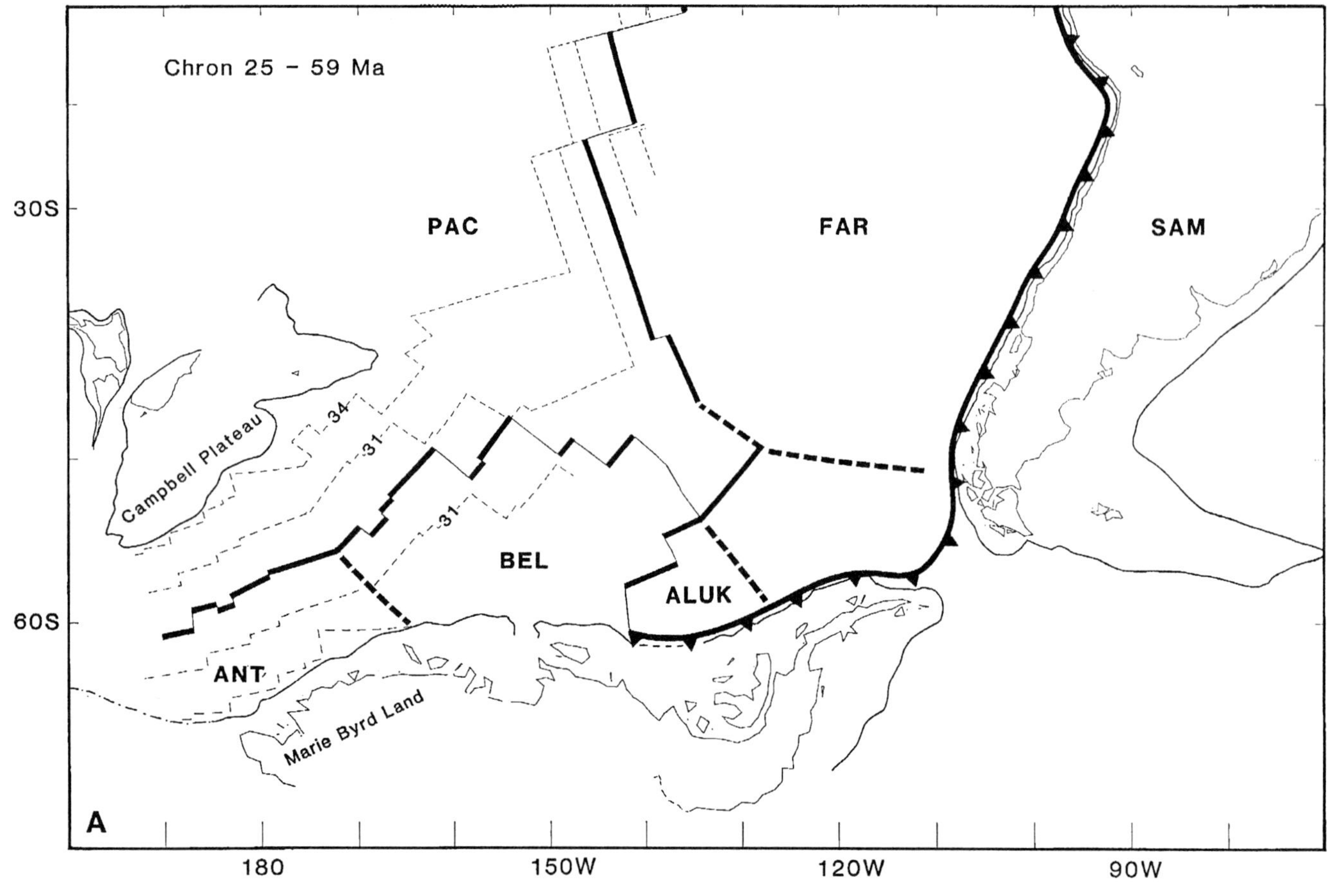

Figure 2—Reconstructions of the South Pacific at A) Chron 25 (59 Ma), B) Chron 21 (50 Ma), and C) present-day chart of the South Pacific that outlines the two main changes in the plate geometry since the Late Cretaceous: between Chrons 25 and 21, and between Chrons 7 and 5 (after Mayes et al., 1990). Thick line segments represent the spreading ridge segments. Undocumented but suspected plate boundaries are represented by a thick dashed line. Saw-toothed lines represent the trenches; thin lines with open saw-teeth show the extinct trenches. Thin dashed lines represent the isochrons still present in the Southern Pacific, and the numbers refer to the corresponding chrons (Chrons 34, 31, 25, 21, 7 and 5; respectively: 84, 68, 59, 50, 26 and 11 Ma). The stippled area shows the portion of the Pacific plate that transferred onto the Antarctic plate between Chrons 25 and 21. ANT = Antarctic plate, BEL = Bellingshausen plate, FAR = Farallon plate, PAC = Pacific plate, SAM = South American plate.

charted around Antarctica because of the lack of shipboard data, also appear as lineations in the data. Isolated features such as seamounts and submarine ridges produce strong signatures in the vertical deflection profiles. However, because many of these features are probably narrower than the spacing of Geosat profiles, they are not well resolved by this technique. For the sake of clarity, the major seamounts and submarine plateaus in Figure 3 are outlined by their bathymetric contours rather than by the scattered peaks and troughs observed in the vertical deflection.

A New Tectonic Chart for the Southern Ocean

Figure 6 combines the structural information interpreted from the vertical deflection profiles with a compilation of magnetic picks and bathymetric contours. The sources for the magnetic anomaly data have been acknowledged in a previous section. The bathymetric contours have been digitized from the GEBCO chart series (Johnson et al., 1980; Falconer and Tharp, 1981; Hayes and Vogel, 1981; LaBrecque and Rabinowitz, 1981; Fisher et al., 1982; Mammerickx and Cande, 1982; Monahan et al., 1982). We have also incorporated recent detailed mapping of the Astrid Ridge (Bergh, 1987), the Conrad Rise (Driscoll et al., 1985, unpublished manuscript) and the Kerguelen Plateau (Schlich et al., 1987).

Fracture zones are the most numerous tectonic features of the Southern Ocean seen in the lineation map (Figures 3, 4). The most prominent fracture zones in the South Atlantic are the Falkland-Agulhas Fracture Zone and the Bullard Fracture Zone east of the South Sandwich Trench. The en échelon Du Toit, Bain, and Prince Edward Fracture Zones offset the Southwest Indian Ridge by more than 800 km. The Kerguelen Fracture Zone separates the Crozet and Enderby Basins. The George V, Tasman, and Balleny Fracture Zones connect the Antarctic and south Australian margins. The main fracture zones in the South Pacific are the large-offset Udintsev, Tharp, Heezen, and Tula Fracture Zones, and the Hero and Shackleton fracture zones in the Drake Passage. All of these major fracture zones produce one or more prominent lineations in the vertical deflection.

The main information contained in the interpretation of

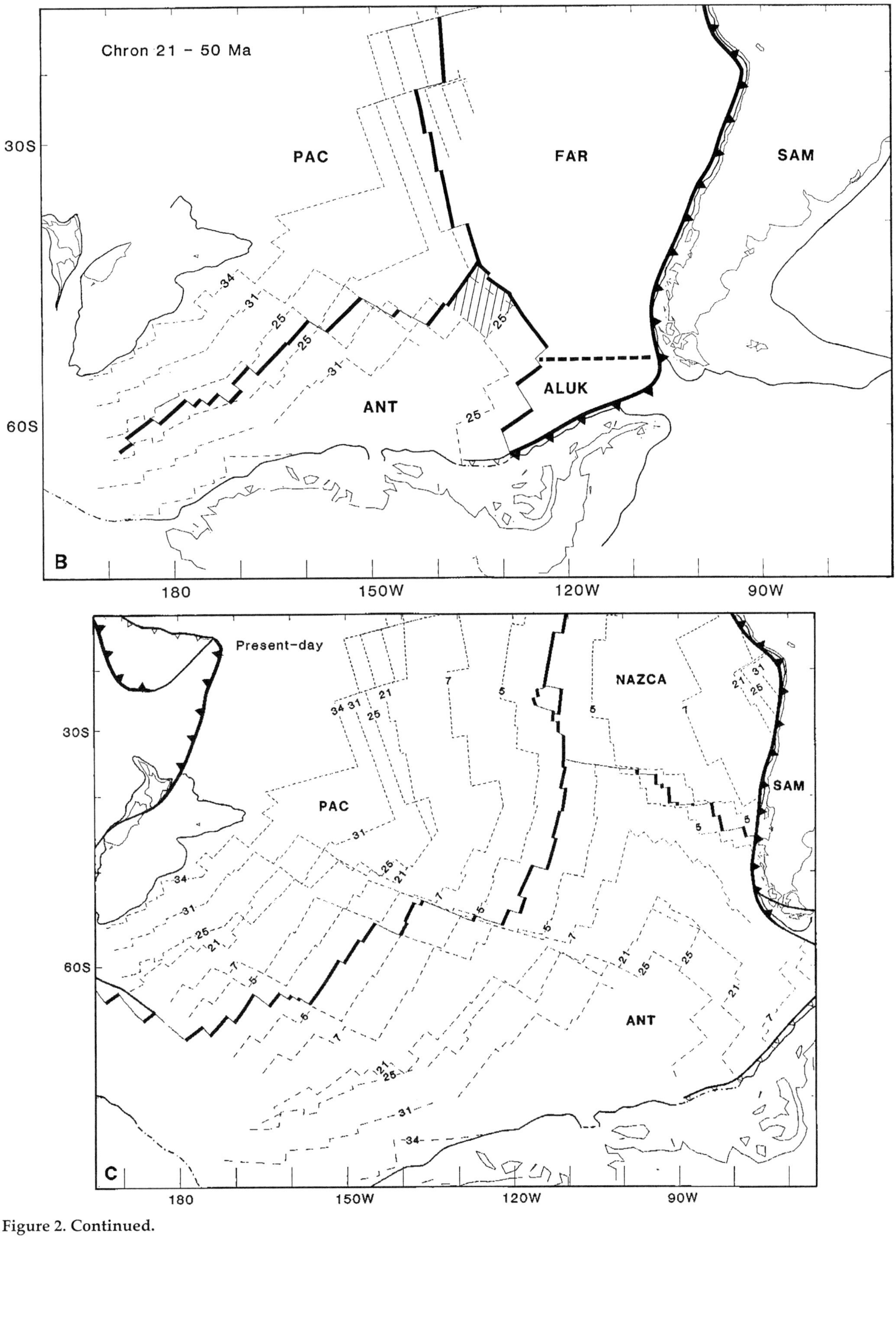

Figure 2. Continued.

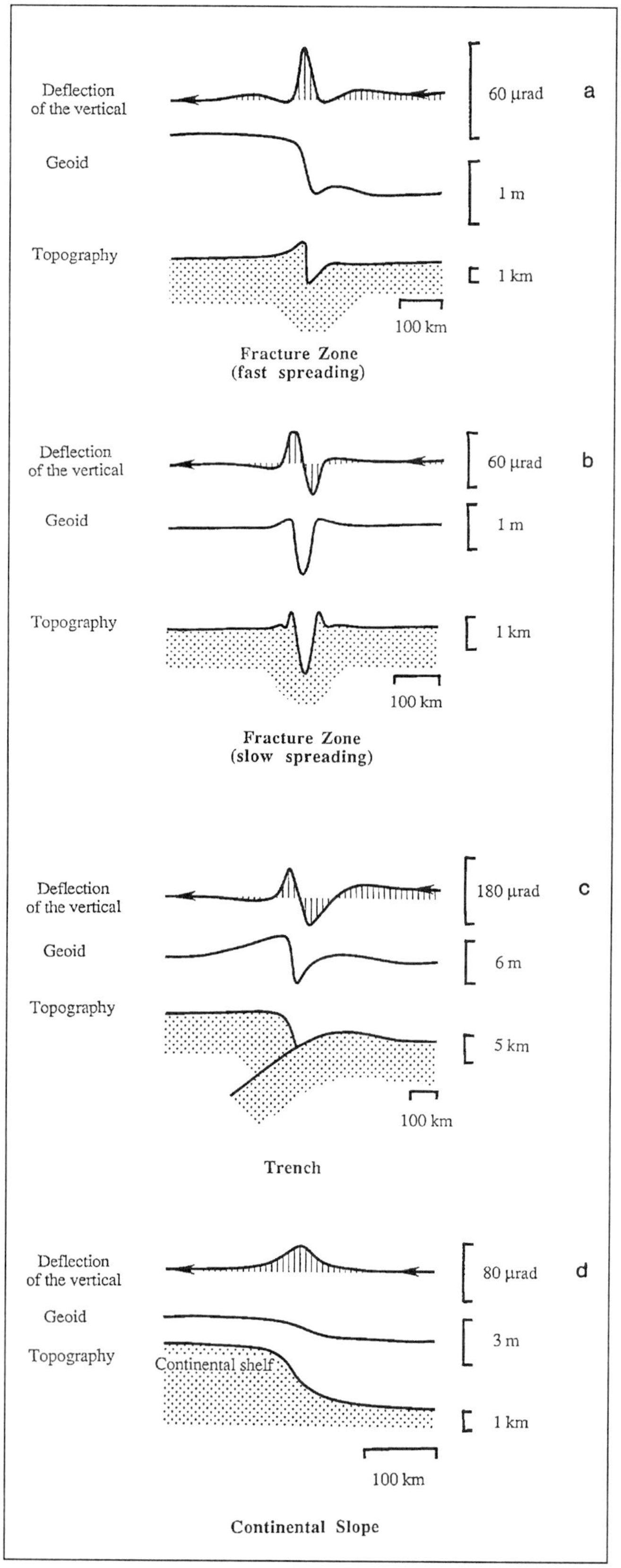

Figure 4—Examples of signatures in the vertical deflection profiles associated to lineated fabric of the ocean floor: a) and b) fracture zones, c) a trench, and d) a continental margin.

the vertical deflection profiles with respect to fracture zone lineations concerns the inferred spreading directions. Even when the sea-floor spreading history appears to be uniform such as in the South Atlantic, subtle changes in the Geosat lineations can be observed (Cande et al., 1988). In other instances, fracture zone lineations record continuous changes in spreading direction such as in the South Pacific. On the Antarctic plate in the Pacific (90°W to 180°, 50°S to 70°S), the orientations of the fracture zones shift progressively from southeast-northwest to almost east-west. As a result, the large offset Heezen and Tharp Fracture Zones converge toward each other as they approach the Pacific-Antarctic Ridge. The change of motion at Chron 6 along the South American-Antarctic Ridge (Barker and Lawver, 1988) is seen as a series of parallel, arcuate lineations in the Weddell Sea Basin. A more abrupt change of motion is indicated in the African-Antarctic Basin where the fracture zones north of the Astrid Fracture Zone display two sharp bends, first toward the west and then toward the east, corresponding to two changes of motion of opposite senses along the Southwest Indian Ridge. These changes occurred between Chrons 32 and 24 (Royer et al., 1988). The juxtaposition of fracture zone trends are evidence for reorganization in the plate boundaries. For instance, south of Conrad Rise in the Indian Ocean, two tectonic trends are observed. The north-northeast–south-southwest oriented lineations reflect the direction of motion between Antarctica and Africa from the Late Jurassic to Late Cretaceous while the southwest-northeast fracture zone in the Crozet Basin records the drift of India away from Antarctica in the Late Cretaceous and Paleogene. This change corresponds to the mid-Cretaceous reorganization of the plate boundaries in the Indian Ocean. In the South Atlantic, south of the Bouvet Triple Junction, the fracture zone lineations related to the Southwest Indian Ridge intersect the fracture zones from the South American-Antarctic Ridge (Sclater et al., 1976; Barker and Lawver, 1988); the cusps defined by the fracture zone lineations may be interpreted as the trace left by the Bouvet Triple Junction on the Antarctic plate.

In addition to spreading direction, the amplitudes and spacings of fracture zone lineations (Figure 3) depend somewhat on spreading rate. Fracture zones that formed at slow spreading ridges (< 35 mm/a half rate) are characterized by a deep topographic trough and correspondingly high amplitude vertical deflection signatures (Figure 4B). In contrast, fracture zones that formed at medium or fast spreading ridges (> 35 mm/a) have much lower-amplitude topographic and vertical deflection signatures; along these fracture zones the step caused by the age offset in the sea floor dominates (Figure 4A). The spacing of fracture zones also depends somewhat on spreading rate; fracture zones that formed at slow spreading ridges are more closely spaced than fracture zones that formed at fast spreading ridges (Sandwell, 1986). The overall result of these two rate-dependent processes is that vertical deflection lineations are more abundant and prominent on sea floor that formed at low spreading rates (e.g., South Atlantic, Figure 5) than they are on sea floor that formed at higher spreading rates (e.g., South Pacific, Figure 5). Thus an examination of the vertical deflection profiles can reveal important changes in spreading rates (Small and Sandwell, 1989).

For example, in the South Atlantic, the vertical deflection data shows very few lineations over the sea floor generated during the Cretaceous Magnetic Quiet Period or during the Late Cretaceous (Argentina Basin). When the spreading rates dropped in the late Paleocene (Chron 31), new fracture zones appeared, although some of them do not extend beyond Chron 20 when the spreading rates increased slightly (Cande

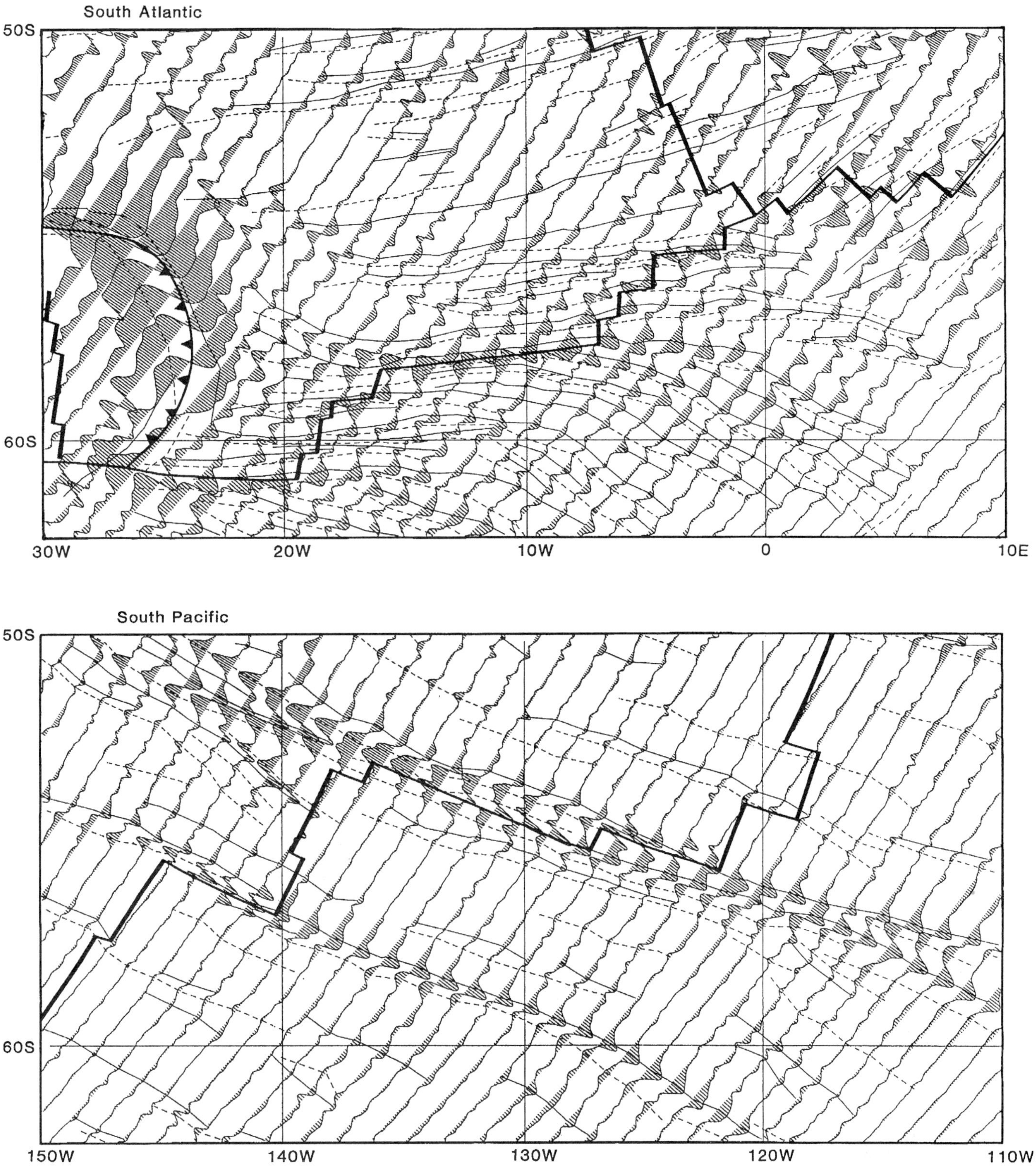

Figure 5—Interpreted vertical deflection profiles in the South Atlantic (top) and the South Pacific (bottom). Vertical deflection is plotted at right angle to the descending ground tracks (i.e., going northeast to southwest), every two data points (~ 7 km), positive to the north. Amplitude scale is 30 μrad (microradians) per degree of longitude. Plain and dashed lines show profile-to-profile correlations of the peaks and troughs in the vertical deflection data. Plates boundaries in the South Atlantic are the South Atlantic Ridge, the Southwest Indian Ridge, and the South American-Antarctic Ridge which connect at the Bouvet triple Junction (at 54°S, 0°E); to the west are the South Sandwich Trench and backarc spreading center. In the South Pacific, the Pacific-Antarctic Ridge is offset from north to south by the Heezen and Tharp Fracture Zones (Eltanin Fracture Zone System) and by the Udintsev Fracture Zone.

et al., 1988). Similar disruptions in the fracture zone pattern occur in the Southwest Pacific where they clearly outline the increase in the sea-floor spreading velocity at Chron 5 (Mayes et al., 1990). Although sea-floor spreading between Wilkes Land (Antarctica) and Australia was initiated in the Late Cretaceous at very slow spreading rates (Cande and Mutter, 1982), fracture zones can clearly be observed only at Chron 18 when the spreading rate drastically increased from 10 to 22 mm/a (half rate). The apparent absence of fracture zones during the slow spreading phase may be caused by the sediment coverage in the vicinity of the margins; however, the magnetic lineations in this area are continuous and are parallel to the coastlines. Another change in the spreading regime may have occurred in the Late Cretaceous in the Weddell Sea Basin (~ 70°S) where abrupt changes in the roughness of the vertical deflection profiles are observed (Haxby, 1988).

Other lineations observed in the Geosat altimeter data can be related to spreading ridges and topographic features produced on active and passive margins. Although spreading centers produce typical signatures in the deflection of the vertical profiles, the spacing of the profiles do not generally permit an accurate mapping of the spreading ridge segments. There are two exceptions south of 45°S where ridge segments can be sampled over several profiles: in the African-Antarctic (Southwest Indian Ridge) and the Australian-Antarctic (Southeast Indian Ridge) basins. Nevertheless, the vertical deflection data can be used to complement identifications of spreading axes that have previously been identified from bathymetric, magnetic, and earthquake epicenter data. The Southwest Indian Ridge is a very slow spreading ridge (< 10 mm/a) characterized by a deep inner valley (Bergh, 1986, unpublished manuscript; LaBrecque and Rabinowitz, 1981), whereas the Southeast Indian Ridge, spreading at medium rate (~30 mm/a), shows a central rise (e.g., Weissel and Hayes, 1972; Vogt et al., 1983). The faster Pacific-Antarctic ridge has smoother topography (e.g., Molnar et al., 1975) which produces only a small signal in the geoid (Sandwell and McAdoo, 1988). Because of the segmentation of the South Atlantic Ridge south of 45°S and its oblique orientation with respect to the Geosat profiles, the vertical deflection signature produced by the South Atlantic Ridge is difficult to distinguish from the fracture zone signal. A detailed analysis of vertical deflection signatures over spreading ridges is presented in Small and Sandwell (1989). Some previously identified extinct spreading centers in the Scotia Sea and the South Pacific can also be mapped from the Geosat profiles. They are seen as major lineations at a right angle to the fracture zone pattern. In the South Pacific, an extinct spreading center can be seen at about 60°S, 100°W on the Antarctic plate (Mayes et al., 1990); the new location of the spreading center is clearly visible on the Pacific plate at 48°S, 146°W. The abandonment of the spreading center corresponds to the plate boundary reorganization that took place between Chrons 25 and 21 in the South Pacific (Figures 2A, 2B).

Trenches are easily identified in the vertical deflection data by their large signatures (e.g., Figure 5). Even fossil trenches such as the one west of the Antarctic Peninsula (Figure 2C) produce a large signature. Troughs associated with presently active or old plate boundaries such as the Hudson Trough in the southeast Pacific, the Falkland Trough south of the Falkland Plateau, and the trough southeast of South Orkney Island in the Weddell Sea are visible in the Geosat data. Also conspicuous are the rifted margins of submarine plateaus: the Kerguelen Plateau which broke apart from Broken Ridge in the middle Eocene (Mutter and Cande, 1983; Coffin et al., 1986; Royer and Sandwell, 1989), and the Campbell Plateau which split off Marie Byrd Land in the Late Cretaceous (Christoffel and Falconer, 1972; Mayes et al., 1990). The continental shelf breaks of Antarctica are identifiable from the Astrid Ridge to George V Land south of Tasmania. Unfortunately, the Antarctic continental margins along the Ross Sea and Marie Byrd Land are south of the coverage of the Geosat orbits (72° of latitude). The new ERS1 satellite, whose orbits will reach 81° of latitude, will cover these remote places. South of Australia, correlations between the vertical deflection data with the continent ocean boundary deduced from magnetic (König, 1980, 1987) and seismic (Talwani et al., 1979; Veevers, 1986, 1987) evidence permit the definition of a criterion for recognition of the structural limit on the Geosat data (Royer and Sandwell, 1989). Such criterion has permitted the mapping of the conjugate continent-ocean boundary off Wilkes Land and is in agreement with interpretations of the few seismic lines available in this remote area (Eittreim and Smith, 1987).

Several intriguing features which are conspicuous in the Geosat data are not yet understood. West of the Kerguelen Plateau and a few degrees southeast of the Kerguelen Fracture Zone, there is a small visible ridge (~ 68°S, 62°E) striking southwest-northeast which may correspond to the Early Cretaceous spreading center between India and Antarctica. In the southernmost part of the Weddell Sea Basin, the vertical deflection data outline a series of small and parallel lineations which cannot be extended to the fracture zone lineations lying farther north. Haxby (1988) speculated that they define an early direction of motion between East Antarctica and South America.

CONCLUSION

The high-density Geosat data set permits the interpretation of structural trends on the ocean floor. By combining this information with the sparse ship-track data set describing the ages of the sea floor, we have derived a tectonic fabric map of the ocean floor (Figure 6, oversize enclosure). The arrival of satellite altimeters, along with the accumulation of shipboard collected data during the last three decades, has allowed major improvements to be made in the mapping and understanding of the Southern Ocean. Some areas such as the southernmost part of the Weddell Sea, the basins off Marie Byrd Land and the Ross Sea, and the margins of West Antarctica still remain poorly charted. Based on this combined information, improved and detailed reconstructions have recently been proposed for the evolution since the Late Cretaceous of the South Atlantic (Cande et al., 1988; Shaw and Cande, 1990), the Indian Ocean (Royer et al., 1988; Royer and Sandwell, 1989), and the South Pacific (Mayes et al., 1990). Models for the early opening of the Southern Ocean after the breakup of Gondwana in the Late Jurassic need to be revised in the light of the new tectonic constraints in the Weddell Sea, the African-Antarctic Basin, and the Enderby Basin.

ACKNOWLEDGMENTS

We thank John LaBrecque and an anonymous reviewer for their constructive comments on the manuscript. This work has been supported by the sponsors of the Paleoceanographic Mapping Project at the Institute for Geophysics, the University of Texas at Austin, and by NSF Grants OCE 86-17193 and DPP 86-15307. D. Sandwell was partially supported by the NASA Geodynamics Program (NAG 5-787) and the Topex/Poseidon Science Program (NASA/JPL 958122). UTIG contribution no. 765.

REFERENCES

Barker, P. F., 1972, A spreading center in the east Scotia Sea, Earth Planetary Science Letter, v. 15, p. 123-132.

Barker, P. F., 1979, The history of ridge-crest offset at the Falkland-Agulhas Fracture Zone from a small-circle geophysical profile, Geophysical Journal Royal Astronomical Society, v. 59, p.131-145.

Barker, P. F., 1982, The Cenozoic subduction history of the Pacific margin of the Antarctic Peninsula: Journal Geological Society of London, v. 139, p. 787-801.

Barker, P. F. and J. Burrell, 1977, The opening of the Drake Passage, Marine Geology, v. 25, p. 15-34.

Barker, P. F. and I. W. D. Dalziel, 1983, Progress in geodynamics in the Scotia arc region, *in* S. J. R. Cabré, ed., Geodynamics of the eastern Pacific region, Caribbean and Scotia arcs: American Geophysics Union Geodynamic Series, v. 9, p. 137-170.

Barker, P. F. and R. A. Jahn, 1980, A marine geophysical reconnaissance of the Weddell Sea: Geophysical Journal Royal Astronomical Society, v. 63, p. 271-283.

Barker, P. F. and L. A. Lawver, 1988, South American-Antarctic plate motion over the past 50 Myr, and the evolution of the South American-Antarctic Ridge: Geophysical Journal Royal Astronomical Society, v. 94, p. 377-386.

Bergh, H. W., 1977, Mesozoic seafloor off Dronning Maud Land, Antarctica: Nature, v. 269, p. 686-687.

Bergh, H. W., 1986, Seafloor spreading symmetry between Africa and Antarctica: South African Journal of Antarctic Research, v. 16, p. 9-12.

Bergh, H. W. 1987, Underlying fracture zone nature of Astrid Ridge off Antarctica's Queen Maud Land, Journal of Geophysical Research, v. 92, p. 475-484.

Bergh, H. W. and D. M. Barrett, 1980, Agulhas magnetic bight, Nature, v. 287, p. 591-595.

Bergh, H. W. and I. O. Norton, 1976, Prince Edward fracture zone and the evolution of the Mozambique Basin: Journal of Geophysical Research, v. 81, p. 5221-5239.

British Antarctic Survey, 1985, Tectonic map of the Scotia arc (scale 1:3000000), BAS (Misc.) 3: Cambridge, British Antarctic Survey.

Cande, S. C. and R. B. Leslie, 1986, Late Cenozoic tectonics of the southern Chile Trench: Journal of Geophysical Research, v. 91, p. 471-496.

Cande, S. C. and J. C. Mutter, 1982, A revised identification of the oldest sea-floor spreading anomalies between Australia and Antarctica: Earth Planetary Science Letter, v. 58, p. 151-160.

Cande, S. C., E. M. Herron and B. R. Hall, 1982, The early Cenozoic history of the southeast Pacific: Earth Planetary Science Letter, v. 57, p. 63-74.

Cande, S. C., J. L. LaBrecque and W. F. Haxby, 1988, Plate kinematics of the South Atlantic: chron 34 to present, Journal of Geophysical Research, v. 93, p. 13,479-13,492.

Christoffel, D. A. and R. K. H. Falconer, 1972, Marine magnetic measurements in the southwest Pacific Ocean and the identification of new tectonic features, *in* D. E. Hayes, ed., Antarctic Oceanology II: The Australian-New Zealand sector: American Geophysical Union Antarctic Research Series, v. 19, p. 197-209.

Coffin, M. F., H. L. Davies and W. F. Haxby, 1986, Structure of the Kerguelen Plateau province from SEASAT altimetry and seismic reflection data: Nature, v. 324, p. 134-136.

Davey, F. J., 1977, Marine seismic measurements in the New Zealand region: New Zealand Journal of Geophysical Research, v. 20, p. 719-777.

Eittreim, S. L. and G. L. Smith, 1987, Seismic sequences and their distribution on the Wilkes Land margin, *in* S. L. Eittreim and M. A. Hampton, eds., the Antarctic Continental Margin: geology and geophysics of offshore Wilkes Land: Circum-Pacific Council Energy and Mineral Resources, Earth Science Series, v. 5A, Houston, p. 15–43.

Falconer, R. H. K. and M. Tharp, 1981, General Bathymetric Chart of the Oceans (GEBCO), sheet 5·14: Canadian Hydrographic Service, Ottawa, Canada.

Fisher, R. L. and J. G. Sclater, 1983, Tectonic evolution of the Southwest Indian Ridge since the Mid-Cretaceous: plate motions and stability of the pole of Antarctica/Africa for at least 80 Ma: Geophysical Journal Royal Astronomical Society, v. 73, p. 553-576.

Fisher, R.L., M. Z. Jantsch and R. L. Comer, 1982, General Bathymetric Chart of the Oceans (GEBCO), sheet 5·9: Canadian Hydrographic Service, Ottawa, Canada.

Gahagan, L. M., J.-Y. Royer, C. R. Scotese, D. T. Sandwell, K. Winn, R. L. Tomlins, M. I. Ross, J. S. Newman, D. Müller, C. L. Mayes, L. A. Lawver and C. E. Huebeck, 1988, Tectonic fabric map of the ocean basins from satellite altimetry data: Tectonophysics, v. 155, p. 1-26.

Goodlad, S. W., A. K. Martin and C. J. H. Hartnady, 1982, Mesozoic magnetic anomalies in the southern Natal Valley: Nature, v. 295, p. 686-688.

Grunow, A. M., D. V. Kent and I. W. D. Dalziel, 1987, Mesozoic evolution of West Antarctica and the Weddell Sea Basin: new paleomagnetic constraints: Earth Planetary Science Letter, v. 86, p. 16-26.

Haxby, W. F., 1985, Gravity field of World's Oceans (color map): Lamont-Doherty Geological Observatory of Columbia University, Palisades, NY.

Haxby, W. F., 1988, Organization of oblique sea floor spreading into discrete, uniformly spaced ridge segments: evidence from the Geosat altimeter data in the Weddell Sea: American Geophysical Union, EOS Transactions, v. 69, p. 1155.

Haxby, W. F., G. D. Karner, J. L. LaBrecque and J. K. Weissel, 1983, Digital images of combined oceanic and continental data sets and their use in tectonic studies: American Geophysical Union, EOS Transactions, v. 64, p. 995-1004.

Hayes, D. E. and M. Vogel, 1981, General Bathymetric Chart of the Oceans (GEBCO), sheet 5·13: Canadian Hydrographic Service, Ottawa, Canada.

Heezen, B. C. and M. Tharp, 1965, Physiographic diagram of the Indian Ocean (with descriptive sheet): GSA, Boulder, CO.

Herron, E. M., 1972, Sea-floor spreading and the Cenozoic history of the east-central Pacific: GSA Bulletin, v. 83, p. 1671-1692.

Herron, E. M. and B. E. Tucholke, 1976, Sea-floor magnetic patterns and basement structure in the southeastern Pacific, *in* C. D. Hollister, C. Craddock et al., eds., Initial Reports of the Deep Sea Drilling Project: v. 35, Washington, (U.S. Government Printing Office), p. 263-278.

Hill, I. A. and P. F. Barker, 1980, Evidence for Miocene back-arc spreading in the central Scotia Sea, Geophysical Journal Royal Astronomical Society, v. 63, p. 427-460.

Johns Hopkins Applied Physics Laboratory Technical Digest, 1987, vol. 8.

Johnson, G. L., J.-R. Vanney, D. J. Drewry and G. de Q. Robin, 1980, General Bathymetric Chart of the Ocean (GEBCO), sheet 5·18: Canadian Hydrographic Service, Ottawa, Canada.

König, M., 1980, Geophysical investigations of the southern continental margin of Australia and the conjugate sector of East Antarctica. PhD. dissertation, Columbia University, New York, NY, 337 p.

König, M.,1987, Geophysical data from the continental margin off Wilkes Land, Antarctica. Implications for breakup and dispersal of Australia-Antarctica, *in* S. L. Eittreim and M. A. Hampton, eds., the Antarctic Continental Margin: geology and geophysics of offshore Wilkes Land: Circum-Pacific Council Energy and Mineral Resources, Earth Science Series, v. 5A, Houston, p. 117-146.

LaBrecque, J. L. and P. F. Barker, 1981, The age of the Weddell Sea Basin, Nature, v. 290, p. 489-492.

LaBrecque, J. L. and S. C. Cande, 1986, Total intensity magnetic anomaly profiles, *in* J. L. LaBrecque, ed., Ocean Margin Drilling Program Atlas 13: South Atlantic Ocean and adjacent Antarctic continental margin: Marine Science Institute, Woods Hole, sheets 9-11.

LaBrecque, J. L. and D. E. Hayes, 1979, Seafloor spreading history of the Agulhas Basin, Earth Planetary Science Letter, v. 45, p. 411-428.

LaBrecque, J. L. and M. Keller, 1982, A geophysical study of the Indo-Atlantic Basin, *in* C. Craddock, ed., Antarctic Geosciences part 2: Scotia Arc: Madison, University of Wisconsin Press, v. 4, p. 387-396.

LaBrecque, J. L. and P. D. Rabinowitz, 1977, Magnetic anomalies bordering the continental margin of Argentina: AAPG Map Series, Tulsa, OK.

LaBrecque, J. L. and P. D. Rabinowitz, 1981, General Bathymetric Chart of the Ocean (GEBCO), sheet 5·16: Canadian Hydrographic Service, Ottawa, Canada.

Lawver, L. A. and H. J. B. Dick, 1983, The America-Antarctic Ridge: Journal of Geophysical Research, v. 88, p. 8193-8202.

Lawver, L. A., J. G. Sclater and L. Meinke, 1985, Mesozoic and Cenozoic reconstructions of the South Atlantic: Tectonophysics, v. 114, p. 233-254.

Lawver, L. A., J.-Y. Royer, D. T. Sandwell and C. R. Scotese, in press, Evolution of the Antarctic continental margins, *in* M. R. A. Thomson, ed., Proceedings of the 5th International Antarctic Earth Sciences Symposium, Cambridge, U. K.

Mammerickx, J. and S. C. Cande, 1982, General Bathymetric Chart of the Ocean (GEBCO), sheet 5·15: Canadian Hydrographic Service, Ottawa, Canada.

Markl, R.G., 1974, Evidence for the breakup of Eastern Gondwanaland by the Early Cretaceous: Nature, v. 251, p. 196-200.

Markl, R.G., 1978, Further evidence for the Early Cretaceous breakup of Gondwanaland off Southwestern Australia: Marine Geology, v. 26, p. 41-48.

Martin, A. K., S. W. Goodlad, C. J. H. Hartnady and A. du Plessis, 1982, Cretaceous paleopositions of the Falkland Plateau relative to southern Africa using Mesozoic seafloor spreading anomalies: Geophysical Journal Royal Astronomical Society, v. 71, p. 567-579.

Mayes, C. M., D. T. Sandwell and L. A. Lawver, 1990, Tectonic history and new isochron chart of the South Pacific: Journal of Geophysical Research, v. 95, p. 8543-8567.

McKenzie, D. P. and J. G. Sclater, 1971, The evolution of the Indian Ocean since the Late Cretaceous, Geophysical Journal Royal Astronomical Society, v. 25, p. 437-528.

Molnar, P., T. Atwater, J. Mamerickx and S. M. Smith, 1975, Magnetic anomalies, bathymetry, and the tectonic evolution of the South Pacific since the Late Cretaceous: Geophysical Journal Royal Astronomical Society, v. 40, p. 483-520.

Monahan, D., R.H. K. Falconer and M. Tharp, 1982, General Bathymetric Chart of the Oceans (GEBCO), sheet 5·10: Canadian Hydrographic Service, Ottawa, Canada.

Mutter, J. C. and S. C. Cande, 1983, The early opening between Broken Ridge and Kerguelen Plateau: Earth Planetary Science Letter, v. 65, p. 369-376.

Norton, I. O. and J. G. Sclater, 1979, A model for the evolution of the Indian Ocean and the breakup of Gondwanaland: Journal of Geophysical Research, v. 84, p. 6803-6830.

Patriat, P., 1979, L'océan Indien occidental: la dorsale ouest-indienne, Mémoire du Muséum National d'Histoire Naturelle, v. 43, p. 49-5.

Patriat, P., 1987, Reconstitution de l'évolution du système de dorsales de l'océan Indien par les méthodes de la cinématique des plaques: Territoire des Terres Australes et Antarctiques Françaises, Paris, 308 p.

Patriat, P., J. Ségoufin, J. Goslin and P. Beuzart, 1985, Relative positions of Africa and Antarctica in the Upper Cretaceous: evidence for a non-stationary behaviour of fracture zones: Earth Planetary Science Letter, v. 75, p. 204-214.

Rabinowitz, P. D. and J. L. LaBrecque, 1979, The Mesozoic South Atlantic Ocean and evolution of its continental margins: Journal of Geophysical Research, v. 84, p. 5973-6002.

Royer, J.-Y. and D. T. Sandwell, 1989, Evolution of the eastern Indian Ocean since the Late Cretaceous: Constraints from GEOSAT altimetry: Journal of Geophysical Research, v. 94, p. 13,755-13,782.

Royer, J.-Y., J. G. Sclater and D. T. Sandwell, 1989, A preliminary tectonic fabric chart for the Indian Ocean, Proceedings of the Indian Academy of Sciences, Earth and Planetary Sciences, v. 98, p. 7-24.

Royer, J.-Y., P. Patriat, H. Bergh and C. R. Scotese, 1988, Evolution of the Southwest Indian Ridge from the Late Cretaceous (anomaly 34) to the Middle Eocene (anomaly 20): Tectonophysics, v. 155, p. 235-260.

Sandwell, D. T., 1984, A detailed view of the South Pacific geoid from satellite altimetry: Journal of Geophysical Research, v. 89, p. 1089-1104.

Sandwell, D. T., 1986, Thermal stress and the spacings of transform faults: Journal of Geophysical Research, v. 91, p. 6405-6417.

Sandwell, D.T. and D. C. McAdoo, 1988, Marine gravity of the Southern Ocean and Antarctic Margin from GEOSAT: tectonic implications, Journal of Geophysical Research, v. 93, p. 10,389-10,396.

Schlich, R., 1975, Structure et âge de l'océan Indien occidental: Mémoire hors série de la Société Géologique de France, no. 6, 103 p.

Schlich, R., 1982, The Indian Ocean: aseismic ridges, spreading centers and basins, *in* A. E. Nairn and F. G. Stheli, eds., The Ocean Basins and Margins: the Indian Ocean: New York, Plenum Press, v. 6, p. 51-147.

Schlich, R., M. F. Coffin, M. Munschy, H. M. J. Stagg, Z. G. Li and K. Revill, 1987, Bathymetric chart of the Kerguelen Plateau: Joint publication of the Bureau of Mineral Resource, Canberra, Australia, and the Institut de Physique du Globe, Strasbourg, France.

Sclater, J. G., R. L. Fisher, P. Patriat, C. Tapscott and B. Parsons, 1981, Eocene to recent development of the Southwest Indian Ridge, a consequence of the evolution of the Indian Ocean triple junction: Geophysical Journal Royal Astronomical Society, v. 64, p. 587-604.

Sclater, J. G., C. Bowin, R. Hey, H. Hoskins, J. Peirce, J. Philipps and C. Tapscott, 1976, The Bouvet triple junction, Journal of Geophysical Research, v. 81, p. 1857-1869.

Ségoufin, J., 1978, Anomalies magnétiques mésozoïques dans le bassin de Mozambique, Comptes Rendus de l'Académie des Sciences, Paris, 287, p. 109-112.

Ségoufin, J. and P. Patriat, 1980, Existence d'anomalies mésozoïques dans le bassin de Somalie. Implications pour les relations Afrique-Antarctique-Madagascar, Comptes Rendus de l'Académie des Sciences, Paris, v. 291, p. 85-88.

Shaw, P. R. and S. C. Cande, 1990, High-resolution inversion for South Atlantic plate kinematics using joint altimeter and magnetic anomaly data: Journal of Geophysical Research, v. 95, p. 2625–2644.

Simpson, E. S. W., J. G. Sclater, B. Parsons, I. O. Norton and L. Meinke, 1979, Mesozoic magnetic lineations in the Mozambique Basin: Earth Planetary Science Letter, v. 43, p. 260-264.

Small, C. and D. T. Sandwell, 1989, An abrupt change in ridge-axis gravity with spreading rate, Journal of Geophysical Research, v. 94, p. 17,383–17,392.

Stock, J. and P. Molnar, 1987, Revised history of early Tertiary plate motion in the south-west Pacific: Nature, v. 325, p. 495-499.

Talwani, M., J. Mutter, R. Houtz and M. König, 1979, The crustal structure and evolution of the area underlying the magnetic quiet zone on the margin south of Australia, *in* J. S. Watkins, L. Montadert and P. W. Dickerson, eds., Geological and geophysical investigations of the continental margins: AAPG Memoir 29, p. 151-175.

Tapscott, C., P. Patriat, R. L. Fisher, J. G. Sclater, H. Hoskins and B. Parsons, 1980, The Indian Ocean triple junction: Journal of Geophysical Research, v. 85, p. 4723-4739.

Udintsev, G. B., R. L. Fisher, V. F. Kanaev, A. S. Laughton, E. W. W. Simpson and D. I. Zhiv, 1975, Geological-geophysical atlas of the Indian Ocean: Academy of Sciences of the U.S.S.R., Moscow, 151 p.

Veevers, J. J., 1987, The conjugate continental margins of Antarctica and Australia, *in* S. L. Eittreim and M. A. Hampton, eds., The Antarctic Continental Margin: geology and geophysics of offshore Wilkes Land: Circum-Pacific Council Energy and Mineral Resources, Earth Science Series, v. 5A, Houston, p. 45-73.

Veevers, J. J., 1986, Breakup of Australia and Antarctica estimated as mid-Cretaceous (95±5 Ma) from magnetic and seismic data at the continental margin: Earth Planetary Science Letter, v. 77, p. 91-99.

Veevers, J. J., J. W. Tayton, B. D. Johnson and L. Hansen, 1985, Magnetic expression of the continent-ocean boundary between the western margin of Australia and the Eastern Indian Ocean: Journal of Geophysics, v. 56, p. 106-120.

Vogt, P. R., N. Z. Cherkis and G. A. Morgan, 1983, Project Investigator I: evolution of the Australia-Antarctic discordance deduced from a detailed aeromagnetic survey, *in* R. L. Oliver, P. R. James and J. B. Lago, eds., Antarctic Earth Science: Proceeding of the IV International Symposium on Antarctic Earth Science: Canberra, Australian Academy Press, p. 608-613.

Weissel, J.K. and D. E. Hayes, 1972, Magnetic anomalies in the Southeast Indian Ocean, *in* D.E. Hayes, ed., Antarctic Oceanology II: The Australian-New Zealand sector: American Geophysical Union Antarctic Research Series, v. 19, p. 165-196.

Weissel, J.K., D.E. Hayes and E. M. Herron, 1977, Plate tectonics synthesis, the displacements between Australia, New Zealand, and Antarctica since the Late Cretaceous: Marine Geology, v. 25, p. 231-277.

Chapter 8

The Stratigraphy, Setting and Hydrocarbon Potential of the Mesozoic Sedimentary Basins of the Antarctic Peninsula

David I. M. Macdonald
Peter J. Butterworth*
British Antarctic Survey
Natural Environment Research Council
Cambridge, United Kingdom

*Present address: The Robertson Group plc, Llandudno, Gwynedd, Wales, United Kingdom

ABSTRACT

The Antarctic Peninsula is a relatively accessible area of the continent, a fact which has stimulated interest in its hydrocarbon potential. This chapter uses known stratigraphic information to provide general constraints on the hydrocarbon potential of the Mesozoic basins of the Antarctic Peninsula.

The peninsula lies on a medium-sized block of continental crust. It is one of a mosaic of crustal blocks forming West Antarctica which underwent a complex tectonic evolution during Gondwana breakup. It was the site of an active volcanic arc above easterly subducting proto-Pacific ocean floor throughout the Mesozoic and part of the Cenozoic.

As a result the exposed Mesozoic basins display a complex stratigraphy, reflecting local tectonic and volcanic events. No units can be correlated between any two basins, but there are a few general trends. Almost all basins are post-Oxfordian; their fill is entirely clastic, and largely derived from the Antarctic Peninsula volcanic arc. Most basins were affected by a period of arc expansion in Late Jurassic or Early Cretaceous times, which manifests itself as inputs of lava or coarse volcaniclastic sediment. Berriasian and older mudstones are generally finer-grained and darker than mudstones from post-Berriasian strata. Deformation is variable, but rarely penetrative.

This stratigraphic information provides the basis for general constraints on the hydrocarbon potential. Organic geochemistry shows that Berriasian and older mudstones from the backarc region are the best potential oil source rocks; all other mudstones tend to be lean and gas-prone. Reservoir and seal facies tend to be better in deep marine (generally older) strata. Reservoir quality is generally poor due to breakdown of labile volcanic grains, but younger sandstones tend to be more quartz-rich.

The only significant prospective basin is the Larsen Basin, east of the Antarctic Peninsula. Geological evidence suggests that the best plays would involve deep strata. The major problem in this basin is the very difficult access, even by Antarctic standards.

INTRODUCTION

The Antarctic Peninsula (AP) (Figure 1) is the surface expression of a medium-sized, discrete block of continental crust; it is one of five continental blocks which make up West Antarctica (AP block, Figure 2). The AP block is 2130 km long, up to 850 km wide, covers an area of 1,300,000 km^2, and has a varied geology, largely due to active-margin processes operating from Paleozoic to Recent times.

The present-day Antarctic Peninsula is a chain of rugged mountains (up to 4190 m high) running almost the entire length of the AP block. It represents the deeply eroded roots of a Mesozoic volcanic arc, formed on a Paleozoic continental margin. Flanking the peninsula to east and west are extensive Mesozoic arc-related basins (Figure 1). These are the only upper Mesozoic sedimentary basins exposed on Antarctica. Since the Antarctic Peninsula is one of the more accessible areas of the continent, this has led to considerable interest in the hydrocarbon potential of the area.

The purpose of this chapter is to review the current knowledge of the setting and stratigraphy of the Mesozoic sedimentary basins of the Antarctic Peninsula crustal block, and to comment on known constraints on their prospectivity for hydrocarbons. It should be noted that there are extensive outcrops of highly deformed sedimentary rock in various basement and accretionary units in the Antarctic Peninsula area. These are not considered in this chapter except insofar as they form the basement to Mesozoic basins.

Previous Appraisals

Most hydrocarbon appraisals of the Antarctic have focused on the deep basins which lie between crustal blocks or off the continental margins (St. John, 1986). Major appraisals of Antarctic hydrocarbon potential are given by Behrendt (1983) and St. John (1986); other references can be found in Macdonald et al. (1988). Most previous appraisals are based on the few seismic lines which have been shot around the Antarctic margins.

In contrast, the Mesozoic basins to be reviewed here are integral parts of the continental crust. They are known mainly from outcrop studies, with subsidiary information from gravity and aeromagnetic surveys. Most outcrops of Mesozoic rock are either parts of overmature, inverted basins, or of basins which were never hydrocarbon-prone. Detailed knowledge of these basins is, however, vital for any future exploration of unexposed basins in the region.

Various authors have commented specifically on the possible prospective potential of the Weddell Sea, which lies to the east of the Antarctic Peninsula. Davey (1985) and St. John (1986) both considered it to be one of the potentially best areas around the continent. Deuser (1971) compared it with the Lake Maracaibo region of Venezuela and Ivanov (1989) draws parallels with the Gulf of Mexico. Both Deuser (1971) and Ivanov (1989) argued the case for the area to be highly prospective by analogy, but neither parallel is likely to stand up to detailed examination.

The only detailed geological study of the hydrocarbon potential of the area was carried out by Macdonald et al. (1988), who concluded that the Larsen Basin (northeast of the Antarctic Peninsula; Figure 1) should have moderate potential for oil generated from Upper Jurassic source rocks and reservoired in Cretaceous and Tertiary clastic rocks.

TECTONIC SETTING OF THE ANTARCTIC PENINSULA

Reconstructions of West Antarctica

From the very earliest reconstructions of the former supercontinent of Gondwana, it was realized that the Antarctic Peninsula constituted a special problem. Treating the whole Antarctic continent as a single rigid entity resulted either in impossible overlap of the peninsula onto continental crust of the Falkland Plateau (Norton and Sclater, 1979), or to unacceptable rotation of the main part of the continent relative to the other southern continents. When it was realized that East and West Antarctica had probably moved independently of one another during at least part of their geological histories, some of the overlap problem was removed (Dalziel and Elliot, 1982). However, it was not until the recognition that West Antarctica is composed of a mosaic of five separate blocks of continental crust, that a possible resolution of the overlap problem presented itself. These crustal blocks were originally identified on a purely topographic basis, by using seismic techniques and radio-echo sounding to delineate major low areas of sub-ice topography. Recent geological fieldwork and aeromagnetic surveying

Figure 1—Geological sketch map highlighting the known outcrop and presumed extent of the Mesozoic sedimentary basins of the Antarctic Peninsula crustal block, and the main localities discussed in the text. Dashed line marks the boundary of the AP block. Note that sedimentary rock within accretionary complexes are not considered in this chapter.

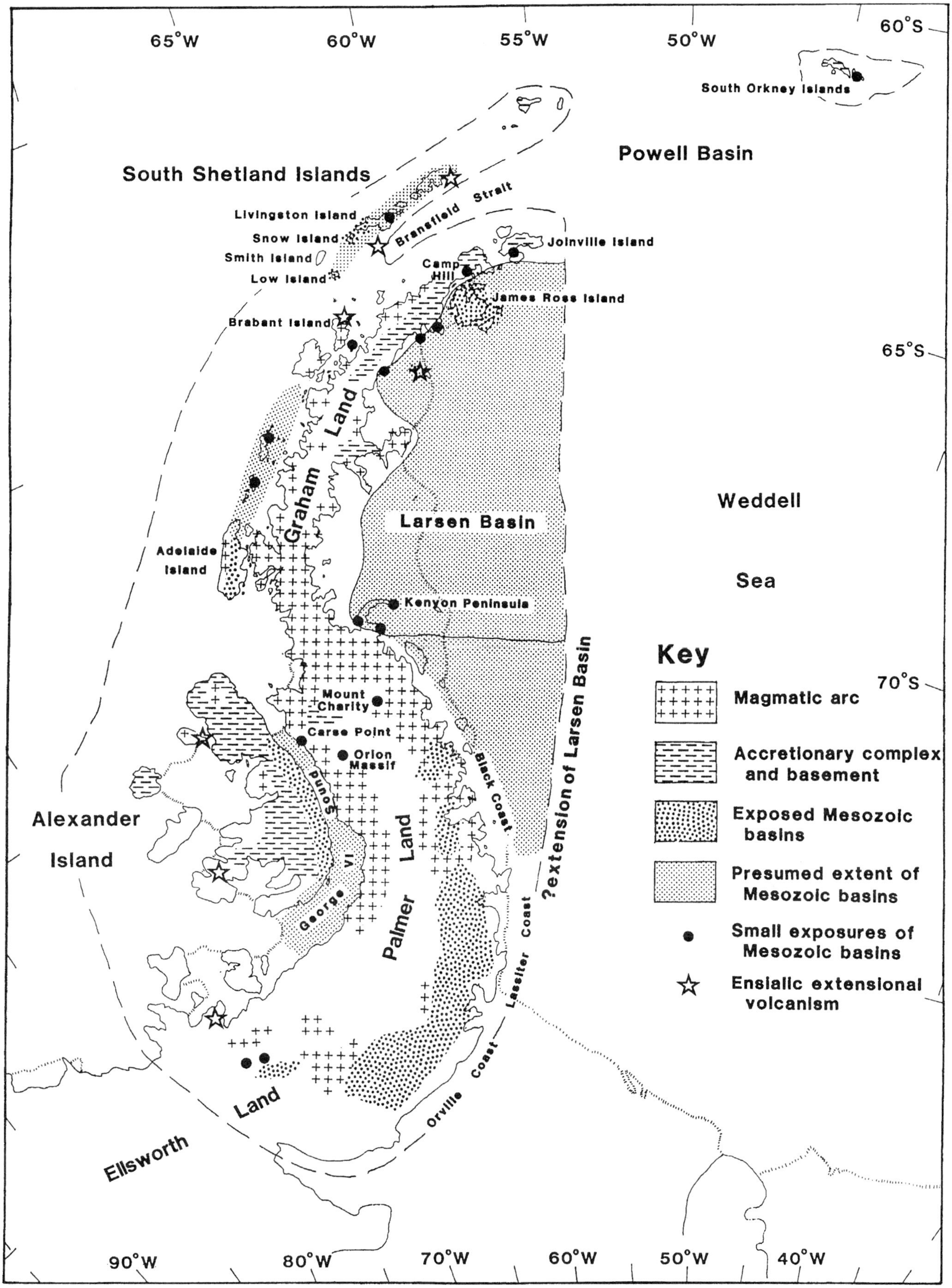
65°W
60°W
55°W
50°W
60°S
65°S
70°S
South Orkney Islands
Powell Basin
South Shetland Islands
Bransfield Strait
Livingston Island
Snow Island
Smith Island
Low Island
Joinville Island
Camp Hill
James Ross Island
Brabant Island
Graham Land
Larsen Basin
Weddell
Sea
Adelaide Island
Kenyon Peninsula
Mount Charity
Carse Point
Orion Massif
George VI Sound
Black Coast
?extension of Larsen Basin
Alexander Island
Palmer Land
Lassiter Coast
Orville Coast
Ellsworth Land
Key
Magmatic arc
Accretionary complex and basement
Exposed Mesozoic basins
Presumed extent of Mesozoic basins
Small exposures of Mesozoic basins
Ensialic extensional volcanism
90°W
80°W
70°W
60°W
50°W
40°W

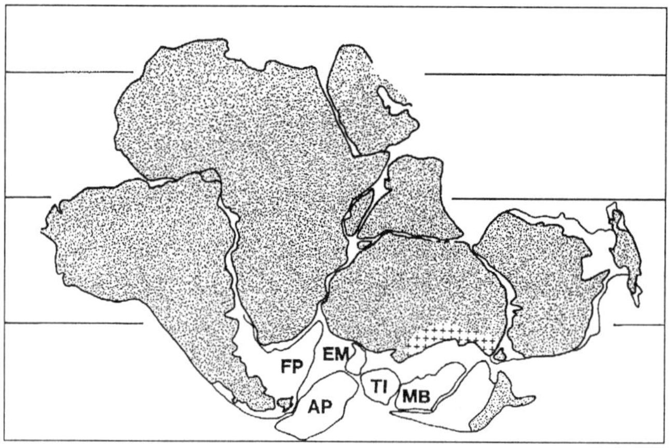

Figure 2—Reconstruction of Gondwana at 180 Ma (after Lawver and Scotese, 1987) showing the microplates of West Antarctica. AP: Antarctic Peninsula; EM: Ellsworth Mountains; MB: Marie Byrd Land; TI: Thurston Island. Falklands Plateau (FP), and the Transantarctic Mountains (crosses) are also shown. The area of the Weddell Sea lies between AP, EM, and FP.

(Storey et al., 1988a) has shown that each block has some distinct geological elements, although the extent of this is the subject of current debate.

Figure 2 shows a recent reconstruction of the continental blocks of West Antarctica (Lawver and Scotese, 1987), which effectively deals with the overlap problem. A consequence of reconstructions of this type is that some of the crustal blocks must have moved independently, during at least part of their history. Recent paleomagnetic results (Grunow et al., 1987) suggest that the blocks are all of local derivation, moved from their original position during the breakup of Gondwana and reassembled under the influence of active-margin processes.

Evolution of the Weddell Sea

The tectonic evolution of the Weddell Sea region is poorly understood for pre-Cretaceous times. There is uncertainty in dating of marine magnetic anomalies in the southern Weddell Sea, and adverse pack-ice conditions prevent marine geophysical work in the western Weddell Sea.

The opening of the Weddell Sea was originally thought to have begun about 165 Ma (La Brecque and Barker, 1981), but more recent work shows that the earliest datable anomaly is probably of Early Cretaceous age (ca. 135 Ma; Barker et al., in press). The nature of the early phase of crustal stretching is unclear. There is, however, evidence that the earliest phase of opening occurred in Jurassic times. Within-plate magmatism affected much of the interior of West Antarctica (Storey et al., 1988b) as well as the Transantarctic Mountains and other regions of Gondwana (Kyle et al., 1981) at about 180 Ma. This can be related to a lithosphere melting event associated with the breakup of Gondwana (Dalziel et al., 1987; Storey et al., 1988b). This event may be recorded by the transition from oceanic to continental crust at the southern edge of the Weddell Sea, where a buried basement ridge is associated with seaward-dipping sub–basement reflectors. The structure has been interpreted as the earliest phase of rifting or oblique rifting of a continental margin of Middle Jurassic or younger age (Hinz and Krause, 1982; Hinz and Block, 1984; Kristoffersen and Haugland, 1986). Recent results from ODP Leg 113 (Barker et al., 1988) tend to support this view of the age of the margin; at Site 692, on the eastern margin of the Weddell Sea, the basement reflector is overlain by 1200 m of sediment, the top of which is Tithonian-Berriasian in age (Doyle et al., in press). Taken together with the regional evidence of Storey et al. (1988b), this suggests that earliest opening of the Weddell Sea may be Middle Jurassic (ca. 180 Ma).

This view of the southern margin of the Weddell Sea as representing Middle Jurassic breakup is challenged by paleomagnetic evidence, which also bears on the former disposition of the crustal blocks, discussed in the previous section. This evidence suggests that the Ellsworth-Whitmore Mountains crustal block was situated in lower paleolatitudes (47°S), in what is now the eastern Weddell Sea, until 122 Ma when it moved into its present position (Grunow et al., 1987; Storey et al., 1988a).

Whichever view turns out to be correct, it is almost certain that the Weddell Sea area was subject to stretching and subsidence in Middle-Late Jurassic times, and that the main phase of opening of the Weddell Sea was during the Early Cretaceous.

Subduction-Related Processes

The AP block now forms part of the Antarctic Plate, and lies close to its most complex boundary (Tectonic Map of the Scotia Arc, 1985). At the northern end of the peninsula, the Drake Plate is being subducted very slowly southeastward under the South Shetland Islands (Barker, 1982), which forms a microplate separated from the Antarctica Peninsula by backarc spreading in Bransfield Strait (Figure 1). The only currently active volcano in the area is associated with this margin. This is a relic of almost continuous subduction which affected the area through most of Mesozoic and Tertiary times.

Subduction ceased by progressive ridge crest–trench collision along a series of fracture zones at high angle to the peninsula. This process began at 50 Ma with collision of a ridge crest off southern Alexander Island; subduction ceased along most of the rest of the peninsula between about 25 and 10 Ma (Barker, 1982).

Variations in subduction rates, and age of ocean crust being subducted, have led to the development of three overlapping tectonic regimes: accretionary, magmatic, and extensional (Storey and Garrett, 1985). Most of the Antarctic Peninsula crust was—"formed by accretionary and magmatic processes, and modified to its present shape by extension" (Storey and Garrett, 1985, p. 5). The three regimes documented by Storey and Garrett (1985) are diachronous across the area, and have given rise to a complex geological history, which has produced seven major strato-tectonic elements (discussed in the next section).

Schmidt and Rowley (1986) suggested the need for dextral transform faulting between East and West Antarctica to accommodate known plate movements. Their model was revised by Storey and Nell (1988), who pointed out the important role of strike-slip tectonics in the evolution of the whole peninsula. These effects span at least Middle Jurassic to Tertiary times (Storey and Nell, 1988; Nell et al., 1989; Nell and Storey, in press).

Discussion

Despite current uncertainties in understanding the tectonic evolution of the area, four points can be made, all potentially relevant to the hydrocarbon potential of the Antarctic Peninsula Mesozoic basins:

1. The Weddell Sea probably formed during Gondwana breakup, and much of the eastern margin of the AP block may have behaved as a passive margin to the Weddell Sea.
2. Subduction-related tectonics have been the major influence in shaping the Antarctic Peninsula.
3. Strike-slip tectonics have played an important role in the area.
4. The long history of subduction was brought to a close during Tertiary times by progressive ridge crest-trench collision.

GEOLOGY OF THE ANTARCTIC PENINSULA

The Antarctic Peninsula is divided into two major segments: Graham Land and Palmer Land (Figure 1). There is a strong topographic distinction between them, which to a certain extent reflects distinctions in the geology (Wyeth, 1977; Renner et al., 1985). Graham Land lies north of 69°S, and is a relatively narrow peninsula with many small offshore islands. Palmer Land is somewhat higher, and much broader, with a single large island (Alexander Island) to the west. The division between them is transitional over about 150 km. The southwest end of the Antarctic Peninsula block forms part of Ellsworth Land, but is topographically and geologically contiguous with Palmer Land. Despite the obvious differences, there are enough common elements to allow the geology of the whole peninsula to be discussed together.

In Storey and Garrett's (1985) model for the crustal growth of the Antarctic Peninsula, they envisage three overlapping tectonic processes creating seven major strato-tectonic elements: basement, accretionary complex, magmatic arc, forearc sedimentary basin, backarc sedimentary basin, intra-arc extensional zones, and ensialic extensional volcanic rocks (Figure 1). The relationship of the sedimentary basins to the main tectonic elements is well shown by the Tectonic Map of the Scotia Arc (1985), which also includes regional cross sections. The sedimentary basins are the subject of this chapter, but all the other tectonic elements are important in three ways:

1. Basement, accretionary complex, and (to a lesser extent) magmatic arc rocks are liable to underlie the Antarctic Peninsula sedimentary basins.
2. All three, but especially magmatic arc rocks, may have been a source of sediment.
3. The two types of extension may both have contributed to elevated heat flow and increased thermal gradients.

These are discussed in the next three sections.

Pre-Mesozoic Units

Most of the major sedimentary basins in the area are probably floored by pre-Mesozoic gneissic basement, or by rocks formed in late Paleozoic–Mesozoic accretionary complexes. Some intra-arc basins may be floored by magmatic arc rocks.

Gneissic Basement

Recent work has demonstrated the existence of pre-Mesozoic basement in the Antarctic Peninsula. In Graham Land, Milne and Millar (1989) have found granite orthogneiss of early Paleozoic age (410 ± 15 and 426 ± 12 Ma). Evidence from Palmer Land is poorer, but Paleozoic ages have been reported (Harrison and Loske, 1988; Harrison and Piercy, in press).

The evidence reviewed by Milne and Millar (1989) suggests that although there are few outcrops, Paleozoic orthogneiss may underlie much of the peninsula. They also present strong evidence of a major metamorphic event in the mid-Carboniferous, which may have been associated with plutonism to the east of the Antarctic Peninsula.

Accretionary Complexes

Major units of deformed siliciclastic turbidites, commonly with subsidiary units of chert, pillow lavas, impure metalimestones, and tuff, are found throughout the Antarctic Peninsula (Storey and Garrett, 1985). They go by various stratigraphic names in different areas, but all are unfossiliferous, or only sparsely fossiliferous, and few have well-constrained ages.

Two units are particularly widespread in the region. Large areas of northern Graham Land are underlain by the Trinity Peninsula Group, probably deposited in Permian-Triassic times, incorporating material derived from the older Paleozoic basement (Smellie, 1987; Milne and Millar, 1989). The sequence was deformed in Late Triassic or Early Jurassic times (Pankhurst, 1983). The other major exposure of accretionary complex rocks is in Alexander Island, where the LeMay Group forms most of the western two-thirds of the island. Known ages range from Early Jurassic (Thomson and Tranter, 1986) to Early Cretaceous (Burn, 1984), and the whole unit may well span Late Triassic-Paleogene times, representing a major accretionary episode on the western margin of the peninsula (Tranter, 1987).

Sediment Source

Both the gneissic basement and accretionary prism rocks discussed earlier may have been sediment sources for Mesozoic basins, but the magmatic arc rocks are much more important in volumetric terms.

Arc-Related Rocks

The present-day Antarctic Peninsula largely corresponds to the Mesozoic arc (Thomson, 1982a). The arc terrane comprises volcanics of the Antarctic Peninsula Volcanic Group (APVG), intruded by cogenetic plutons. Much of the arc activity was coeval with sedimentation.

In Graham Land there were apparently two major phases of arc activity: 180-160 and 130-80 Ma (Pankhurst, 1982). The end of the second pulse was diachronous across the arc, finishing at about 80 Ma in eastern Graham Land, but continuing well into Tertiary times in the west. This reflects a westward migration of arc activity with time (Pankhurst, 1982).

Timing of events in Palmer Land is less well-defined, but volcanism and plutonism span Early Jurassic until mid-Cretaceous times (Meneilly et al., 1987). During Late Jurassic to earliest Cretaceous times there was extension in the arc; calc-alkaline igneous activity declined, and there was extensive emplacement of basic hypabyssal intrusions, at least in western Palmer Land (Meneilly et al., 1987; Piercy and Harrison, in press). Intermediate–acid plutonism resumed in mid-Cretaceous times, accompanied by deformation, uplift, and volcanism; there is no evidence for any magmatic activity younger than 90 Ma in Palmer Land (Piercy and Harrison, in press). As in Graham Land, the arc stepped westward, with a major calc-alkaline intrusion into the

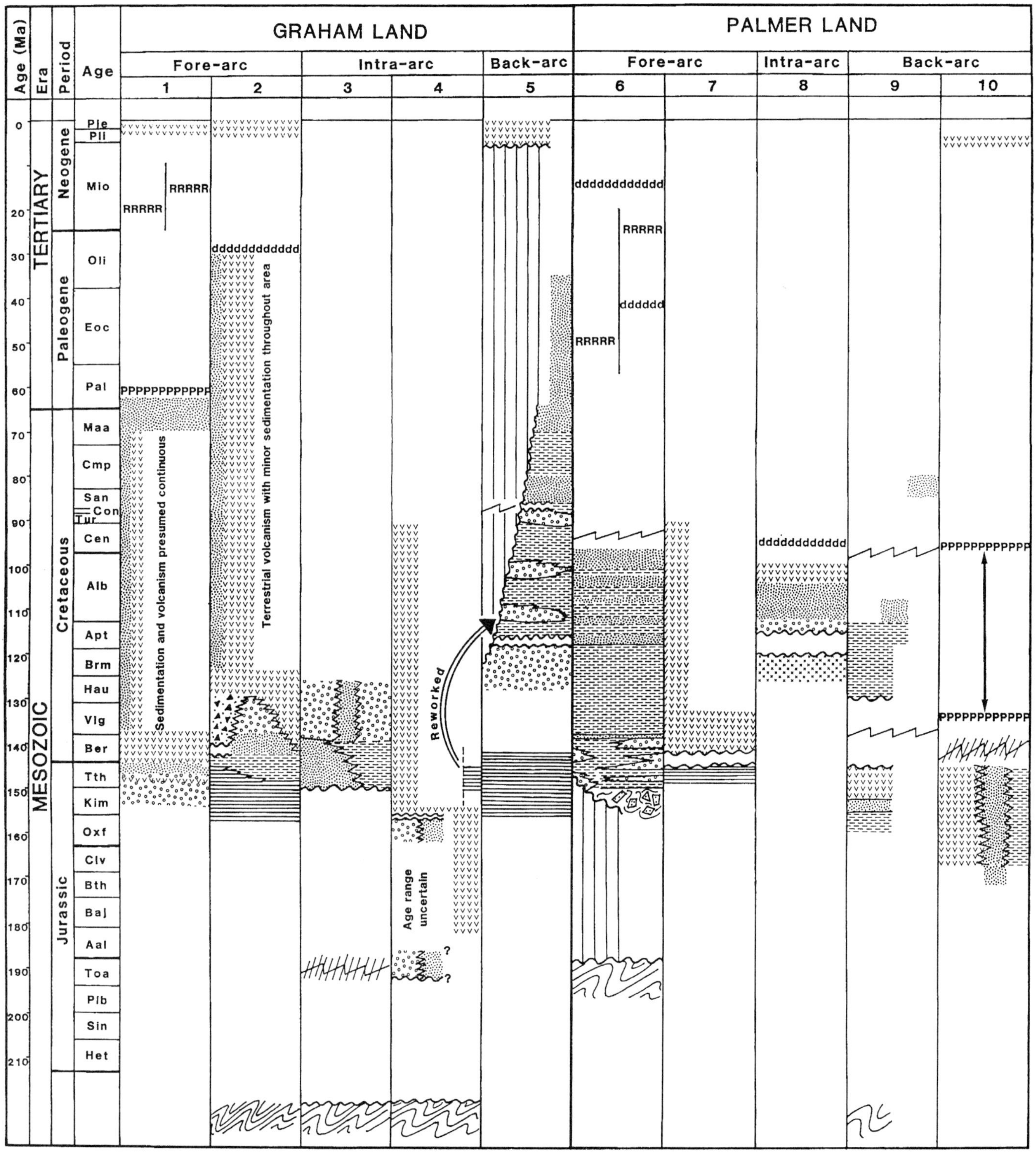

Figure 3—Summary chart of the stratigraphy of the Mesozoic sedimentary basins of the Antarctic Peninsula.

accretionary complex at 46 Ma (Pankhurst, 1982), and 40-60 Ma volcanism in Alexander Island (Burn, 1981).

Rocks from the volcanic arc terrain were transported into the Mesozoic basins by two main mechanisms:

1. Direct airfall: Tuff beds are found in all the basins, however, direct airfall is volumetrically negligible (e.g., Pirrie, 1989).

2. Surface transport of "epiclastic" detritus: Most sedimentary rocks contain a variety of volcanic, plutonic and basement clast types, indicating erosion and mixing in the arc before transport to basins.

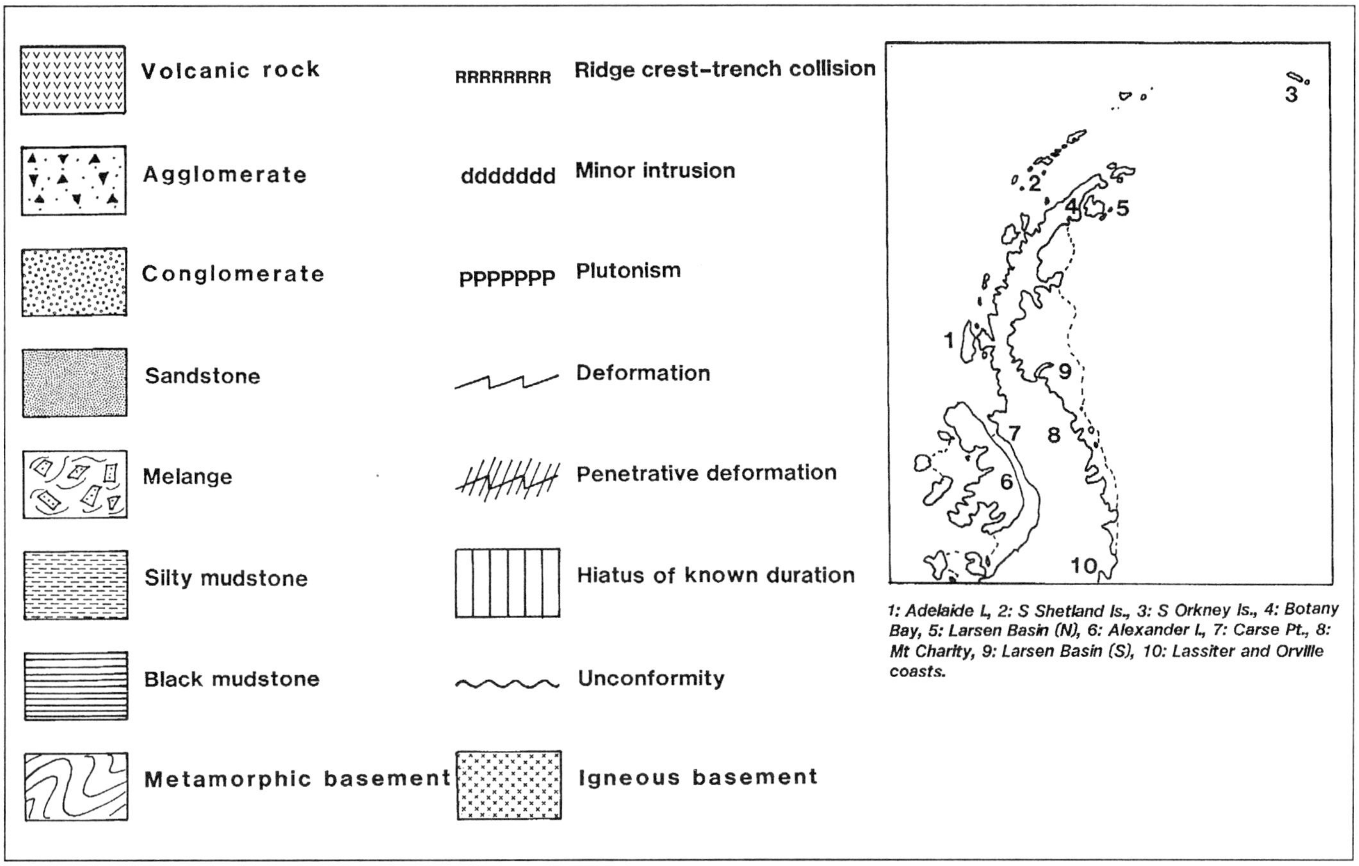

Figure 3 Continued.

Extensional Effects

Important extensional episodes occurred in Palmer Land in Late Jurassic and Tertiary times, and in Graham Land during the Early Cretaceous and Tertiary (Storey and Garrett, 1985; Maslanyj, 1988). The effects of extension are variably expressed as block faulting (Edwards, 1980; Crabtree et al., 1985), mafic hypabyssal intrusion (Piercy and Harrison, in press), and emplacement of large mafic bodies in a linear belt along the west side of the AP block (Renner et al., 1985; Jones and Maslanyj, in press). Storey and Garrett (1985) suggest that Bransfield Strait is the latest and best-developed of these intra-arc extensional features.

Neogene alkaline volcanic rocks are present at a number of localities in the Antarctic Peninsula area. These represent extension after cessation of subduction (Smellie et al., 1988) and are quite distinct from the intra-arc extensional rocks, which were coeval with subduction.

Broadly speaking, the Jurassic–earliest Cretaceous extensional episodes probably correspond to the formation of the major Mesozoic basins (perhaps in a transtensional regime). Tertiary extension is important in elevating thermal gradients, and thus hastening maturation. The Tertiary extensional basins of the Antarctic Peninsula may also provide analogues for the likely sedimentation and maturation processes in the Mesozoic basins. Compressional events took place in most basins at a variety of times (Figure 3). In contrast to the extensional effects there are no peninsula-wide compressional events. The local deformation history of each basin is discussed below.

STRATIGRAPHY OF THE MESOZOIC BASINS

Three factors are crucial to understanding the Mesozoic stratigraphy of the Antarctic Peninsula: igneous rocks of the volcanic arc were the main source of sediment; the evolution of the arc strongly affected sedimentation patterns; and tectonic processes were the major control on basin development.

Although the distinction between forearc, arc, and backarc is not always clear-cut, they are useful divisions. This, coupled with the split of the Antarctic Peninsula into Graham Land and Palmer Land, provides the basis for a convenient sixfold division of the Mesozoic basins. Only the rocks of Ellsworth Land do not fit this division, and are discussed in a separate section. Stratigraphic information for Graham Land and Palmer Land is summarized in Figure 3.

Graham Land: Forearc

The forearc region of Graham Land is one of the most complex and least understood parts of the Antarctic Peninsula. This is partly a reflection of disruption by Tertiary plutonism during westward migration of the volcanic arc, and partly a function of the lack of modern sedimentological and stratigraphic work in the area.

Extent of Sedimentation

Extensive deposits of sedimentary rock are found on Adelaide Island (Dewar, 1970), the South Shetland Islands (Smellie et al., 1984), and on many of the smaller islands between them (Figure l) (Smellie et al., 1985). Sedimentary rocks occur as far east as the western fringes of the Antarctic Peninsula (Smellie et al., 1985).

Neither gravity nor aeromagnetic maps show any significant sedimentary basins west of Graham Land (Renner et al., 1985). In the case of the aeromagnetic map this is almost certainly due to the high-amplitude West Coast Magnetic Anomaly (WCMA) which represents Tertiary extension (Garrett and Storey, 1987), and obscures any Mesozoic magnetic signature. Gravity and aeromagnetic observations do not extend as far west as the shelf break, although there are seismic results from this region (Kimura, 1982; Anderson, et al., 1990), which indicate 2.5 sec or more of Cenozoic sediment at the continental margin. Kimura (1982) claimed to recognize a pre-Miocene arc-trench system west of Adelaide Island, which may date back to Cretaceous times. His results are equivocal, but seem to indicate significant buried topography which could correspond to pre-Cenozoic basins.

Basement

Accretionary complex rocks of the Trinity Peninsula Group are widespread in western Graham Land (Smellie, 1987); high levels of this group have yielded a fauna of Triassic age in the area (Thomson, 1975a).

There is a similar unit on the South Shetland Islands (Miers Bluff Formation; Smellie et al., 1984) thought to be of ?Carboniferous–Triassic age. Smith Island, nearby, is entirely formed of blueschists, which can be no older than 135 Ma (Tanner et al., 1982) . The evidence suggests that a paired metamorphic belt (Smellie and Clarkson, 1975), representing a broad Paleozoic—early Mesozoic accretionary complex, underlies most of the islands and continental shelf west of Graham Land.

Stratigraphy

Sedimentation began in the area in Oxfordian-Kimmeridgian times, continuing into the early Miocene in the South Shetland Islands. Most sedimentation is intimately associated with active arc volcanism but, due to the paucity of fossils, there are major problems with dating sedimentary sequences.

On Adelaide Island, Dewar (1970) recognized an upward-coarsening then fining succession of tuff, tuffaceous sandstone and granulestone, and water-lain volcaniclastic conglomerate over 1000 m thick. This is capped by a laterally variable sequence of brecciated lava (up to 800 m), with minor sedimentary interbeds. The base of the succession has yielded a molluscan fauna of Kimmeridgian age (Thomson, 1972; P. Doyle, 1988, personal communication). Farther to the south, the supposedly identical unit (Dewar, 1970) has yielded a Maastrichtian-Paleocene flora (Jefferson, 1980). This suggests that volcanism and volcaniclastic sedimentation on Adelaide Island were continuous through latest Jurassic and Cretaceous times. Sedimentation was clearly in a marginal marine environment, proximal to volcanic centers; there is no evidence of separate, well-defined sedimentary basins. The sedimentary sequence is cut by calc-alkaline plutons dated at 60 ± 3 and 62 ± 2 Ma (Pankhurst, 1982) and ridge crest-trench collision occurred between 20 and 15 Ma (Barker, 1982). The volcanic succession and later plutons are cut by alkaline lamprophyre dykes at one locality; by analogy with other occurrences in the area this probably represents Plio-Pleistocene extension (Smellie et al., 1988).

On the smaller islands north of Adelaide Island, the same picture seems to hold, of intimately associated sedimentary and volcanic rocks, cut by later plutons (Smellie et al., 1985; J. W. Thomson, 1988, unpublished BAS data). On Brabant Island there is a thick sequence of alkaline volcanic rocks less than 1 Ma old (Smellie et al., 1988).

On the South Shetland Islands, the exposed sedimentary rocks are younger to the east. At the western end of the chain, sediments are uppermost Oxfordian-Valanginian; in the central part they are Upper Cretaceous; and towards the east, Maastrichtian to lowest Miocene (Smellie et al. 1984; Birkenmajer and Zastawniak, 1989). This eastward change in the age of onset of sedimentation almost exactly parallels the apparent diachronism in the onset of plutonism documented by Pankhurst and Smellie (1983). Thin-bedded dark mudstone and tuffaceous mudstones, interbedded with thin, graded sandstones and tuffs of Oxfordian-Tithonian age, crop out on Snow, Low, and Livingston islands (Thomson, 1982b; Smellie et al., 1984). On Livingston Island these rocks are overlain (perhaps with a slight hiatus) by a mixed succession of fossiliferous shales, volcaniclastic sandstones, and pebble conglomerates of Berriasian-Valanginian age (Smellie et al., 1980). This unit is laterally equivalent to, and overlain by, terrestrial pyroclastic rocks with interbedded lavas and local marine bands. The stratigraphic evidence suggests that open marine conditions in Late Jurassic times gave way to marginal marine and terrestrial conditions proximal to a volcanic center during the Cretaceous. The Cretaceous and Tertiary successions of the rest of the archipelago are formed of similar interbeds of lavas, pyroclastic rocks and locally derived sediment (Smellie et al., 1984). As in other areas west of Graham Land, Plio-Pleistocene alkaline volcanic rocks related to Tertiary extension are widespread (Smellie et al., 1988).

Discussion

Most of the Mesozoic deposits west of Graham Land represent Late Jurassic-Cretaceous, coarse, volcaniclastic sedimentation proximal to volcanic centers, probably laid down in small basins, with little interrelationship. It is debatable whether this represents a true forearc area, but the description is probably apt as the locus of magmatic activity lay to the east until Late Cretaceous times (Pankhurst, 1982). The distinction between forearc and arc seems ill-defined. The transition seen in western Livingston Island, from open marine–marginal marine–proximal volcanic, is probably typical of the process which went on throughout the area in the Late Jurassic, with marine sedimentation swamped by expansion and encroachment of the volcanic arc.

It is possible that a well-defined Mesozoic forearc basin lies to the west of the western archipelago of Graham Land on the continental shelf, but proof of this must await much more detailed seismic work.

Graham Land: Intra-Arc

Extent of Sedimentation

A series of intra-arc basins are exposed in northeast Graham Land, Joinville Island, and on the South Orkney Islands (which have been moved relatively oceanwards from their original position by creation of Powell Basin during Tertiary times; Figure 1). The sedimentology of these deposits suggests they may be of limited local extent (Elliot and Wells, 1982; Farquharson, 1984), and it is unlikely that they are all the remains of a single large basin. The small thickness of these deposits, coupled with their low density and magnetic susceptibility contrast with the underlying strata means that they are effectively invisible to any geophysical techniques used in the area to date.

Available evidence points to the existence of a series of small intra-arc basins in northern Graham Land. No similar deposits have yet been found in southern Graham Land.

Basement

At all the localities known, intra-arc basin rocks rest with pronounced angular unconformity on polyphase-deformed accretionary complex strata. Paleozoic gneisses are also likely to form basement.

Stratigraphy

The sedimentary rocks comprise sandstone and conglomerate up to 200 m thick, derived from the immediately underlying accretionary complex rocks; analysis of clast provenance proves these to be of very local derivation (Elliot and Wells, 1982). Deposition was in alluvial fans or fan deltas marginal to flood plains or lacustrine deposits (Elliot and Wells, 1982; Farquharson, 1982, 1984). Only in the case of the Spence Harbour Conglomerate and Gibbon Bay Shale in the South Orkney Islands is there evidence of any marine sedimentation (Thomson, 1981).

At most localities, the upper levels of the succession are interbedded with tuff, and pass upwards, apparently conformably, into andesites of the Antarctic Peninsula Volcanic Group (Farquharson, 1984).

The Spence Harbour Conglomerate and associated rocks in the South Orkney Islands span Tithonian-Neocomian times, and are well dated from a molluscan fauna in interbedded marine deposits (Thomson, 1981). All other areas contain only terrestrial floras. Recent work at Camp Hill suggests an age of latest Early Jurassic or earliest Middle Jurassic for this flora (P. M. Rees, 1988, personal communication). Recent radiometric dating of garnets from a sill intruding the base of the overlying volcanic succession at one of the localities (152 ± 8 Ma) and of comagmatic garnets within the sedimentary succession at another (156 ± 6 Ma) suggests that volcanism and at least the late stages of sedimentation were Late Jurassic (Millar et al., in press). None of these radiometric dates contradict Rees' paleobotanical dates, and there is clearly more work needed in this area.

Discussion

The sedimentary successions in both of the dated areas, and all the undated ones, are remarkably similar. All probably represent largely terrestrial sedimentation in small, fault-bounded basins, prior to the onset of volcanism. These basins were probably of limited extent and duration, and might be of any age ranging from Early Jurassic to Early Cretaceous.

Graham Land: Backarc

Extent of Sedimentation

Most of James Ross Island and the surrounding island group is formed of sedimentary rock, which also crops out on many of the small nunataks and peninsulas farther south, at least as far as Kenyon Peninsula in northeast Palmer Land (Figure 1). Aeromagnetic investigations reveal a major sedimentary basin east of Graham Land, with its western boundary roughly coincident with the coastline (Renner et al., 1985; Garrett and Storey, 1987). This basin, termed the Larsen Basin by Macdonald et al. (1988), probably extends as far east as the shelf break (Paterlini et al., 1984), and reaches at least 70°S. The southern boundary is, however, ill-defined and the basin may stretch much farther south, east of the backarc area of eastern Palmer Land (Figure 1).

Limited geophysical evidence (reviewed by Macdonald et al., 1988) indicates that the sedimentary succession is about 5 km thick. Since the Larsen Basin covers an area of at least 200,000 km^2, it could potentially contain 1,000,000 km^3 of fill.

Basement

By analogy with eastern Graham Land, accretionary complex rocks and Paleozoic orthogneisses are the most likely rocks to form the basement of the Larsen Basin. Magnetotelluric studies in the northern part of the basin tend to support this conclusion (Fournier et al., 1980). Aeromagnetic results are also consistent with this interpretation, but since magnetic susceptibility contrast is so low, large variations in depth to basement (e.g., caused by block faulting) would be effectively invisible to this technique (Macdonald et al., 1988).

Stratigraphy

Most of the evidence presented here comes from the 5 km-thick stratigraphic sequence exposed on James Ross Island and islands to the east. A partial column put together from scattered outcrops in the Kenyon Peninsula area is consistent with this. The exposed stratigraphic succession consists of five major sedimentary sequences, separated by unconformities or hiatuses.

1. The oldest rocks in the area are a series of organic-rich black mudstones and ash layers of Kimmeridgian-Berriasian age, known as the Nordenskjöld Formation (Farquharson, 1983a; Whitham and Doyle, 1989). These crop out at scattered localities along the western basin margin, and are also found as reworked clasts in younger sedimentary rocks. They represent marine sedimentation remote from any major source of clastic detritus, but at a time when the volcanic arc had a subaerial edifice (Macdonald et al., 1988; Millar et al., in press).

2. There is an apparent hiatus in sedimentation from late Berriasian–Valanginian times. This separates the youngest part of the Nordenskjöld Formation (Whitham and Doyle, 1989) from coarse volcaniclastic conglomerates exposed south of James Ross Island which contain a reworked Hauterivian fauna (Thomson and Farquharson, 1984). On the island itself the oldest strata are Barremian conglomerates. These are at least 500 m thick, and were deposited in a submarine fan. Relationships with the rest of the Cretaceous succession are unclear (Ineson et al., 1986), but presumed to be unconformable (Whitham and Marshall, 1988, their figure 5).

3. From Aptian until Turonian times, sedimentation was dominated by coarse clastic input, with 1700 m of thick, laterally variable, channelized conglomerates and coarse sandstones encased in silty mudstone (Ineson, 1989). Most of the clasts are volcanic, plutonic, or accretionary complex material, derived from the arc terrain, but a high proportion of clasts in the Aptian-Albian age conglomerates are Nordenskjöld Formation debris (Ineson et al., 1986); this includes three giant slide blocks of Nordenskjöld Formation, clearly derived from the arc margin (Ineson, 1985). Deposition was in a deep

submarine fan-slope apron complex (Ineson, 1989).

4. Sediments of Coniacian age are locally unconformable on the deep marine deposits. Coniacian-Santonian strata comprise about 400 m of coarse sandstones with interbedded mudstones and represent sedimentation in a shallow marine environment with local deposition by fan deltas (Whitham et al., 1987; Whitham and Ineson, in press). This change from deep to relatively shallow or marginal marine sedimentation seems to have been basinwide, as the same change is found in the Kenyon Peninsula area (Macdonald, 1985). The change is accompanied by a decrease in the marine fauna and a great increase in the fossil terrestrial flora.

5. Santonian-Oligocene strata, which conformably overlie the fan-delta unit, are dominantly bioturbated fine sandstone, siltstone and mudstone, with subordinate coarse sandstone and rare conglomerate (Pirrie, 1989; Macellari, 1988a; Sadler, 1988). There is some reworked Cretaceous material in Upper Cretaceous rocks. The Cretaceous part of this sequence represents shallow marine deposition on a shelf with waning tectonic activity (Pirrie, 1989). Tertiary strata, exposed on islands east of James Ross Island, represent various shelf and deltaic environments (Macellari, 1988a), with a major unconformity separating Eocene from Paleocene strata (Sadler, 1988).

6. All of the Cretaceous units are overlain with pronounced overstep by the alkaline volcanic rocks of the James Ross Island Volcanic Group (Nelson, 1975). This unit is probably related to late extension in the area (Smellie et al., 1988).

Structure

The greater part of the Larsen Basin fill is apparently undeformed, with only a gentle easterly tilt. Previous suggestions of a broad open warping (Bibby, 1966) have been discounted, but Pirrie and Riding (1988) document oblique-slip normal and reverse faults, some of which may be synsedimentary.

It is only close to the basin margin that there has been any significant deformation: each of the unconformity-bound sequences mentioned above dips progressively less steeply than the underlying one. Evidence of local strike-slip deformation has also been documented from the Nordenskjöld Formation (Whitham and Storey, 1989). Whitham and Marshall (1988) used vitrinite reflectance data to demonstrate that none of the Cretaceous sediments on northern James Ross Island has ever been deeply buried, and that the structure represents progressive deformation at the basin margin.

Discussion

The Nordenskjöld Formation was probably deposited at an early stage in the development of the Larsen Basin, and probably over a wider area than the well-defined basin which developed in the Early Cretaceous. There is uncertainty as to how much of the Larsen Basin is underlain by Nordenskjöld Formation (Macdonald et al., 1988).

Sometime during the Valanginian-Hauterivian, there was major uplift of the arc terrain along a fault-bounded basin margin, and deep marine fan sedimentation was initiated. It appears that the major change to fan delta and subsequent shallow marine sedimentation was a result of partial basin inversion in the Late Cretaceous.

There are no post-Oligocene sediments exposed in the basin, but on the nearby South Orkney microcontinent, King and Barker (1988) recognized a pervasive early Miocene breakup unconformity. They attributed this to uplift and block faulting during formation of the Powell Basin; a similar event may well have affected northern parts of the Larsen Basin.

Palmer Land: Forearc

The forearc region west of Palmer Land is exceptionally broad, encompassing the whole width of Alexander Island. The forearc is separated from the arc terrain of Palmer Land by George VI Sound, a Tertiary transtensional feature (Storey and Nell, 1988; Maslanyj, 1988), which obscures the exact relationship of arc to forearc.

Extent of Sedimentation

The bulk of Alexander Island is formed of accretionary complex rocks (LeMay Group: Tranter, 1987) but the east side of the island comprises a perched forearc basin (Fossil Bluff Group: Butterworth et al., 1988) which crops out in a strip 250 km long by up to 30 km wide (Figure 1). The aeromagnetic map shows a flat magnetic signature over Alexander Island and the continental shelf, reflecting the generally low magnetic susceptibilities both of the deformed sedimentary rocks of the accretionary complex, and the forearc basin rocks (Renner et al., 1985). Density contrasts are also low, hence any sedimentary basins would be invisible to gravity surveys.

LeMay Group rocks include deposits of trench-slope basins (Tranter, 1987). In central Alexander Island these deposits are thoroughly deformed. There is nothing, however, to preclude the existence of undeformed trench-slope basins farther to the west. Any such basins would most likely be of Tertiary age.

Basement

At the western margin of the basin, forearc basin rocks are unconformable on, and faulted against, the accretionary complex (Edwards, 1980). Recent field work has demonstrated that this boundary is complex, and reflects local emergence of the accretionary prism (Nell and Storey, in press; Macdonald, 1987). Away from the western margin, the base of the sedimentary succession in not seen, but accretionary complex probably forms the basin floor.

The LeMay Group is generally unfossiliferous, but is clearly the youngest accretionary complex exposed in the Antarctic Peninsula region, with a good Liassic, possibly Sinemurian, age from the central eastern part (Thomson and Tranter, 1986). Farther west, a possible Cretaceous age has been recorded (Burn, 1984) and the extreme western edge may be as young as Tertiary (P.A.R. Nell and B.K. Holdsworth, 1989, personal communication).

Stratigraphy

Most work to date has concentrated on the northern part of the Fossil Bluff Group basin, where the oldest strata are exposed. The succession gets younger towards the south, accompanied by a facies change from marine to terrestrial (A.C.M. Moncrieff, 1989, personal communication). In the northern area the minimum thickness of sediment is 4300 m. The basin fill in this northern area has been divided into four formations (Butterworth et al., 1988), representing progressive basin shallowing from Kimmeridgian-Albian times (Butterworth and Macdonald, in press).

The oldest known strata in the basin are part of a huge Kimmeridgian slump unit, at least 440 m thick, comprising folded and rafted blocks of sedimentary and (subordinate) volcanic rock in chaotic, laterally discontinuous units. The unit has been interpreted as a slope-collapse deposit (Macdonald and Butterworth, 1986).

The overlying 2.2 km comprise a highly variable sequence of conglomerates, sandstones, and mudstones of Tithonian to Berriasian age (Crame and Howlett, 1988). At the type area of this formation, there are four channeled conglomerate complexes up to 170 m thick, and traceable laterally for at least 20 km. These reflect major channels in a complex submarine fan system (Butterworth, 1985). Paleocurrent and provenance studies clearly indicate derivation from the volcanic arc to the east (Horne, 1968; Butterworth, 1988). Away from this major depocenter, facies are much more mudstone-rich, and on the western side of the basin, equivalent strata of Tithonian age rest unconformably on the accretionary complex. Conglomerates with sandstone clasts, derived from the west, directly overlie the unconformity.

The submarine fan complex is overlain by a 1000 m thick, monotonous sequence of mudstone and siltstone with subordinate thin, fine sandstone interbeds of Valanginian-Aptian age. This unit is characterized by slumps and synsedimentary melanges up to 120 m thick, swarms of sandstone dykes (Taylor, 1982), synsedimentary faults, and local angular unconformities. The unit was probably deposited in a slope-outer shelf environment which was undergoing active tectonism during deposition (Butterworth and Macdonald, in press).

Transitionally overlying the mudstone unit is a sequence of bioturbated, cross-bedded medium sandstones, probably deposited in a shelf environment. This unit has been comparatively little-studied as yet, but it probably interdigitates with a paralic–delta plain facies farther south, where extensive fossil forests of Aptian-Albian age have been found (Jefferson, 1982).

No sediments younger than Albian have been recorded. The sequence is intruded by a single large andesite dyke, dated by K-Ar at 41 ± 1.5 Ma (I.L. Millar, 1988, personal communication), and by several camptonite (alkaline) dykes (Horne and Thomson, 1967) dated at 15 ± 1 Ma (Rex, 1970). Elsewhere in Alexander Island there are extensive deposits of ?Upper Cretaceous-Paleogene andesitic volcanic rocks (Bell, 1974) and Neogene alkaline volcanic rocks (Smellie et al., 1988). These are probably related to arc activity after westward migration, and post-subduction extension, respectively. The evidence of the two sorts of dykes intruding the Fossil Bluff Group suggests that it may have been overlain by volcanic rocks at one time.

Structure

The Fossil Bluff Group is deformed by a series of north-northwest-trending open anticlines related to major east-northeast-directed thrusts. The abundant sandstone dykes which cut the group also trend east-northeast.

This evidence of structures oblique to the basin led Nell and Storey (in press; see also Storey and Nell, 1988) to suggest deformation in a transpressional regime related to dextral strike-slip faulting. Timing of this deformation is poorly constrained, but was between the end of sedimentation at approximately 100 Ma and intrusion of the andesite dyke at 41 Ma. The fact that there was unlithified sediment capable of intrusion as sedimentary dykes suggests early deformation.

Strike-slip activity continued in the region with creation of George VI Sound by Tertiary transtension (Maslanyj, 1988). Eastern marginal facies of the Fossil Bluff Group are probably downfaulted, a conclusion supported by geophysical evidence of thick, nonmagnetic sequences on the floor of the sound (Maslanyj, 1988; Storey and Nell, 1988).

It is possible that the forearc basin was initiated as a transtensional pull-apart (Storey and Nell, 1988). However, from sedimentological evidence of vertical stacking of conglomerate channels in the type section (Butterworth, 1988), there was either no strike-slip movement along the eastern basin margin from Tithonian-Berriasian times, or the channels were located south of a releasing step.

All of the basin fill has been metamorphosed to zeolite facies, and there is abundant development of laumontite in all of the sandstones (Horne, 1968). Timing of this event is uncertain.

Discussion

This large sedimentary basin clearly demonstrates the interplay of tectonic and sedimentary controls on sedimentation. The main points to note are the presence of large bodies of coarse clastic material completely enclosed within mudstone, and the possible early formed structures.

Fossil trench-slope basins are located on the accretionary complex of central Alexander Island (Tranter, 1987). Younger, less-deformed examples may occur farther west.

Palmer Land: Arc Margin/Intra-Arc

Extent of Sedimentation

Arc margin sedimentary rocks crop out at Carse Point (Figure 1), flanking the forearc basin margin, and may be a part of a more extensive basin-fill sequence. Intra-arc basin sedimentation is restricted to small exposures in the areas of Mount Charity and the Orion Massif (Figure 1); the sedimentology of both suggests limited extent.

Basement

The basement to the arc margin sediments at Carse Point is unknown, although there is evidence of older (?lower Paleozoic) basement in western Palmer Land (Harrison and Loske, 1988). Intra-arc basin sediments at Mount Charity rest with a pronounced unconformity on a little-deformed granite complex, dated at 120 Ma (R.J. Pankhurst, 1989, personal communication).

Stratigraphy

At Carse Point, arc margin sedimentary rocks comprise >60 m of fossiliferous black pyritic mudstones and shales with subordinate thin-bedded sandstones and conglomerate lenses (Culshaw, 1975). The fauna indicates a Late Jurassic (? Tithonian) age (Thomson, 1975b). Clast composition and petrography suggest a volcanic source terrain. Deposition was clearly within a quiet-water marine environment. The sequence is (?) unconformably overlain by arc volcanics, and is intruded by an altered microdiorite sill (Culshaw, 1975).

The sedimentary rocks of the restricted intra-arc basin at Mount Charity comprise approximately 150 m of red, cross-bedded and parallel-laminated coarse sandstone and pebbly sandstone/conglomerate, with subordinate mudstone (K.D. Holmes, 1967, unpublished B.A.S. data). Clast composition suggests derivation from the arc basement. Deposition was clearly terrestrial, and most probably reflects an alluvial fan complex. The sequence fines up and becomes interbedded with tuffs, and is conformably overlain by volcanics. The whole sequence is cut by a basic dyke; as there is no volcanism post 90 Ma in Palmer Land (S.M. Harrison, 1989, personal communication), it is most probable that this intra-arc basin is of Aptian-Albian age. A poorly exposed 20 m thick sequence of mudstones on Orion Massif (Skinner, 1973) represents intra-arc sedimentation of unknown age.

Discussion

The transition from open marine conditions to a major episode of volcanism which occurred during Late Jurassic times at Carse Point is comparable to that observed on Livingston Island in the South Shetland Islands.

The intra-arc sediments at Mount Charity most probably represent local terrestrial sedimentation in a basin of limited extent and duration. Although this is the only area of intra-arc sedimentation discovered in Palmer Land, extensive block faulting occurs in western Palmer Land (Meneilly et al., 1987). It is possible that this process was responsible for the creation of other, similar small basins in the area.

Palmer Land: Backarc

The backarc area of Palmer Land is much more complex than that of Graham Land. It is likely that the Larsen Basin, or some correlative extends southward along the eastern shelf of Palmer Land (Figure 1). Macdonald et al. (1988) put an arbitrary southern limit of 70°S on the basin, but it could reach at least 73°S. South of this latitude, Middle-Upper Jurassic rocks, predating the Larsen Basin, crop out close to the east coast of the peninsula (Rowley et al., 1983), which is almost coincident with the margin of the Antarctic Peninsula microplate (Lawver and Scotese, 1987).

The outcrop geology of eastern Palmer Land (inland of any Larsen Basin correlative) records sedimentation, volcanism, deformation, and plutonism (Figure 3). This complex history, coupled with the remoteness of the area, means that the geology of this region is less well-known than any other part of the Antarctic Peninsula.

Extent of Sedimentation

Sedimentary rocks crop out over an area extending more than 600 km along the Black, Lassiter, and Orville coasts of eastern Palmer Land and Ellsworth Land (Laudon et al., 1983) (Figure 1). It is possible that the basin extended farther north into the central and northern Black Coast area, where a sequence of metasedimentary rocks (Mount Hill Formation) crops out. This correlation was first suggested by Singleton (1980) and supported by the collection of a sparse fauna and flora (Meneilly et al., 1987). More recent work shows different styles of deformation in the two areas (Storey et al., 1987).

Basement

The metasedimentary rocks of the northern Black Coast area were thought to be underlain by a pre-late Paleozoic metamorphic complex (Singleton, 1980), although more recent work has shown that this "complex" is probably deformed plutons of the Mesozoic magmatic arc (180-200 Ma; Meneilly et al., 1987). It is likely, however, that Paleozoic basement underlies much of the area (Storey et al., 1989).

Stratigraphy

Most work to date has concentrated on the thick backarc basin sediments exposed on the Lassiter and Orville Coasts (Figure 1). The succession comprises black and gray slate, siltstone, and mudstone with subordinate sandstones and coal; conglomerate is rare (Rowley et al., 1983). Neither the base nor the top are exposed. The greatest continuously exposed stratigraphic thickness measured is 830 m, but based on lithology and structure, Laudon et al. (1983) estimate that the basin fill is at least several kilometers thick. Simplistically, there appears to be a transition in depositional environments from a terrestrial setting near the axis of the peninsula (magmatic arc) into a truly marine environment farther south. Nonmarine sedimentary rocks, containing abundant plant fossils and thin coal beds, intertongue with calc-alkaline silicic to intermediate air-fall and ash-flow tuffs, lava flows, and volcanic breccias at the basin margin (Rowley et al., 1983). The nonmarine deposits pass laterally through shallow marine, possibly deltaic rocks, into open-marine deposits (Laudon et al., 1983).

On the basis of locally abundant marine faunas, most of these sediments have been assigned a Late Jurassic age (Thomson, 1983), although in eastern Ellsworth Land, Middle Jurassic faunas have been described (Quilty, 1982). The inferred location of the Jurassic paleo-shoreline trended parallel to the axis of the Antarctic Peninsula (Thomson, 1980; Rowley et al., 1983). Provenance studies suggest that the active magmatic arc was the primary source of sediment, with significant amounts of metamorphic material, which may represent erosion of the arc basement (Laudon et al., 1983).

Structure

Mafic dykes and steep extensional shear zones throughout northern Palmer Land are related to Jurassic extension of the arc and formation of the backarc basin (Meneilly et al., 1987). The whole backarc basin fill was subsequently tightly folded on axes trending parallel to the coast (Rowley et al., 1983). Local thrust faults display a south or southeast vergence, with a slaty cleavage parallel to fold axes (Laudon et al., 1983). The age of the main period of deformation is Late Jurassic or Early Cretaceous, between the time of sedimentation and prior to emplacement of Early Cretaceous plutons, which cut the folds and thrusts (Laudon et al., 1983; Rowley et al., 1983). Magmatic arc rocks are affected by heterogeneous ductile shear zones involving east-directed thrusting, extensional normal faulting and strike-slip deformation (Meneilly et al., 1987).

Discussion

Of most relevance here is the predominance of thick sequences of mudstones, and in particular the occurrence of thin coal horizons and abundant plant/carbonaceous detritus. However, this must be put into context with the extensive episode of post-depositional deformation, which is most probably the result of backarc basin closure (Meneilly et al., 1987). The extensive tectonic deformation of these rocks is in marked contrast to the backarc area of the northern Antarctic Peninsula (Larsen Basin), and provides evidence that these two regions have undergone very different basin evolution and inversion.

Ellsworth Land

The area southwest of Palmer Land forms the farthest southwest extremity of the AP block. Sedimentary rocks are not well exposed in the area, but they clearly do not fit conveniently into the tectonic divisions used for the description of the stratigraphy of the rest of the Antarctic Peninsula.

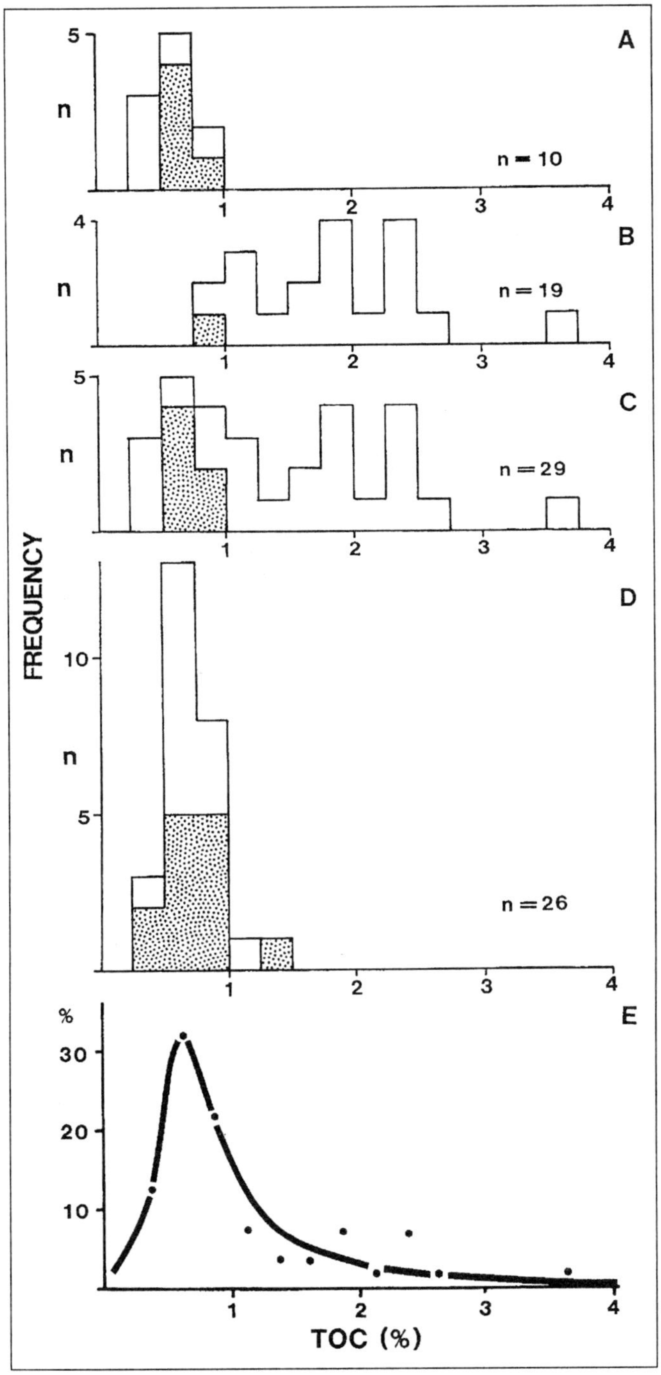

Figure 4—Distribution of total organic carbon (TOC) values in A: Jurassic-Berriasian, forearc; B: Jurassic-Berriasian, backarc; C: Jurassic-Berriasian, total of all areas; D: post-Berriasian, total data; E: percentage frequency of total of all samples. Shading indicates silty mudstone samples. Note that there are no data from the Jurassic backarc basin of Palmer Land.

Extent of Sedimentation

Sedimentary rocks are found at a series of scattered nunataks at 74-75°S and 70-75°W (Figure 1). Jurassic rocks of the Latady Formation crop out in the area, at the extreme western end of their range.

Basement

No basement rocks are known from the area.

Stratigraphy

In the southern part of the area, the Latady Formation is probably Middle-Upper Jurassic (Gee, 1989; Laudon, in press). Just to the north are three small exposures of dark mudstone, siltstone and fine sandstone informally named the Erewhon Beds (Laudon, in press). The thickest exposed sequence is only 28 m, but the rocks are significant in containing a probable Late Permian flora (Gee, 1989). Sandstones are lithic arenites; they contain 30-70% volcanic material, and fall into the transitional arc and recycled orogenic field of a quartz-feldspar-lithics (QFL) plot (Laudon, in press).

Farther north there is a single exposure of pre-Cretaceous quartzite, 300 m thick. The age of this sequence is unknown, but by analogy with the Ellsworth Mountains (Figure 2) it could be Devonian (Laudon, in press).

Structure

The Permian rocks have a moderate dip to the south-southeast at all three localities. There is no penetrative fabric. The quartzite beds have been sheared and hornfels developed by the intrusion of Cretaceous granite plutons (Laudon, in press).

Discussion

Although the exposures are very small, and none of the sections is very thick, these rocks are the only undeformed pre-Jurassic strata on the AP block. The affinities of these rocks are probably with the Cambrian-Permian succession of the Ellsworth-Whitmore Mountains block (Figure 2), in particular with the Permian Polarstar Formation (Collinson et al., in press).

STRATIGRAPHIC SUMMARY

The stratigraphy of the Antarctic Peninsula Mesozoic basins appears very variable at first glance (Figure 3). This is directly relevant to the hydrocarbon potential of the area. Although these basins formed on a single crustal block, all of which was presumably undergoing a similar tectonic history, there are few clues in the stratigraphy of any one basin to the likely stratigraphy of any other. In no case is detailed correlation between basins possible. The stricture of Green (1984) to all explorationists in frontier areas, that small geological changes can result in large changes in prospectivity, should be borne in mind. However, there are a number of general points which can be made about the stratigraphy of the Mesozoic basins.

1. There are few exposed sedimentary rocks older than latest Oxfordian or Kimmeridgian. The only exceptions to this are in Ellsworth Land (where some Permian strata occur, and the Latady Formation is Callovian in part), and in northeast Graham Land where the Botany Bay Group must be older than 152 ± 8 Ma and may be as old as Early Jurassic.

2. In many areas there seems to have been an expansion of the volcanic arc in latest Jurassic or earliest Cretaceous times. Upper Jurassic–Berriasian mudstones with an open marine fauna are overlain by lavas or coarse volcaniclastic sediments at a number of localities.

3. Jurassic and earliest Cretaceous mudstones are generally finer-grained and darker than mudstones from post-

Table 1. Summary of Pyrolysis data for Antarctic Peninsula mudstone samples. Total ranges of values for each parameter are shown.

	Sample area and age	n	TOC (%)	P_1 (ppm)	P_2 (ppm)	H. Index $\frac{P_2}{TOC}$ (mg/g)	Production Index $\frac{P_1}{P_1 + P_2}$	T_{max} (°C)	Lithification Group[1]
Forearc	Cretaceous (post-Berriasian)	8	0.50-0.88	6-19	77-381	9-59	0.021-0.163	358-499[2]	B
	Jurassic-Berriasian	6	0.42-0.92	4-26	12-324	2-77	0.018-0.250	348-498[2]	B
Backarc	Cretaceous (post-Berriasian)	11	0.44-1.13	*	*	*	*	*	B
	Jurassic-Berriasian	11	1.10-3.50	20-543	2122-10807	174-309	0.009-0.091	432[3] 439-454[2]	A
	Jurassic-Berriasian	2	1.50-2.44	180-874	1380-2140	87-92	0.290-0.330	439-443[3]	B

Notes: 1. See text for explanation of induration levels.
2. Analysis performed at Imperial College on non-standard Rock-Eval equipment.
3. Analysis performed at Robertson Group on standard Rock-Eval equipment.
* Data not available due to lab problems, should be run soon.

Berriasian strata. The only exception to this is Alexander Island, where all mudstones from Kimmeridgian–Albian are apparently similar.

4. Only in the Jurassic Latady Formation (Palmer Land backarc) is there widespread penetrative deformation. Although inverted, deformation is only locally penetrative in all other basins.

5. All arenites and rudites in all the basins are strongly volcaniclastic, with the exception of basement-derived, intra-arc sequences in Graham Land and the upper levels of the Larsen Basin fill.

6. There are no Mesozoic carbonates in any basin.

PETROLEUM GEOLOGY

The stratigraphic information presented here provides the basis for a series of constraints on the petroleum geology, which is discussed in general terms, rather than concentrating on any one area.

Seeps and Shows

No oil or gas seeps have been reported from the AP block. The only known occurrence in the area comes from the Bransfield Strait, where thermogenic hydrocarbons were discovered in a core of unconsolidated Recent sediments (Whiticar et al., 1985). The 8.6 m-long core smelled strongly petroliferous, with significant gas concentrations below 4.0 m; methane yields were up to 150 ppb, with ethane to pentane reaching 200 ppb (Whiticar et al., 1985).

Backarc extension in the Bransfield Strait only began at 4 Ma, with formation of a deep basin floored by transitional oceanic crust by 1.6 Ma (Barker and Dalziel, 1983). The sediment cover is thin, with 0.7-1.0 s maximum development (Jeffers and Anderson, 1990). The early maturation of hydrocarbons in such a shallow sequence implies that there could be significant losses before the formation of structural traps. Although the Bransfield Strait is a Neogene structure, it might provide an analog for some of the Mesozoic basins.

Source Rocks

Mudstone samples have been analyzed from various units in the area. Results from the Larsen Basin (around 50% of data presented here) have previously been published by Macdonald et al. (1988). All other data are new.

In a regional field survey of this type, it is necessary to collect samples from wherever they are exposed, rather than following a statistical sampling plan. Many basins either contain volcanic rocks or are intruded by later plutons; some are best exposed at their deformed margins. As a result the burial signature of maturation is commonly obscured by a later regional dynamothermal pattern. There is a very broad spread of relative maturities. Since degree of maturity affects total organic carbon (TOC), pyrolysis, and elemental ratios, the geochemical parameters discussed here are variably biased, making meaningful comparison difficult.

Where appropriate, samples are divided into two broad groups reflecting the degree of metamorphism. **Group A** represents unlithified or poorly lithified sediments: dominantly samples from the Larsen Basin. **Group B** represents well-indurated or metamorphosed samples. These vary from well-lithified rocks from the Larsen Basin margin, through the low-grade regionally metamorphosed rocks of the Fossil Bluff Group to samples of hornfels on the arc margins.

Total Organic Carbon (TOC)

Fifty-five samples were analyzed (Figure 4). TOC values range from 0.31-3.50% and exhibit a log-normal distribution with a mode of about 0.5-0.75%.

Splitting the data reveals interesting trends:

1. Mudstones of Jurassic-Berriasian age have a distinct separation in TOC between forearc and backarc areas (Figure 4). This could be due partly to clastic dilution, as more of the forearc samples are silty mudstones. However, the almost complete lack of overlap between the two distributions suggests that this is not the case. Berriasian strata are included with the Jurassic as the natural break in many sections occurs in late Berriasian times, and Valanginian strata are commonly missing.

2. No such areal separation occurs in Cretaceous (post-Berriasian) mudstones.

3. Cretaceous (post-Berriasian) mudstones have a much lower TOC than Jurassic-Berriasian ones. Again, clastic dilution may play a part, but the fact that fine mudstones also have low TOC implies some other cause.

Pyrolysis

Thirty-eight samples were analyzed by Rock-Eval pyrolysis. The results are presented in Table 1 as a series of ranges for the various parameters.

There is no significant difference between the Jurassic and Cretaceous of the forearc or Cretaceous of the backarc. P_1 values are generally low (<30 ppm) and there are no free hydrocarbons. Potential yields (P_2) are also low (<400 ppm). The low P_2 and moderate TOC values combine to give very low hydrogen indices (<80 mg/g) which indicate high maturity. This is confirmed by high values of production index; a few values over 0.15 could be due either to hydrocarbon contamination during cutting or to very high maturity. T_{max} in this study is not directly comparable to values normally obtained on commercially available Rock-Eval equipment, but values up to 400°C indicate high maturity, equivalent to dry gas generation.

In contrast, the Jurassic-Berriasian backarc rocks have relatively high P_1, indicating some free hydrocarbons. This is confirmed by visual examination which shows bright yellow or yellow-orange fluorescing palynomorphs under UV. Potential yields are all above 2000 ppm, and reach 11,000 ppm. Taken in conjunction with hydrogen indices over 150, this shows that these rocks have good potential to generate oil. Production Indices are low, which indicates relative immaturity, as does the tight grouping of T_{max} values below 455°C (Table 1).

The high P_1 and P_2 peaks of the Jurassic-Berriasian backarc rocks are partly a function of their lower maturity as compared to forearc and Cretaceous backarc samples. This reflects the fact that most were collected from reworked, poorly lithified material (Lithological Group A) in the Larsen Basin. However, the two samples which were collected from the deformed margin of the basin (Lithological Group B) suggest that the main difference is original source rock quality.

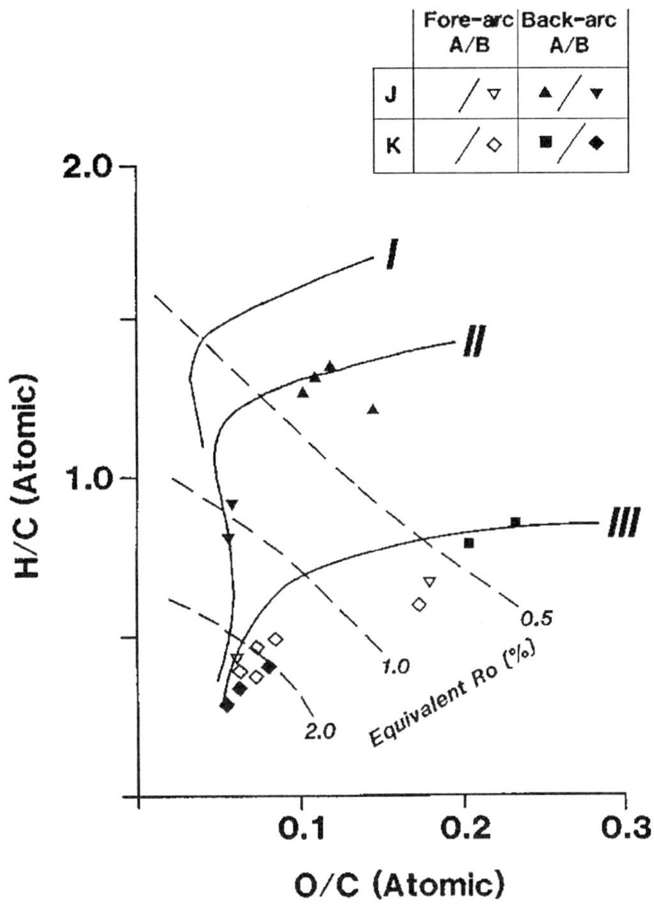

Figure 5—Van Krevelen diagram showing elemental ratios of Antarctic Peninsula mudstones, divided into four groups according to location and age (J: Jurassic-Berriasian; K: post-Berriasian). Each group is further subdivided according to lithification state (A and B); see text for details.

Kerogen Types

Elemental analysis was performed on kerogen isolates from 18 samples. This confirms the contrast between the Jurassic-Berriasian of the backarc areas (Type II), and all other samples which fall on a Type III trend (Figure 5). There is a wide spread of maturity, reflecting the variety of settings referred to above.

The contrast in kerogen types is largely confirmed by visual examination. Samples from the Jurassic-Berriasian of the backarc are dominantly composed of amorphous algal matter, waxy kerogen (spores and cuticle) and woody material; there are subsidiary, and highly variable proportions of spores, phytoplankton, and allochthonous coaly material. In complete contrast there is little or no amorphous organic matter in post-Berriasian backarc samples, and assemblages are dominated by humic macerals and terrestrial spores. The situation is not so clear-cut in the forearc, where low-grade regional metamorphism has destroyed much of the microfossil assemblage, leaving a residue of amorphous organic matter, some of which may be marine. However, the presence of abundant plant macerals, and the position on a Type III trend, suggest the dominance of terrestrial organic material.

Maturity

Pyrolysis and elemental analysis suggests that rocks in Group A (largely from little-deformed parts of the Larsen Basin) are immature, while Group B (Larsen Basin margins, and all other areas) are mature to over-mature. Vitrinite reflectance measurements were made on kerogen isolates of 16 samples and compared with results from 105 samples used by Whitham and Marshall (1988) in their detailed study of the margin of the Larsen Basin on James Ross Island.

Samples from forearc and intra-arc areas have R_o of 1.83-3.71% (Figure 6). The backarc samples all come from the deformed margin of the Larsen Basin, but can be split into two groups: those on the Antarctic Peninsula mainland, which are more highly deformed, and commonly affected by plutons, vary from 0.56-4.0%. A single sample from the eastern edge of the deformed zone on James Ross Island was R_o = 0.39%. This compares well with the work of Whitham and Marshall (1988) who found constant values of about 0.4% in the deformed zone on James Ross Island. At the transition out of the marginal deformed zone, R_o values declined over about 700 m of section to about 0.27%.

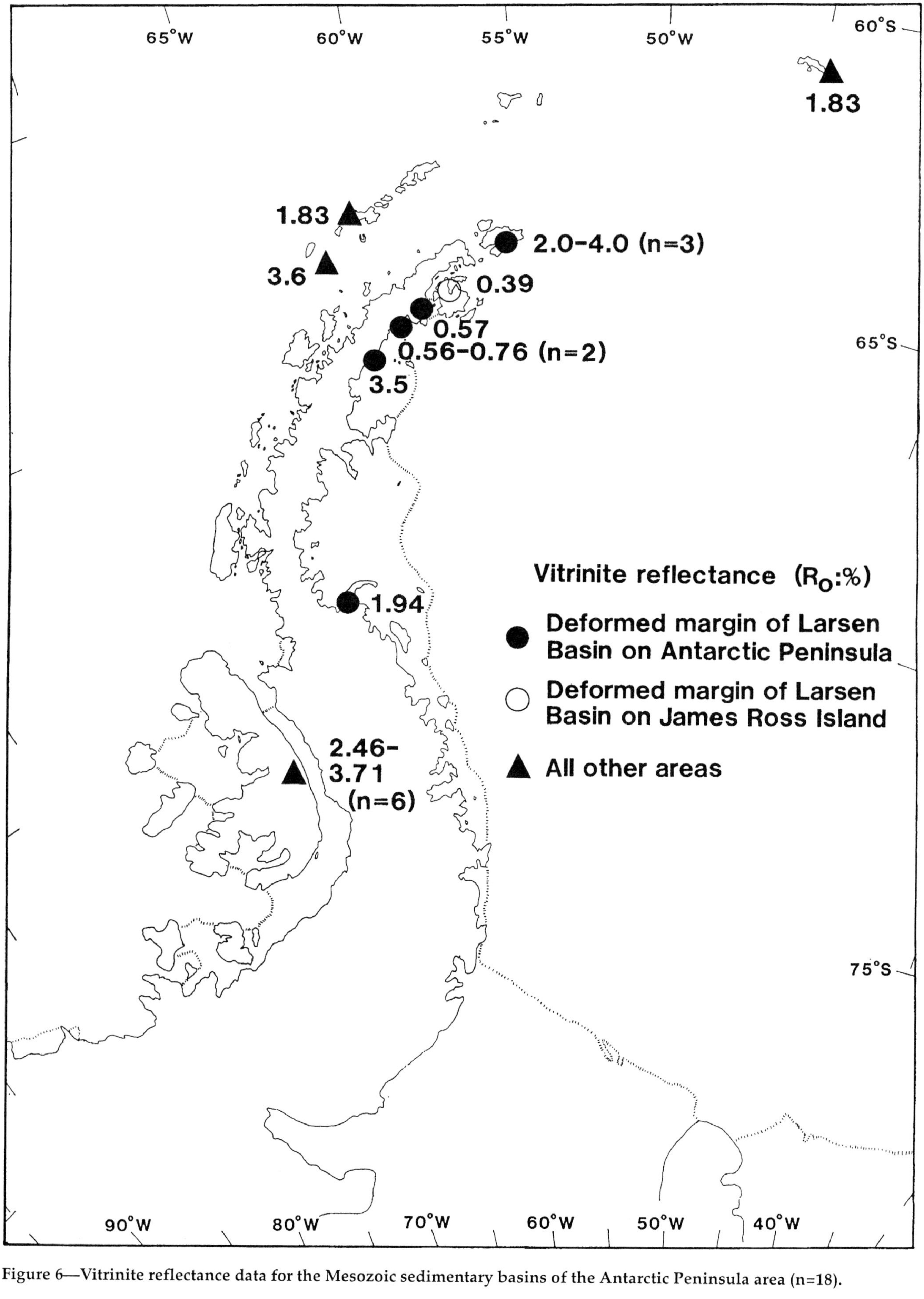

Figure 6—Vitrinite reflectance data for the Mesozoic sedimentary basins of the Antarctic Peninsula area (n=18).

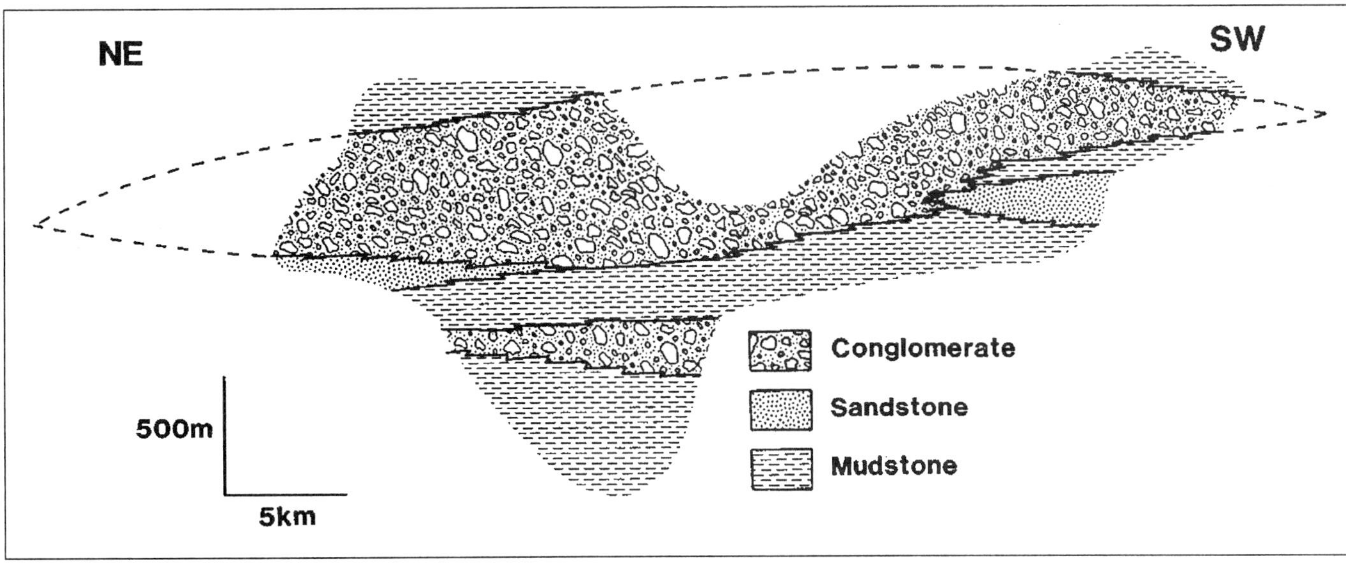

Figure 7—Schematic representation of the relationships between a lenticular body of channelled conglomerate and encasing sandstone and mudstone facies from the Aptian of James Ross Island (modified from Ineson, 1989).

Whitham and Marshall (1988) interpreted this pattern to reflect progressive deformation on the basin margin.

Conclusions

Although the bulk of the samples were mature or overmature, these results provide pointers for the likely source quality of Mesozoic mudstones generally.

1. Due to the degree of maturity of most of these field samples TOC and potential yield should be considered *minimum* values.
2. About 90% of the AP mudstones measured contain more than the minimum 0.5% TOC to be considered as potential source rocks.
3. There is a significant difference in TOC between the Jurassic-Berriasian mudstones of the backarc area (>1% TOC) and all other mudstones (mostly <1% TOC).
4. Only the Jurassic-Berriasian mudstones have potential yields sufficient to be considered good source rocks.
5. The Jurassic-Berriasian backarc samples contain Type II (oil-prone) kerogens; all other samples contain Type III dominantly gas-prone material.

Using the simple fourfold division in time (Jurassic + Berriasian vs. post-Berriasian) and space (forearc vs. backarc), only the Jurassic samples from the backarc are significant oil source rocks. These samples all come from the Nordenskjöld Formation (Farquharson 1983a; Whitham and Doyle 1989). The contrast in source rock quality between the Nordenskjöld Formation and equivalent-age strata in the forearc area could be due to the AP volcanic arc acting as a barrier separating the enclosed early Weddell Sea and open marine conditions in the proto-Pacific. This is a different interpretation to that of Farquharson (1983a), who considered Late Jurassic anoxia to be due to a mid-water oxygen minimum.

The decline of TOC and potential yield into the Cretaceous could be a function of arc expansion leading to greater clastic dilution. The concomitant change from Type II to Type III kerogen would then be a function of encroachment of a more extensive forested land area. This model implies that more marine-kerogen-rich mudstones may have been deposited farther away from the Cretaceous volcanic arc.

Reservoir

All of the Mesozoic sedimentary basins considered here are at least partly inverted; most are overmature. As such, they are not directly relevant to any assessment. Assessment is also complicated by the outcrop pattern, which is biased toward the proximal parts of basins. However, there are several general points which can be made, which may provide pointers to likely facies patterns in unexposed Mesozoic basins.

Facies

Most data come from the two well-exposed large basins: the Fossil Bluff Group (FBG) Basin of Alexander Island and the Larsen Basin of northeast Graham Land. Although timing of events is different, the sedimentological trends are similar. The fill of both basins is about 4-6 km thick, comprising regressive mega-sequences, from deep marine fan to shallow marine/terrestrial facies (Macdonald et al., 1988; Butterworth and Macdonald, in press).

In the deep marine parts of both basins there are thick inner fan channels, filled by clast-supported conglomerate, encased in slope mudstone (Figure 7). Both successions are relatively deficient in sandy facies, where seen in outcrop. It is reasonable to argue that the exposed strata represent the inner parts of high-efficiency fan systems (cf. Mutti, 1985); this implies that most of the sand bypassed these areas and was deposited as detached lobes in the basin. The width of the FBG channels is 15-20 km and they are up to 170 m thick (Butterworth, 1988). Those in the Larsen Basin are thought to be of a similar size (Figure 7) (Ineson, 1989). Consequently, any channel-fill reservoir could contain anything from a few to tens of km^3 of conglomerate, depending on channel length. The FBG examples are at least 5 km long, although they pinch out rapidly into the basin. Sandstone lobes would probably be of the same order of magnitude. Sedimentation within both the FBG and Larsen Basins is inferred to have been controlled by tectonism along the faulted arc-basin margin (Butterworth, 1988; Ineson, 1989) and as such each channel–lobe system is probably constrained in position relative to the margin (cf. Macdonald, 1986). Within the FBG, channel-fills are vertically stacked and overlap, with discrete fan complexes separated in space by interchannel mudstones (Butterworth, 1988). If such stacked complexes occurred in an unexposed basin, they would constitute an attractive multiple target. Within the Larsen Basin, it is possible that there may be a series of separate fan systems in elongate "axial" troughs parallel to the basin margin, as the equivocal paleocurrent information may suggest (Ineson, 1989).

The evolution of both the FBG and Larsen Basins culminated in the development of broad areas of shallow marine sedimentation with paralic/deltaic complexes (Macellari, 1988a; Pirrie, 1989; A.C.M. Moncrieff, 1989, personal communication). These deposits are much more sandstone-rich than the lower part of either succession. The deposits of the FBG Basin contain relatively thick mudstones at this level (A.C.M. Moncrieff, 1989, personal communication), but equivalent deposits in the Larsen Basin are sandier (Pirrie, 1989).

There is relatively little useful information from the other basins. It is interesting to note that in all the smaller basins examined here, sedimentary rocks are commonly interbedded with volcaniclastic and volcanic rocks and are sand-deficient. The Botany Bay Group intra-arc deposits of Graham Land show the type of basin which would develop on uplifted accretionary complex or metamorphic basement prior to the local onset of volcanism. Such intermontane-type basins are likely to contain the only significant quantities of nonvolcanic-derived clastic material in the areas of the Mesozoic basin. The Latady Formation of western Palmer Land illustrates the importance of tectonic control in all the AP Mesozoic basins. Throughout Late Jurassic times the position of the paleoshoreline seems to have been relatively constant (Laudon et al., 1983). As a result, the basin fill comprises three broad facies belts of volcanic rock, coarse sedimentary rock, and mudstone.

Petrography

Clast composition and diagenesis is a significant factor in reservoir assessment. Again, most information comes from the FBG and Larsen Basins.

Within the FBG, there is a clear time-dependant shift in the composition of conglomerate clasts, from dominantly volcanic to dominantly plutonic. This is mirrored by a similar petrofacies shift from volcaniclastic lithic to feldspathic sandstones (Figure 8) (Butterworth, 1988), reflecting a change from an undissected to a dissected arc terrane between Kimmeridgian and Valanginian times. Within the Aptian-Albian-age strata, sandstones are predominantly arkosic arenites. There is some evidence of quartz being more abundant in the very shallow marine/terrestrial sandstones which constitute the youngest part of the basin fill (Horne, 1968). The FBG is well-cemented, with abundant isolated calcareous concretions, and semi-continuous concretionary bands. The whole sequence has undergone a phase of low-grade metamorphism up to zeolite grade. This has produced mottled, laumontitized sandstones with no intergranular porosity.

Within the Larsen Basin, sandstones are arkosic and lithic arenites, with subordinate sublitharenites and quartz arenites (Pirrie, in press). There is a fundamental time-dependent change in the proportions of framework grains (Elliot and Trautman, 1982; Farquharson, 1983b; Macdonald et al., 1988; Pirrie, in press). In localities within 30 km of the basin margin in the James Ross Island area there is a shift from Hauterivian-Aptian arkosic arenites to sublitharenites and lithic arenites in the Campanian-Maastrichtian. In more distal localities there is a shift from Maastrichtian times (lithic and sublitharenites) to quartz arenites in Eocene-Oligocene-age strata (Macdonald et al., 1988; Pirrie, in press). Macdonald et al. (1988) pointed out that in Maastrichtian strata it appeared that deposits farther from the basin margin were more quartz-rich than their proximal counterparts. Although such proximal-distal comparisons are not possible for strata of other ages, they suggested that this might be a general relationship in the rest of the succession. More recent work by Pirrie (in press) suggests a series of unroofing events punctuated by volcanic pulses. Hence the view of the petrographic evolution presented by Macdonald et al. (1988) is probably oversimplified. In particular the proximal to distal change could be a function of poor age control.

About 70% of the Larsen Basin succession is unlithified or only partly lithified. Early diagenetic concretions are common (Pirrie, 1987). Carbonates, chlorite and silica are the dominant cements; together with other clays they are responsible for significant occlusion of porosity. Open porosity is generally rare, especially in the volcanic-rich sandstones.

Summary

There are five general points which might be applicable to the reservoir potential of unexposed Mesozoic basins.

1. Basins were tectonically active during deposition; facies patterns are related to marginal faults. As a result major sediment bodies are likely to be stacked.
2. Most basins exhibit an overall shallowing-upward trend with time. Large, coarse, clastic bodies encased in mudstone are likely to be more common in deeper-water facies. Shelf deposits appear to be more sand-rich.
3. All rudites and most arenites are rich in igneous detritus, demonstrating the dominance of the AP volcanic arc as a sediment source throughout Mesozoic times.
4. Petrographic trends indicate arc unroofing through time; only in the youngest deposits of any basin are there significant quartzose sandstones. However, local volcanic pulses can significantly affect the overall trend.
5. The dominance of abundant volcanic lithic grains has led to complex diagenesis and significant porosity reduction. No unit with a high percentage of volcanic grains is likely to constitute an attractive reservoir target.

Seals

Just as reservoir information was biased towards proximal areas, so are data on potential seals. The best sealing lithology in any of the basin is dark organic-rich mudstone of Kimmeridgian-Berriasian age; this is best developed on the backarc side of the peninsula. Most mudstones are somewhat silty, but should be adequate seals.

The major problem is in the Upper Cretaceous of the Larsen Basin, where the background sedimentation was bioturbated muddy sandstone (Pirrie, 1989). Macdonald et al. (1988) suggested that any reservoir of this age in the Larsen Basin would not be sealed. However, they also suggested that mudstones might be better developed in areas farther offshore.

Traps

The likelihood of traps is more difficult to predict than any other element of a potential hydrocarbon system, as it involves three separate factors (reservoir, seal, and structure), each with its own degree of uncertainty. Likely traps fall into two major categories:

Stratigraphic

Stratigraphic evidence presented here shows that large clastic bodies encased in mudstone are common, especially in deep marine deposits (Figure 7). The strong tectonic control in the AP Mesozoic basins has brought about abrupt facies switches which could lead to well-sealed, vertically stacked reservoirs. Sand bypassing of proximal areas may have resulted in large sandstone bodies in basin centers.

Structural

None of the basins studied is undeformed, but structural style is variable. In the Fossil Bluff Group forearc basin, large anticlines, involving the whole stratigraphic succession, trend obliquely across the basin, with variable plunge. These are associated with deep-seated thrusts, formed in a dextral strike-slip regime (Storey and Nell, 1988; Nell and Storey, in press). In any similar basin, folds such as these would make ideal traps.

In the Larsen Basin, deformation is concentrated at the basin margin, with the rest of the succession dipping gently eastward (Macdonald et al., 1988; Whitham and Marshall, 1988). Within the deformed marginal zone, there is evidence of strike-slip deformation affecting strata of Kimmeridgian-Berriasian age (Whitham and Storey, 1989). Reverse faults (Pirrie and Riding, 1988) may reflect basin inversion.

It is probable, given the intense tectonism which has affected the AP throughout Mesozoic times that any sedimentary basin will be at least locally deformed. Hence, structural traps are likely.

Maturation and Timing

Evidence from Bransfield Strait shows that thermogenic hydrocarbons can be generated very early at shallow depths in a basin with high heat flow (Whiticar et al., 1985). It is unlikely that any of the Mesozoic basins considered here was floored by oceanic crust, with the possible exception of the backarc basin of Palmer Land. However, heat flows were probably high.

Thermal modeling by Macdonald et al. (1988, their figure 10) showed that the Nordenskjöld Formation of the Larsen Basin would enter the oil window at about 80 Ma, with peak generation by 60 Ma. Even using the most conservative estimates of thickness and thermal gradient they suggested oil generation by 50 Ma. Realistic burial models implied generation of dry gas from 50 Ma.

An additional factor is the possible effect of ridge crest-trench collision from 50 Ma, along the AP subduction zone (Barker, 1982). De Long et al. (1978) suggested that one of the effects of subducting an ocean spreading center at an island arc would be regional low-grade thermal metamorphism (δT = 100-300°C). However, fission track work in progress has failed to detect temperature elevation above 100°C in northeast Alexander Island since mid-Eocene time (B.C. Storey, 1989, personal communication), despite ridge-crest trench collision in the area at 25 Ma. This is more in accord with the models of De Long et al. (1979) who suggested that abrupt temperature increases are restricted to shallow locations within 40 km of the trench, and only last a few million years. The major problem with early generation is that traps might not have formed. However, evidence presented here of early stratigraphic traps, and early deformation in some basins suggests that this would not be a problem.

CONCLUSIONS

The AP crustal block was the site of an active volcanic arc throughout Mesozoic time. It was in a region undergoing complex tectonics during Gondwana breakup. Exposed

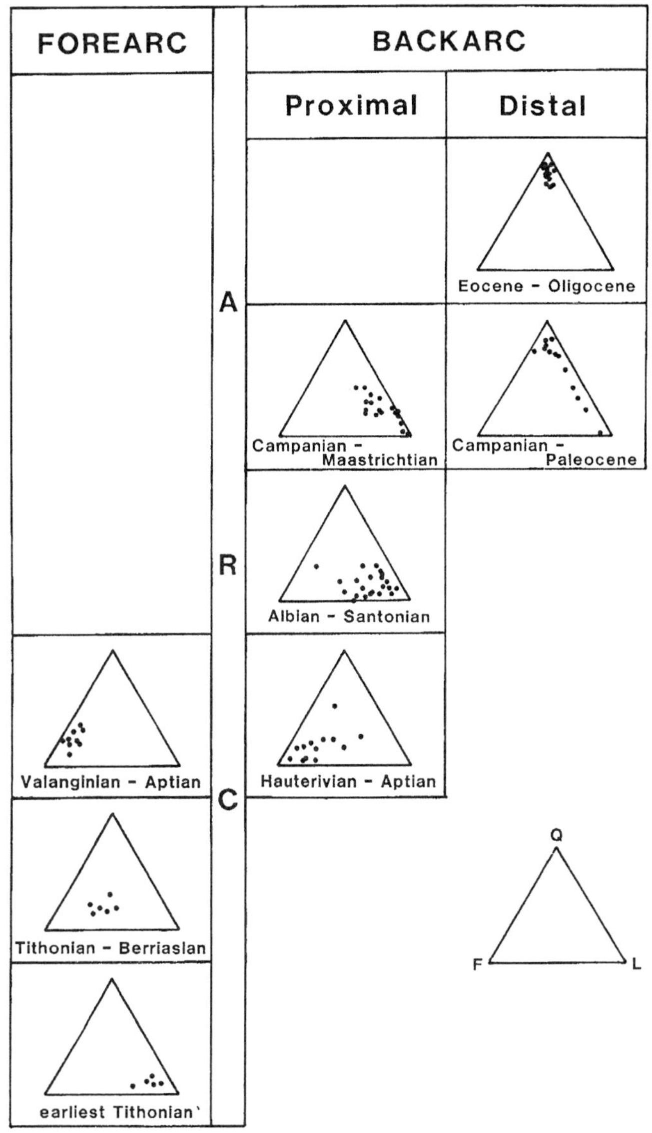

Figure 8—Petrographic evolution of sandstones from the Fossil Bluff Group (from Butterworth, 1988) and Larsen Basin (from various sources collated in Macdonald et al., 1988), showing the relative percentages of quartz (Q), feldspar (F), and lithic fragments (L). Recent work by Pirrie (in press) shows that the Campanian-Maastrichtian data presented here are oversimplified, and do not adequately represent the detail of the petrographic evolution of the Larsen Basin.

Mesozoic basins reflect this in their variable stratigraphy, strongly influenced by tectonics and the predominantly volcaniclastic input (Figures 3, 8).

Over the past 10-15 years, great emphasis has been placed on the use of analog methods in the preliminary assessment of frontier basins (Klemme, 1980). However, such methods require the best possible knowledge of the geological history of the area to be effective. Even small changes in the geology can radically alter the prospect possibilities of a basin (White and Gehman, 1979). This is particularly true of comparisons with southern South America

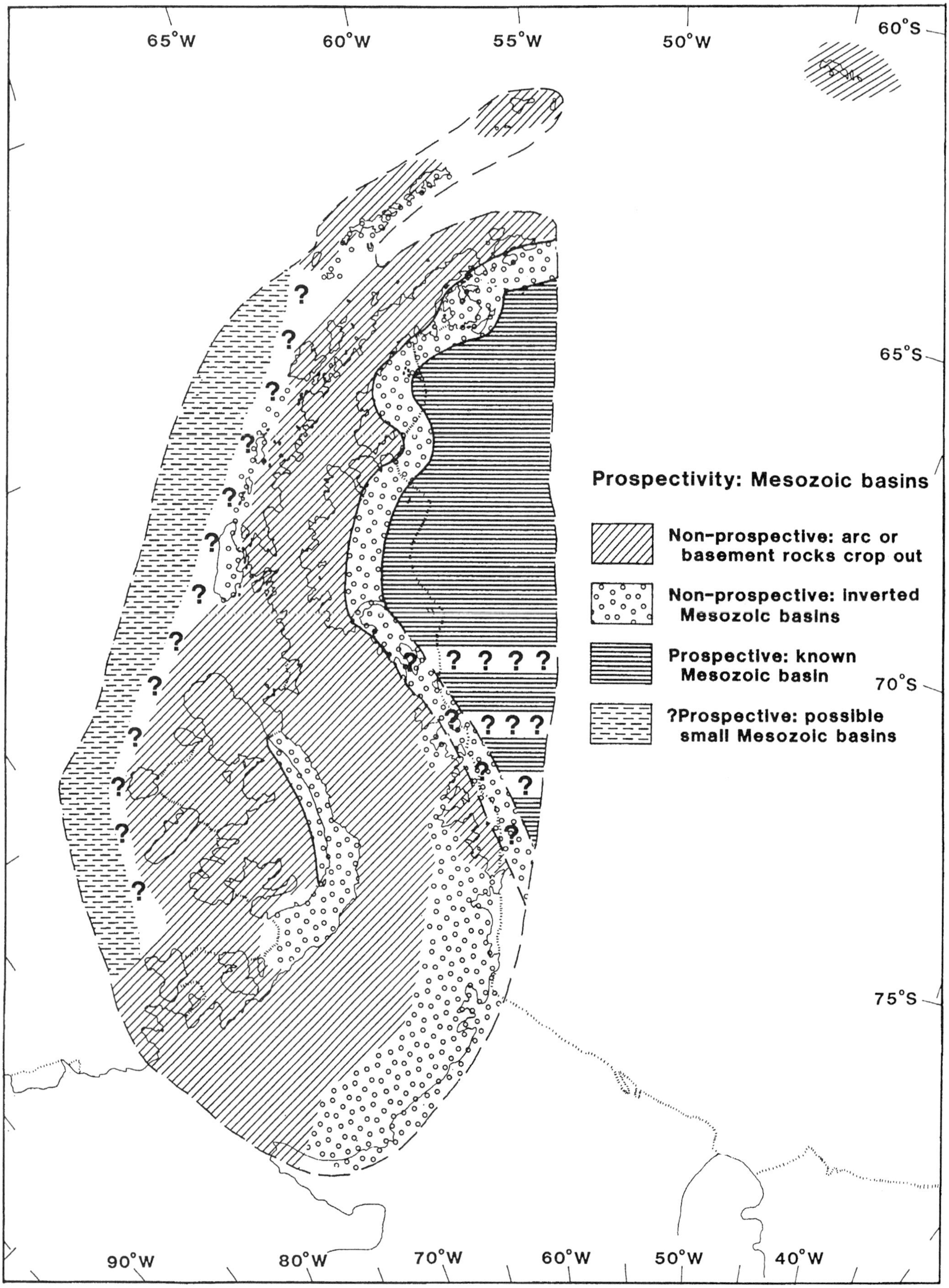

Figure 9—Schematic prospect possibility diagram for the Mesozoic sedimentary basins of the Antarctic Peninsula crustal block.

Table 2. Summary of the hydrocarbon potential of the Larsen Basin showing information from individual elements as exposed, and their likely trend into unexposed areas offshore. ? Basement-derived sandstones reflect the possibility of analogues to the intra arc basins of Graham Land occurring at depth.

	SOURCE		RESERVOIR FACIES		RESERVOIR QUALITY		SEALS	
SUCCESSION	Exposed	Offshore	Exposed	Offshore	Exposed	Offshore	Exposed	Offshore
Sandstone and mudstone (Con-Oli)	Lean, gas-prone	?Improving	Poor-moderate	Same or worse	Poor-moderate	Improving	Poor	Improving
Conglomerate and mudstone (Brm-Tur)	Lean, gas-prone	?Improving	Good	Same or worse	Poor	Improving	Moderate	Improving
Organic-rich mudstone (Kim-Ber)	Good, oil-prone	Same or better	—	—	—	—	Very good	Same
?Basement-derived sandstones (pre-Kim)	—	—	Good	Same	Good	Same	—	—

where the oil-producing Austral (Magallanes) Basin displays an overall Cretaceous shallowing-upward cycle (Malumián et al., 1983) similar to that seen in some AP basins. However, detailed comparisons show large differences. The early stage of basin evolution involved a widespread continental extensional volcanic suite of Jurassic age. This is paraconformably overlain by the transgressive Springhill Formation (of Late Jurassic–Valanginian age; Macellari, 1988b). This unit is the main reservoir horizon in the Austral Basin, and there are no similar deposits in any of the exposed AP basins. The Cretaceous history of the Austral Basin is also significantly different from the AP basins. In Early Cretaceous time the Austral Basin was a marginal basin with a broad eastern (pericratonic) shelf. Mid-Cretaceous closure of the marginal basin led to uplift of the "Paleoandes" along the line of the marginal basin, and establishment of a successor foredeep to the east (Macellari, 1988b). As a result there is a Cenomanian-Coniacian unconformity throughout the Austral Basin. None of this history compares in detail with any of the history of the AP basins.

For the moment, we have to rely on constraints available from the known geology. Figure 9 summarized prospect possibility of the AP crustal block for Mesozoic basins. Seventy percent (900,000 km^2) of the block is nonprospective, with basement, arc rocks, or inverted Mesozoic basins cropping out. It is possible that there may be hidden Mesozoic basins in this area, but outcrop data suggest they would be small and incapable of hosting large fields.

There is little information on the Pacific margin of the block, but it is possible that a forearc basin (equivalent to the Fossil Bluff Group Basin) occurs west of Graham Land. There may also be Mesozoic-Tertiary trench-slope basins on the accretionary prism west of Alexander Island. This whole zone covers about 13% (170,000 km^2) of the AP block. This is the area most sensitive to ridge crest-trench collision since decollement lubricated by high fluid pressure can lead to forward collapse of the accretionary prism and creation of sedimentary basins as the ridge crest approaches (Nell, in press). While it is unlikely that there are any large Mesozoic basins in this region, marine seismic exploration would be relatively easy, compared to other areas of the Antarctic margin.

The only large possible prospect on the block is the Larsen Basin, east of Graham Land, which probably extends down the eastern margin of Palmer Land (Figure 9). This basin has a narrow, deformed western margin, but its relatively undeformed part occupies 17% (220,000 km^2) of the AP block. This is probably the only Mesozoic basin in the area capable of hosting giant or supergiant fields.

Table 2 summarizes our current knowledge of various factors relevant to a potential hydrocarbon system in the Larsen Basin, and their likely changes into unexposed parts of the basin. As can be seen, the best prospects would be at deeper levels, with most parts of the system improving away from the volcanic arc. The major problem would be the extreme difficulty of operating in this region. To date few ships have penetrated the area bounded by 65° and 75°S, 50°W and the margin of the ice shelf, and survived.

Given the remoteness of the area, the short summer season, and the high cost of operating in the Antarctic, exploration is unlikely to proceed without a far better level of background knowledge than is currently available.

ACKNOWLEDGMENTS

A work of this type necessarily draws on data from many people. We would like to thank all our colleagues at BAS for much stimulating discussion. In particular we are grateful to the following people who shared their ideas, or gave permission to quote unpublished data, or read early drafts of the manuscript: Tony Alabaster, Alistair Crame, Peter Doyle, Steve Harrison, Malcolm Hole, Rob Larter, Alastair Milne, Andrew Moncrieff, Philip Nell, Bob Pankhurst, Duncan Pirrie, Bryan Storey, Janet Thomson, Mike Thomson, Andrew Whitham, and Rob Willan. We owe a great debt of gratitude to Bill St. John for his patience and forebearance during the editing process, and to Gill Clarke for typing all 43 versions of the manuscript. The authors are indebted to Malcolm Laird and Carlos Macellari for thorough reviews.

REFERENCES

Anderson, J. B., P. G. Pope and M. A. Thomas, 1990, Evolution and hydrocarbon potential of the northern Antarctic Peninsula continental shelf, this volume.

Barker, P. F., 1982, The Cenozoic subduction history of the Pacific margin of the Antartic Peninsula: ridge crest-trench interactions: Journal Geological Society of London, v. 139, p. 787-801.

Barker, P. F. and I. W. D. Dalziel, 1983, Progress in geodynamics in the Scotia Arc region: American Geophysical Union, Geodynamics Series 9, p. 137-170.

Barker, P. F., J. P. Kennett and Scientific Party, 1988, Weddell Sea palaeoceanography: preliminary results of ODP Leg 113: Palaeogeography, Palaeoclimatology, Palaeoecology, v. 67, p. 75-102.

Barler, P. F., I. W. D. Dalziel and B.C. Storey, in press, The tectonic evolution of the Scotia arc region, *in* R. J. Tingey, ed., Geology of Antarctica: Oxford, Oxford University Press.

Behrendt, J.C., ed., 1983, Petroleum and mineral resources of Antarctica: USGS Circular 909, 75 p.

Bell, C. M., 1974, The geology of parts of Alexander Island: Ph.D. dissertation, University of Birmingham, UK, 125 p.

Bibby, J. S., 1966, The stratigraphy of part of north-east Graham Land and the James Ross Island Group: British Antarctic Survey Scientific Report 53, 37 p.

Birkenmajer, K. and E. Zastawniak, 1989, Late Cretaceous–early Tertiary floras of King George Island, West Antarctica: their stratigraphic distribution and palaeoclimatic significance, *in* J. A. Crame, ed., Origins and evolution of the Antarctic biota: Geological Society of London Special Publication 47, p. 227-240.

Burn, R. W., 1981, Early Tertiary calc-alkaline volcanism on Alexander Island: British Antarctic Survey Bulletin, no. 53, p. 175-193.

Burn, R. W., 1984, The geology of the LeMay Group, Alexander Island: British Antarctic Survey Scientific Report 109, 65 p.

Butterworth, P. J., 1985, Sedimentology of Ablation Valley, Alexander Island: British Antarctic Survey Bulletin, no. 66, p. 73-82.

Butterworth, P. J., 1986, Slope collapse deposits in a Mesozoic marine fore-arc basin, Antarctica (abs.): 12th International Sedimentological Congress, Canberra, p. 193-194.

Butterworth, P. J., 1988, Sedimentology and stratigraphy of the Mesozoic Fossil Bluff Group, Alexander Island, Antarctica: Ph.D. dissertation, Council for National Academic Awards, 351 p.

Butterworth, P. J. and D. I. M. Macdonald, in press, Basin shallowing from the Mesozoic Fossil Bluff Group of Alexander Island and its regional tectonic significance, *in* M. R. A. Thomson, J. A. Crame and J. W. Thomson, eds., Geological evolution of Antarctica: Cambridge, Cambridge University Press.

Butterworth, P. J., J. A. Crame, P. J. Howlett and D.I.M. Macdonald, 1988, Lithostratigraphy of Upper Jurassic–Lower Cretaceous strata of eastern Alexander Island, Antarctica: Cretaceous Research, v. 9, p. 249-264.

Collinson, J. W., C. L. Vavra and J. M. Zawiskie, in press, Sedimentology of the Polarstar Formation, Permian, Ellsworth Mountains, Antarctica: GSA Memoir 170.

Crabtree, R. D., B. C. Storey and C. S. M. Doake, 1985, The structural evolution of George VI Sound, Antarctic Peninsula: Tectonophysics, v. 114, p. 431-442.

Crame, J. A. and P. J. Howlett, 1988, Late Jurassic and Early Cretaceous biostratigraphy of the Fossil Bluff Formation, Alexander Island: British Antarctic Survey Bulletin, no. 78, p. 1-35.

Culshaw, N. G., 1975, The geology of Carse Point, Palmer Land: British Antarctic Survey Bulletin, no. 41 and 42, p. 23-30.

Dalziel, I. W. D. and D. H. Elliot, 1982, West Antarctica: problem child of Gondwanaland: Tectonics, v. 1, p. 3-19.

Dalziel, I. W. D., B. C. Storey, S. W. Garrett, A. M. Grunow, L. D. B. Herrod and R. J. Pankhurst, 1987, Extension tectonics and the fragmentation of Gondwanaland, *in* M. P. Coward, J. F. Dewey and P. L. Hancock, eds., Continental extensional tectonics: Geological Society of London Special Publication 28, p. 433-441.

Davey, F. J., 1985, The Antarctic margin and its possible hydrocarbon potential: Tectonophysics, v. 114, p. 443-470.

DeLong, S.E., P.J. Fox and F.W. McDowell, 1978, Subduction of the Kula Ridge at the Aleutian Trench: GSA Bulletin, v. 89, p. 83-95.

DeLong, S. E., W.M. Schwarz and R.N. Anderson, 1979, Thermal effects of ridge subduction: Earth and Planetary Science Letters, v. 44, p. 239-246.

Deuser, W.G., 1971, Lake Maracaibo and Weddell Sea: comparison in petroleum geology: AAPG Bulletin, v. 55, p. 705-708.

Dewar, G.J., 1970, The geology of Adelaide Island: British Antarctic Survey Scientific Report 57, 66 p.

Doyle, P., J.A. Crame and M.R.A. Thomson, in press, Late Jurassic-Early Cretaceous macrofossils from ODP Leg 113, Site 692B, eastern Weddell Sea: Initial Reports of the Ocean Drilling Program.

Edwards, C. W., 1980, New evidence of major faulting on Alexander Island: British Antarctic Survey Bulletin, no. 49, p. 15-20.

Elliot, D. H. and T. A. Trautman, 1982, Lower Tertiary strata on Seymour Island, Antarctic Peninsula, *in* C. Craddock, ed., Antarctic Geoscience: Madison, University of Wisconsin Press, p. 287-297.

Elliot, D. H., and N. A. Wells, 1982, Mesozoic alluvial fans of the South Orkney Islands, *in* C. Craddock, ed., Antarctic Geoscience: Madison, University of Wisconsin Press, p. 235-244.

Farquharson, G. W., 1982, Late Mesozoic sedimentation in the northern Antarctic Peninsula and its relationship to the southern Andes: Journal Geological Society of London, v. 139, p. 721-728.

Farquharson, G. W., 1983a, The Nordenskjöld Formation of the northern Antarctic Peninsula: an Upper Jurassic radiolarian mudstone and tuff sequence: British Antarctic Survey Bulletin, no. 60, p. 1-22.

Farquharson, G. W., 1983b, Evolution of late Mesozoic sedimentary basins in the northern Antarctic Peninsula, *in* R. L. Oliver, P. R. James and J. B. Jago, eds., Antarctic Earth Science: Cambridge, Cambridge University Press, p. 323-327.

Farquharson, G. W., 1984, Late Mesozoic, non-marine conglomeratic sequences of northern Antarctic Peninsula (The Botany Bay Group): British Antarctic Survey Bulletin, no. 65, p. 1-32.

Fournier, H., M. Keller, J. Demichelli and H. Irigoin, 1980, Prospection magnetotel urica en la Isla Vicecomodoro Marambio, Antartida: Contribuciones Cientificas del Instituto Antartico Argentino, v. p. 37-46.

Garrett, S. W. and B. C. Storey, 1987, Lithospheric extension on the Antarctic Peninsula during Cenozoic subduction, *in* M. P. Coward, J. F. Dewey and P. L. Hancock, eds., Continental extensional tectonics: Geological Society of London Special Publication 28, p. 419-431.

Gee, C. T., 1989, Permian *Glossopteris* and *Elatocladus* megafossil floras from the English Coast, eastern Ellsworth Land Antarctica: Antarctic Science, v. 1, p. 35-44.

Green, A. R., 1984, Future petroleum province exploration: Proceedings of the 11th World Petroleum Congress 2, Chichester, John Wiley and Sons, p. 167-175.

Grunow, A. M., D. V. Kent and I. W. D. Dalziel, 1987, Evolution of the Weddell Sea basin, new palaeomagnetic constraints: Earth and Planetary Science Letters, v. 86, p. 16-26.

Harrison, S. M. and W. P. Loske, 1988, Early Paleozoic U-Pb isotope age for an orthogneiss from north-western Palmer Land, Antarctic Peninsula: British Antarctic Survey Bulletin, no. 81, p. 11-18.

Harrison, S. M. and B. A. Piercy, in press, Basement gneisses in northwestern Palmer Land: further evidence for pre-Mesozoic rocks in Lesser Antarctica, *in* M. R. A. Thomson, J. A. Crame and J. W. Thomson, eds., Geological Evolution of Antarctica: Cambridge, Cambridge University Press.

Hinz, K. and M. Block, 1984, Results of geophysical investigations in the Weddell Sea and in the Ross Sea, Antarctica: Proceedings of the 11th World Petroleum Congress 2, Chichester, John Wiley and Sons, p. 79-91.

Hinz, K. and W. Krause, 1982, The continental margin of Queen Maud Land, Antarctica: seismic sequences, structural elements and geological development: Geologische Jahrbuch E23, p. 17-41.

Horne, R. R., 1968, Petrology and provenance of the Cretaceous sediments of south-eastern Alexander Island: British Antarctic Survey Bulletin, no. 17, p. 73-82.

Horne, R. R. and M. R. A. Thomson, 1967, Post-Aptian camptonite dykes in south-eastern Alexander Island: British Antarctic Survey Bulletin, no. 14, p. 15-24.

Ineson, J. R., 1985, Submarine glide blocks from the Lower Cretaceous of the Antarctic Peninsula: Sedimentology, v. 32, p. 659-670.

Ineson, J. R., 1989, Coarse-grained submarine fan and slope apron deposits in a Cretaceous back-arc basin, Antarctica: Sedimentology, v. 36, p. 793-819.

Ineson, J. R., J. A. Crame and M. R. A. Thomson, 1986, Lithostratigraphy of the Cretaceous strata of west James Ross Island: Cretaceous Research, v. 7, p. 141-159.

Ivanov, V. L., 1989, Evolution of Antarctic prospective sedimentary basins: Antarctic Science, v. 1, p. 51-56.

Jeffers, J. D. and J. B. Anderson, 1990, Sequence stratigraphy of the Bransfield Basin, Antarctica: implications for tectonic history and hydrocarbon potential, this volume.

Jefferson, J. H., 1980, Angiosperm fossils in supposed Jurassic volcanogenic shales, Antarctica: Nature, v. 285, p. 157-158.

Jefferson, J. H., 1982, Fossil forests from the Lower Cretaceous of Alexander Island, Antarctica: Palaeontology, v. 25, p. 681-708.

Jones, J. A. and M. P. Maslanyj, in press, An aeromagnetic study of southern Palmer Land and eastern Ellsworth Land, *in* M. R. A. Thomson, J. A. Crame and J. W. Thomson, eds., Geological evolution of Antarctica: Cambridge, Cambridge University Press.

Kimura, K., 1982, Geological and geophysical survey in the Bellingshausen Basin, off Antarctica: Antarctic Record, v. 75, p. 12-24.

King, E. C. and P. F. Barker, 1988, The margins of the South Orkney microcontinent: Journal Geological Society of London, v. 145, p. 317-332.

Klemme, H. B., 1980, Petroleum basins–classifications and characteristics: Journal of Petroleum Geology, v. 5, p. 187-287.

Kristoffersen, Y. and V. C. Haugland, 1986, Geophysical evidence for the East Antarctic plate boundary in the Weddell Sea: Nature, v. 322, p. 538-541.

Kyle, P. R., D. H. Elliot and J. F. Sutter, 1981, Jurassic Ferrar Supergroup tholeiites from the Transantarctic Mountains, Antarctica and their relationship to the initial fragmentation of Gondwanaland, *in* M. M. Cresswell and P. Vella, eds., Gondwana Five: Rotterdam, A.A. Balkema, p. 283-287.

LaBrecque, J. L. and P. F. Barker, 1981, The age of the Weddell Basin: Nature, v. 290, p. 489-492.

Laudon, T. S., in press, Petrology of sedimentary rocks from the English Coast, eastern Ellsworth Land, *in* M. R. A. Thomson, J. A. Crame and J. W. Thomson, eds., Geological evolution of Antarctica: Cambridge, Cambridge University Press.

Laudon, T. S., M. R. A. Thomson, P. L. Williams, K. L. Milliken, P. D. Rowley and J. M. Boyles, 1983, The Jurassic Latady Formation, southern Antarctic Peninsula, *in* R. E. Oliver, P. R. James and J. B. Jago, eds., Antarctic earth science: Canberra, Australian Academy of Science and Cambridge University Press, p. 308-314.

Lawver, L. A. and C. R. Scotese, 1987, A revised reconstruction of Gondwana, *in* G. D. McKenzie, ed., Gondwana Six: Structure, tectonics and geophysics: American Geophysical Union Monograph 41, p. 17-23.

Macdonald, D. I. M., 1985, Sedimentological investigations in the Antarctic Peninsula area: British Antarctic Survey Field Report R/1984-85/G4, 24 p.

Macdonald, D. I. M., 1987, Sedimentological studies in northeastern Alexander Island: British Antarctic Survey Field Report R/1986-87/G3, 21 p.

Macdonald, D. I. M., 1986, Proximal to distal sedimentological variation in a linear turbidite trough: implications for the fan model: Sedimentology, v. 33, p. 243-259.

Macdonand, D. I. M. and P. J. Butterworth, 1986, Slope collapse deposits in a Mesozoic marine fore-arc basin, Antarctica (abs.): 12th International Sedimentological Congress, Canberra, p. 193-194.

Macdonald, D. I. M., P. F. Barker, S. W. Garrett, J. R. Ineson, D. Pirrie, B. C. Storey, A. G. Whitham, R. R. F. Kinghorn and J. E. A. Marshall, 1988, A preliminary assessment of the hydrocarbon potential of the Larsen Basin, Antarctica: Marine and Petroleum Geology, v. 5, p. 34-53.

Macellari, C. E., 1988a, Stratigraphy, sedimentology and paleoecology of late Cretaceous/Paleocene shelf-deltaic sediments of Seymour Island (Antarctic Peninsula), *in* R. M. Feldmann and M. O. Woodburne, eds., Geology and paleontology of Seymour Island: GSA Memoir 169, p. 25-53.

Macellari, C. E., 1988b, Cretaceous paleogeography and depositional cycles of western South America. Journal of South American Earth Sciences, v. l, p. 373-418.

Malumián, N., F. E. Nullo and V. A. Ramos, 1983, The Cretaceous of Argentina, Chile, Paraguay and Uruguay, *in* M. Moullade and A. E. M. Nairn, eds., The Phanerozoic geology of the world II, The Mesozoic, B: Amsterdam, Elsevier, p. 265-304.

Maslanyj, M. P., 1988, Gravity and aeromagnetic evidence for the crustal structure of George VI Sound, Antarctic Peninsula: British Antarctic Survey Bulletin, no. 79, p. 1-16.

Meneilly, A. W., S. M. Harrison, B. A. Piercy and B. C. Storey, 1987, Structural evolution of the magmatic arc in northern Palmer Land, Antarctic Peninsula, *in* G. D. McKenzie, ed., Gondwana Six: structure, tectonics and geophysics: American Geophysical Union Geophysical Monograph, v. 40, p. 209-219.

Millar, I. L., A. J. Milne and A. G. Whitham, in press, Implications of Sm-Nd garnet ages for the stratigraphy of northern Graham Land, Antarctic Peninsula: Zentralblatt fur Geologie und Palaontologie.

Milne, A. J. and I. L. Millar, 1989, The significance of mid-Palaeozoic basement in Graham Land, Antarctic Peninsula: Journal Geological Society of London, v. 146, p. 207-210.

Mutti, E., 1985, Turbidite systems and their relations to depositional sequences, *in* G. G. Zuffa, ed., Provenance of arenites: NATO ASI Series, Dordrecht, The Netherlands, Reidel Publishing Company, p. 65-93.

Nell, P. A. R., in press, Deformation in an accretionary melange, Alexander Island, Antarctica, *in* R. J. Knipe, ed.,

Deformation mechanisms, rheology and tectonics: Geological Society of London Special Publication.

Nell, P. A. R., B. K. Kamenov and C. T. Pimpirev, 1989, Field studies on combined Anglo-Bulgarian geological studies in northern Alexander Island: Antarctic Science, v. 1, p. 167-168.

Nell, P. A. R. and B. C. Storey, in press, Strike-slip tectonics within the Antarctic Peninsula forearc, *in* M. R. A. Thomson, J. A. Crame and J.W. Thomson, eds., Geological evolution of Antarctica: Cambridge, Cambridge University Press.

Nelson, P. H. H., 1975, The James Ross Island Volcanic Group of north-east Graham Land: British Antarctic Survey Scientific Report 54, 62 p.

Norton, I. O. and J. G. Sclater, 1979, A model for the evolution of the Indian Ocean and the break-up of Gondwanaland: Journal of Geophysical Research, v. 84, p. 6803-6830.

Pankhurst, R. J., 1982, Rb-Sr geochronology of Graham Land, Antarctica: Journal Geological Society of London, v. 139, p. 701-711.

Pankhurst, R. J., 1983, Rb-Sr constraints on the ages of basement rocks of the Antarctic Peninsula, *in* R. E. Oliver, P. R. James and J. B. Jago, eds., Antarctic earth science: Cambridge, Cambridge University Press, p. 367-371.

Pankhurst, R. J. and J. L. Smellie, 1983, K-Ar geochronology of the South Shetland Islands, Lesser Antarctica: apparent lateral migration of Jurassic to Quaternary island arc volcanism: Earth and Planetary Science Letters, v. 66, p. 214-222.

Paterlini, C. M., J. H. Nunez and M. A. Keller, 1984, Regional aeromagnetic survey over the South Shetland Islands, Bransfield Strait and part of the Antarctic Peninsula: Revista Geologia de Chile, v. 23, p. 3-20.

Piercy, B. A. and S. M. Harrison, in press, Mesozoic metamorphism, deformation and plutonism in the southern Antarctic Peninsula: evidence from north-western Palmer Land, *in* M. R. A. Thomson, J. A. Crame and J. W. Thomson, eds., Geological evolution of Antarctica: Cambridge, Cambridge University Press.

Pirrie, D., 1987, Orientated calcareous concretions from James Ross Island, Antarctica: British Antarctic Survey Bulletin, no. 75, p. 41-50.

Pirrie, D., 1989, Shallow marine sedimentation within an active margin basin, James Ross Island, Antarctica: Sedimentary Geology, v. 63, p. 61-82.

Pirrie, D., in press, Controls on the petrographic evolution of an active margin sedimentary sequence: the Larsen Basin, Antarctica, *in* A. C. Morton and S. P. Todd, eds., Developments in sedimentary provenance studies: Geological Society of London Special Publication.

Pirrie, D. and J. B. Riding, 1988, Sedimentology, palynology and structure of Humps Island, northern Antarctic Peninsula: British Antarctic Survey Bulletin, no. 80, p. 1–19.

Quilty, P. G., 1982, Tectonic and other implications of Middle-Late Jurassic rocks and marine faunas from Ellsworth Land, Antarctica, *in* C. Craddock, ed., Antarctic Geoscience: Madison, University of Wisconsin Press, p. 669-678.

Renner, R. G. B., L. J. S. Sturgeon and S. W. Garrett, 1985, Reconnaissance gravity and aeromagnetic surveys of the Antarctic Peninsula: British Antarctic Survey Scientific Report 110, 50 p.

Rex, D. C., 1970, Age of camptonite dyke from south-east Alexander Island: British Antarctic Survey Bulletin, no. 23, p. 103.

Rowley, P. D., W. R. Vennum, K. S. Kellogg, T. S. Laudon, P. E. Carrara, J. M. Boyles and M. R. A. Thomson, 1983, Geology and plate tectonic setting of the Orville Coast and eastern Ellsworth Land, Antarctica, *in* R. L. Oliver, P. R. James and J. B. Jago, eds., Antarctic earth science: Cambridge, Cambridge University Press, p. 245-250.

Sadler, P. M. 1988, Geometry and stratification of uppermost Cretaceous and Paleogene units on Seymour Island, northern Antarctic Peninsula, *in* R. M. Feldmann and M. O. Woodburne, eds., Geology and paleontology of Seymour Island: GSA Memoir 169, p. 303-320.

Schmidt, D. L. and P. D. Rowley, 1986, Continental rifting and transform faulting along the Jurassic Transantarctic rift, Antarctica: Tectonics, v. 5, p. 279-291.

Singleton, D. G., 1980, The geology of the central Black Coast, Palmer Land: British Antarctic Survey Scientific Report 102, 50 p.

Skinner, A. C., 1973, Geology of north-western Palmer Land between Eureka and Meiklejohn Glaciers: British Antarctic Survey Bulletin, no. 35, p. 1-22.

Smellie, J. L., 1987, Sandstone detrital modes and basinal setting of the Trinity Peninsula Group, northern Graham Land, Antarctic Peninsula: a preliminary survey, *in* G. D. McKenzie, ed., Gondwana Six: Structure, tectonics and geophysics: Washington DC, Geophysical Monograph Series, American Geophysical Union, v. 40, p. 199-207.

Smellie, J. L. and P. D. Clarkson, 1975, Evidence for pre-Jurassic subduction in western Antarctica: Nature, v. 258, p. 701-702.

Smellie, J. L., R. E. S. Davies and M. R. A. Thomson, 1980, Geology of a Mesozoic intra-arc sequence on Byers Peninsula, Livingston Island, South Shetland Islands: British Antarctic Survey Bulletin, no. 50, p. 55-76.

Smellie, J. L., A. B. Moyes, P. D. Marsh and J. W. Thomson, 1985, Field report on the geology of Hugo Island, Quintana Island, Sooty Rock, Betbeder Islands, Rugged Island and parts of the Biscoe and Outcast islands: British Antarctic Survey Field Report R/83-84/G7, 29 p.

Smellie, J. L., R. J. Pankhurst, M. J. Hole and J. W. Thomson, 1988, Age, distribution and eruptive conditions of Late Cenozoic alkaline volcanism in the Antarctic Peninsula and eastern Ellsworth Land: review: British Antarctic Survey Bulletin, no. 80, p. 21–48.

Smellie, J. L., R. J. Pankhurst, M.R.A. Thomson and R.E.S. Davies, 1984, The geology of the South Shetland Islands: VI. Stratigraphy, geochemistry and evolution: British Antarctic Survey Scientific Report 87, 85 p.

St. John, B., 1986, Antarctica—geology and hydrocarbon potential, *in* M. Halbouty, ed., Future petroleum provinces of the world: AAPG Memoir 40, p. 55-100.

Storey, B. C. and S. W. Garrett, 1985, Crustal growth of the Antarctic Peninsula by accretion, magmatism and extension: Geological Magazine, v. 122, p. 5-14.

Storey, B. C. and P. A. R. Nell, 1988, Role of strike-slip faulting in the tectonic evolution of the Antarctic Peninsula: Journal Geological Society of London, v. 145, p. 333-337.

Storey, B, C., H. E. Wever, P. D. Rowley and A. B. Ford, 1987, The geology of the central Black Coast, eastern Palmer Land: British Antarctic Survey Bulletin, no. 77, p. 145-155.

Storey, B. C., I. W. D. Dalziel, S.W. Garrett, A. M. Grunow, R. J. Pankhurst and W. R. Vennum, 1988a, West Antarctica in Gondwanaland: crustal blocks, reconstruction and breakup processes: Tectonophysics, v. 155, p. 381-390.

Storey, B. C., M. J. Hole, R. J. Pankhurst, I. L. Millar and W. Vennum, 1988b, Middle Jurassic within-plate granites in West Antarctica and their bearing on the break-up of Gondwanaland: Journal Geological Society of London, v. 145, p. 999-1007.

Storey, B. C. et al., 1989, Antarctic geoscience transect: AP2: Abstracts of the 28th International Geological Congress, Washington, D.C., v. 3, p. 187.

Tanner, P. W. G., R. J. Pankhurst and G. Hyden, 1982, Radiometric evidence for the age of the subduction complex in the South Orkney and South Shetland Islands, West

Antarctica: Journal Geological Society of London, v. 139, p. 683-690.

Taylor, B. J., 1982, Sedimentary dykes, pipes and related structures in the Mesozoic sediments of south-eastern Alexander Island: British Antarctic Survey Bulletin, no. 41, p. 1-42.

Tectonic Map of the Scotia Arc, 1985, 1:3 000 000. BAS (Misc) 3: Cambridge, British Antarctic Survey.

Thomson, M. R. A., 1972, New discoveries of fossils in the Upper Jurassic Volcanic Group of Adelaide Island: British Antarctic Survey Bulletin, no. 30, p. 95-101.

Thomson, M. R. A., 1975a, New palaeontological and lithological observations on the Legoupil Formation, northwest Antarctic Peninsula: British Antarctic Survey Bulletin, no. 41 and 42, p. 169-185.

Thomson, M. R. A., 1975b, Upper Jurassic Mollusca from Carse Point, Palmer Land: British Antarctic Survey Bulletin, no. 41 and 42, p. 31-42.

Thomson, M. R. A., 1980, Late Jurassic ammonite faunas from the Latady Formation, Orville Coast: U.S. Antarctic Journal, v. 15, p. 28-30.

Thomson, M. R. A., 1981, Late Mesozoic stratigraphy and invertebrate palaeontology of the South Orkney Islands: British Antarctic Survey Bulletin, no. 54, p. 65-84.

Thomson, M. R. A., 1982a, Mesozoic palaeogeography of western Antarctica, *in* C. Craddock, ed., Antarctic geoscience: Madison, University of Wisconsin Press, p. 331-337.

Thomson, M. R. A., 1982b, Late Jurassic fossils from Low Island, South Shetland Islands: British Antarctic Survey Bulletin, no. 56, p. 25-35.

Thomson, M. R. A., 1983, Late Jurassic ammonites from the Orville Coast, Antarctica, *in* R. L. Oliver, P. R. James and J. B. Jago, eds., Antarctic earth science: Cambridge, Cambridge University Press, p. 315-319.

Thomson, M. R. A. and G. W. Farquharson, 1984, Discovery and significance of the ammonite genus *Favrella* in the Antarctic Peninsula area: British Antarctic Survey Bulletin, no. 62, p. 7-14.

Thomson, M. R. A. and T. H. Tranter, 1986, Early Jurassic fossils from central Alexander Island and their geological setting: British Antarctic Survey Bulletin, no. 70, p. 23-39.

Tranter, T. H., 1987, The structural history of the LeMay Group of central Alexander Island, Antarctic Peninsula: British Antarctic Survey Bulletin, no. 77, p. 61-80.

White, D. A. and H. M. Gehman, 1979, Methods of estimating oil and gas resources: AAPG Bulletin, v.63, p.2183-2192.

Whitham, A. G. and P. Doyle, 1989, Stratigraphy of the Upper Jurassic-Lower Cretaceous Nordenskjöld Formation of eastern Graham Land, Antarctica: Journal of South American Earth Sciences, v. 2, p. 371–384.

Whitham, A. G. and J. R. Ineson, in press, Shallow to deep marine sedimentation patterns along a tectonically active back-arc basin margin: Cretaceous of James Ross Island, Antarctica: Journal of Sedimentary Petrology.

Whitham, A. G. and J. E. A. Marshall, 1988, Syndepositional deformation in a Cretaceous succession, James Ross Island, Antarctica. Evidence from vitrinite reflectivity: Geological Magazine, v. 125, p. 583-591.

Whitham, A. G. and B. C. Storey, 1989, Late Jurassic-Early Cretaceous strike-slip deformation in the Nordenskjöld Formation of Graham Land: Antarctic Science, v. 1, p. 269-278.

Whitham, A. G., P. J. Butterworth and D. I. M. Macdonald, 1987, Tectonic significance of fan-delta sediments from the western margin of the Weddell Sea, Antarctica (abs): Bergen, International Symposium on Fan Deltas, p. 91-93.

Whiticar, M. J., E. Suess and H. Wehner, 1985, Thermogenic hydrocarbons in surface sediments of the Bransfield Strait, Antarctic Peninsula: Nature, v. 314, p. 87-90.

Wyeth, R. B., 1977, The physiography and significance of the transition zone between Graham Land and Palmer Land: British Antarctic Survey Bulletin, no. 46, p. 39-58.

Chapter 9

Geophysical Investigations in the Bransfield Strait and in the Bellingshausen Sea—Antarctica

L. A. P. Gambôa
P. R. Maldonado
Petróleo Brasileiro S.A.–PETROBRÁS and
Brazilian Antarctic Program–PROANTAR

ABSTRACT

Marine geophysical investigations off the northern tip of the Antarctic Peninsula by the Brazilian Antarctic Program in 1987 and 1988 have revealed the complex geologic evolution and structure of the Bransfield Basin, and the Bellingshausen continental margin. The Bransfield Basin, within the Bransfield Strait, has an asymmetrical profile with a steeper slope along its northern margin and a conspicuous spreading center closer to the South Shetland Islands. A sedimentary wedge deposited along the southern margin of the basin forms the northern continental margin of the Antarctic Peninsula. Structural features and sedimentary sequences in this wedge show an Atlantic-type margin setting with an older rift sequence and a younger drift sequence.

The Bellingshausen continental margin shows a well-developed continental rise, including a deep-sea fan to the north of Adelaide Island, a steep continental slope and a broad continental shelf. At the outer shelf, clinoforms indicate a prograding shelf to slope environment similar to that of the continental shelf of an Atlantic-type margin. These sediments have prograded above an erosional unconformity, below which tilted and faulted layers are observed and appear to represent an earlier "active" margin setting. A basement high occurs at the eastern limit of the younger passive margin sedimentary wedge, and a closed and buried basin has been discovered to the east of the basement high. The basement high and the closed basin could represent an eroded island arc and a fossil backarc basin, respectively.

INTRODUCTION

In the austral summers of 1987 and 1988, the Brazilian Antarctic Program conducted two marine geophysical expeditions off the Antarctic Peninsula, between Adelaide and Elephant Islands (Figure 1). These surveys were part of a reconnaissance program by PETROBRÁS and the Brazilian Navy, to investigate the structure of the Bransfield Strait and its relation to the South Shetland Trench, and the structure of the Bellingshausen continental margin. The Bransfield Basin is a fairly young rift located between the South Shetland Islands and the Antarctic Peninsula (Figures 1, 2, 3). The Bellingshausen continental margin, south of the South Shetland Islands, has a complex geologic history, changing in the Tertiary from an active margin into a passive margin.

The surveys were conducted aboard the research vessel *Almirante Câmara* of the Brazilian Navy. Using eight high-pressure air guns (540 in.^{3} total pressure) and operating at 4500 psi, 5560 km of seismic data were acquired. The signals were received by a 72-channel streamer and recorded using a DFS-V. Continuous gravity and magnetic measurements were recorded with a Lacoste & Romberg gravimeter and a Geometrics nuclear precession magnetometer. Transit and GPS satellite systems were used for navigation. This chapter reports the results of the interpretation of the seismic data. The gravity and magnetic data, still being processed, will be presented in another article.

The newly acquired seismic data reveal the structure of the Bransfield Basin as well as the seismic sequences that document the rifting history of the strait. The data also show the structure of the Bellingshausen continental margin, where accreted and forearc material is identified beneath younger stacked and prograded deposits formed during the inactive period of the margin. Our seismic investigations support the previously deduced evolutionary history of this region based on magnetic data (Herron and Tucholke, 1976; Herron et al., 1981; Barker, 1982).

REGIONAL GEOLOGIC SETTING

As shown by several authors (e.g. Dalziel, 1984; Dott et al., 1982; Farquharson, 1982; Forsythe, 1982; Harrison et al., 1979; Thomson et al., 1983; Elliot, 1983; Saunders and Tarney, 1982), the geology of the Antarctic Peninsula bears many similarities to that of southernmost South America. Plate reconstructions (e.g., Lawver and Scotese, 1987; Dalziel, 1983; Barker et al., 1976; De Wit, 1977) indicate that an active magmatic arc extended continuously from the Andes to the Antarctic Peninsula. This arc formed the western margin of Gondwana and was active at least since the Triassic (Smellie and Clarkson, 1975; Storey and Garrett, 1985; Thomson et al., 1983). The opening of the Drake Passage and the formation of the Scotia Arc isolated the Antarctic continent after the Oligocene (Barker and Burrell, 1977).

The Antarctic Peninsula is presently surrounded by oceanic crust. The opening of the Weddell and Scotia Seas and the ridge crest interactions along the peninsula's western margin transformed the peninsula into a tectonically unstable strip of continental crust, along which there has been rifting and wrench-fault related block movement.

The evolutionary history of the western margin of the peninsula has previously been inferred primarily from studies of the magnetic anomalies in the adjacent ocean floor (Herron and Tucholke, 1976; Herron et al., 1981; Barker, 1982). These studies indicate that the Bellingshausen continental margin, to the south of the South Shetland Islands, evolved from an active margin with a subduction zone into an inactive margin. The deep-sea trench that is visible in front of the South Shetland Islands, is not present along the Bellingshausen continental margin (Figures 2, 3).

The South Shetland Archipelago is located at the southwestern end of the South Scotia Ridge (Figure 3). Geological and geophysical evidence indicates that the South Shetland Islands were originally part of the Antarctic Peninsula, before the formation of the Bransfield Basin, (Ashcroft, 1972; Davey, 1972; Barker, 1970; Thomson et al., 1983). A magmatic island arc has existed in the region since the Jurassic. Until the formation of the Bransfield Basin, the arc was situated on the Antarctic Peninsula. The Bransfield Basin separates the South Shetland Island, the present location of the volcanic arc, from the Antarctic Peninsula. The age of formation of the Bransfield Basin is unclear. The existing evidence indicates that the basin was formed less than 4 Ma (Barker, 1982; Storey and Garrett, 1985; González-Ferrán, 1985). As we will discuss later, our work indicates that this backarc basin could be correlated with a closed basin south of the Hero Fracture Zone, and thus it could be older than previously estimated.

PRESENT WORK

Bransfield Basin

The Bransfield Basin is a marginal basin located behind the South Shetland Islands and the South Shetland Trench. The South Shetland Trench is bounded to the north by the Shackleton Fracture Zone and to the south by the Hero Fracture Zone (Figure 3). The basin has an asymmetrical profile, with a steeper slope along its northern margin and with its spreading center closer to the South Shetland Arc (Figure 4). The northern and southern limits of the Bransfield Basin are roughly aligned with those of the South Shetland Trench (Shackleton Fracture Zone to the north and Hero Fracture Zone to the south).

The Bransfield spreading center is a conspicuous feature in the deeper parts of the basin (Figure 5), and is extremely well defined on seismic profiles compared to spreading centers described in other marginal basins (e.g., Karig et al., 1978; Mrozowski and Hayes, 1979; Watts, et al., 1977; Hilde and Lee, 1984). The topographic expression of the spreading center varies along its length and crops out subaerially on Deception and Bridgeman Islands. The spreading center is disrupted by a transform fault near the northern end of King George Island (Figure 6).

The deeper areas of the Bransfield Basin, are characterized by a smooth submarine floor (Figures 7, 8). A sequence of sediments showing high-amplitude, semiparallel, horizontal reflections occurs above the oceanic/transitional basement. These reflections show good lateral continuity, suggesting that they represent interbedded turbidites and pelagic layers.

A sedimentary wedge was deposited along the southern margin of the Bransfield Basin, and forms the northern continental margin of the present Antarctic Peninsula (Figures 6, 7). Seismic data show several genetically distinct sedimentary sequences (Figure 7). A region of thin sedimentary cover and outcrops of "acoustic basement" are reported from the peninsula down to the 200 m isobath (Barker and Griffiths, 1972). Because our seismic profiles extended landward only to water depths of about 200 m along the southern margin of the basin, we are unable to speculate about the structure of the area closer to the Antarctic Peninsula. Reflections that dip steeply in a basinward direction are observed at the southern margin of the Bransfield Basin

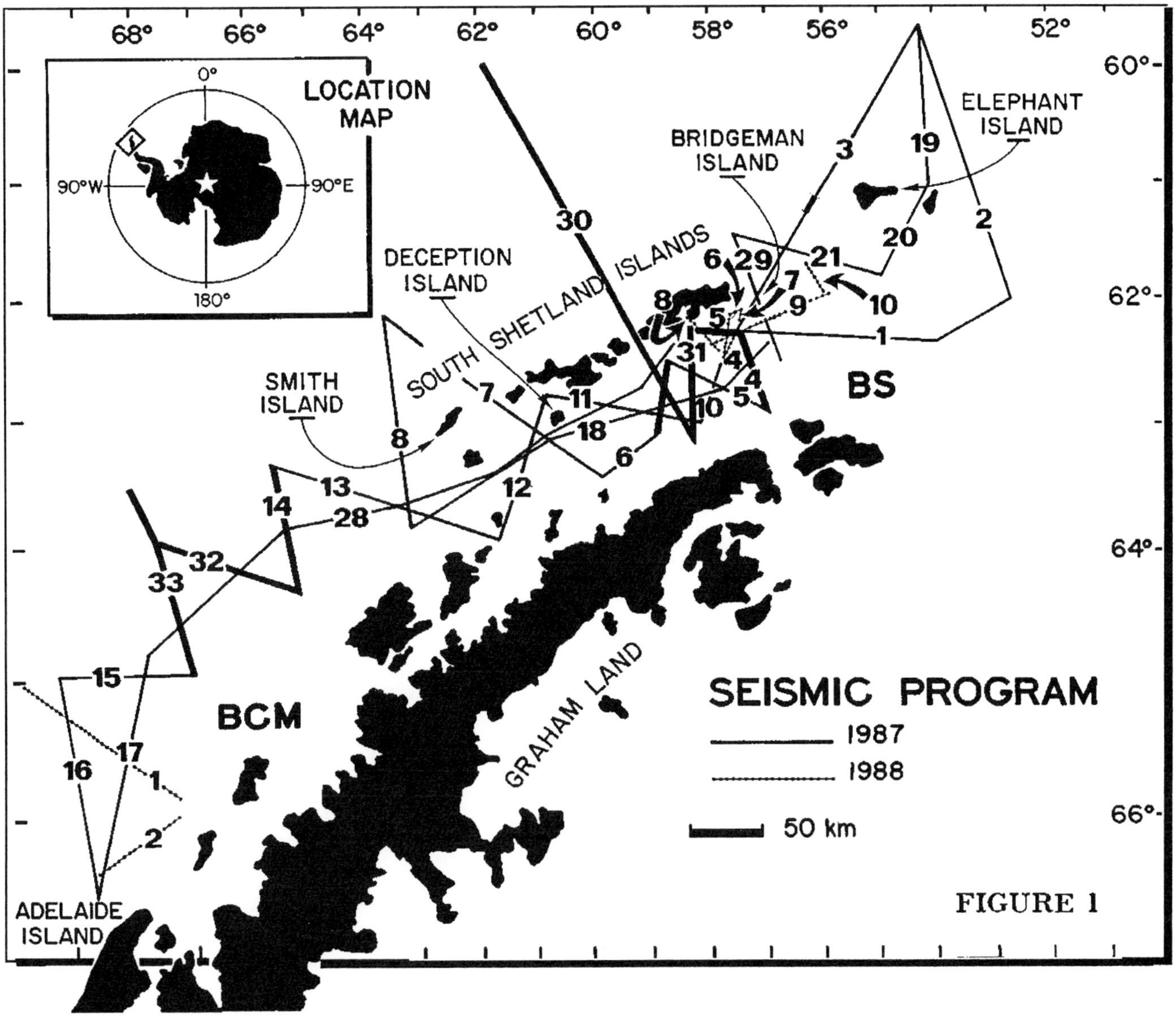

Figure 1—Location of the seismic reflection profiles. Inset shows location of study area. BS = Bransfield Strait; BCM = Bellingshausen continental margin.

(Figure 7). These reflections form a 5 to 15 km wide band (Figure 6) and are interpreted to represent intercalations of volcanic and volcanoclastic layers formed during the initial breakup of the continental mass. This interpretation is comparable to that attributed to the origin of similar "seaward-dipping reflectors" observed by Hinz (1981).

Basinward from the seaward-dipping reflectors, the sedimentary wedge can be divided into two main sequences separated by a regional unconformity. The older sequence shows faulted and folded high-amplitude reflections, which are tilted towards the continent at many localities (Figure 7). Tectonism during and immediately after the deposition of this sequence formed a series of fault blocks in the region between the seaward-dipping reflectors and the basement high at the contact with the "oceanic" crust of the basin (see Figure 7, SPs 900-1400). Our data do not permit mapping of individual blocks, but the structural style (rotated blocks along normal faults) created by the stretching and breakup of the peninsula's northern margin (the "rift" stage of this margin) and formation of the marginal basin (the Bransfield Basin) is clear (see Figure 7).

The regional unconformity separates this older faulted sequence deposited during the rifting stage, from a younger sequence which forms the present morphology of the northern peninsula's continental margin (Figures 7, 8). Except for one small area, located halfway along the basin and which will be discussed later, the younger sequence is not affected by tectonism and the layers are subhorizontal or dip gently basinward. Unconformities do occur within this younger sequence but due to the scale of the present work, they were not studied in detail.

These two main sedimentary sequences correspond to the "rift" and "drift" sequences of the southern margin of the Bransfield Basin, in an arrangement similar to that

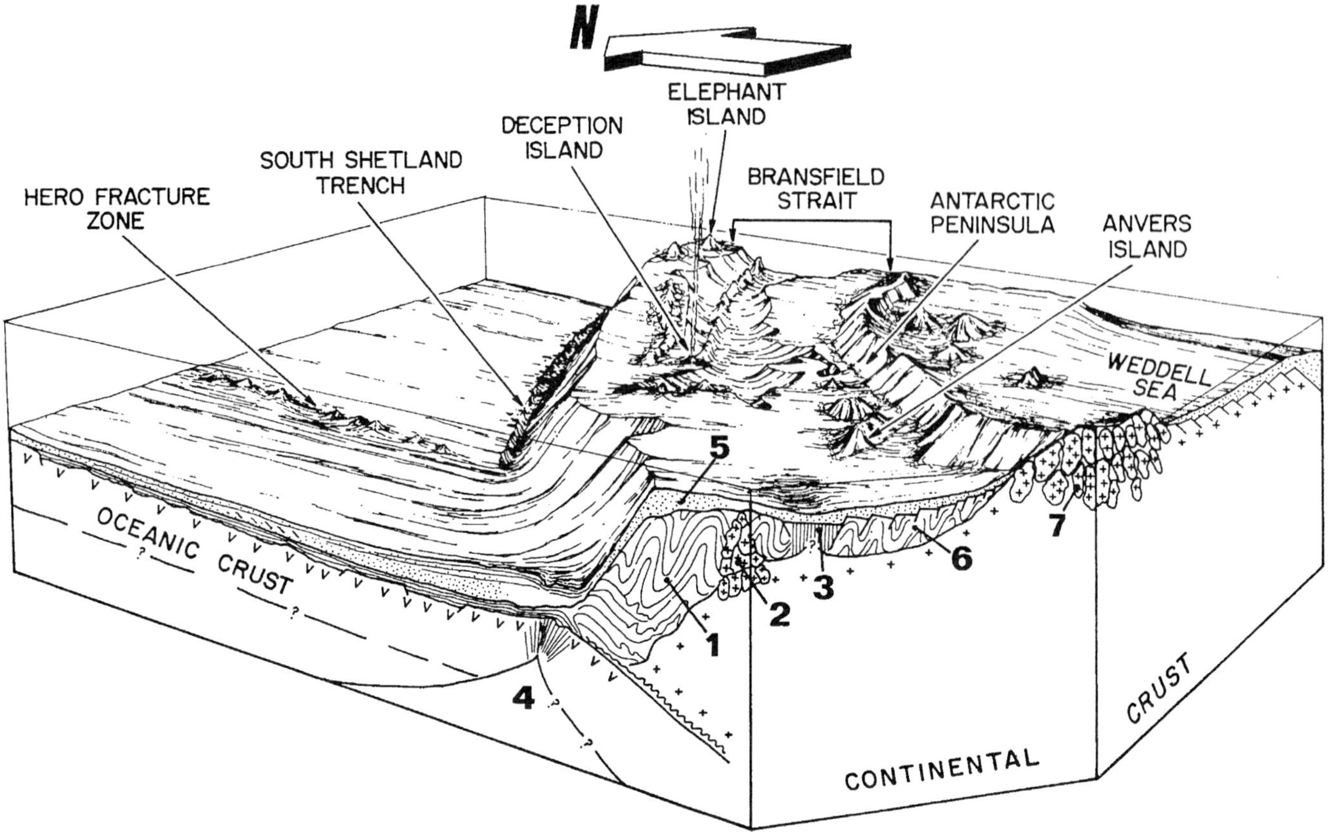

Figure 2—Perspective block diagram of the study area. Note the major change in the margin tectonic style across the Hero Fracture Zone from "passive" margin on the south to "active" margin on the north. (1) forearc material; (2) eroded arc; (3) closed backarc basin; (4) fossil spreading center; (5) sedimentary wedge (Camara Basin); (6) metamorphic rocks; (7) intrusive and extrusive acid rocks.

observed in corresponding sequences described in Atlantic-type margins (for example, in the eastern Brazilian margin; Ojeda, 1982). The occurrence of the "rift" sequence is limited basinward by a basement high. The dimensions of this feature are variable throughout the basin, but its trend is parallel to the spreading center (Figure 6). Basement highs at the boundaries between continental and oceanic crust have been described previously, but their origin is still controversial. Some authors suggest a basaltic accumulation at the beginning of the drift phase of a passive margin (e.g., Rabinowitz and La Brecque, 1977, for passive margins in general; and La Brecque and Zitellini, 1985, and Voggenreiter et al., 1988, for the Red Sea). Other authors prefer a continental origin and envision the high being the last part of the stretched continental crust formed by extensional tectonics (e.g., Tankard and Welsink, 1987, in the Grand Banks, Newfoundland; and Cochran, 1983 and Coleman and McGuire, 1988, for the Red Sea). Our analysis does not permit a choice between these hypotheses; however, our data clearly show that this basement high is situated at the boundary between continental and oceanic crust. The regional unconformity, the upper limit of the "rift" sequence, also terminates at this basement high (Figures 7, 8). These observations suggest a continental nature for the basement high at this boundary between the continental and the oceanic/transitional crust.

The Bransfield spreading center is interrupted by a transform fault halfway along its length, near the northern end of the King George Island. The axis of the ridge is displaced by about 10 km (Figure 6). Anomalous features are observed on both the southern and northern margins of the basin, suggesting that the transform fault may extend itself in the form of a fracture zone at both ends of the transform. In contrast to other areas, the "drift" sequence of the southern margin is cut by several normal faults and by volcanic features, indicating intense tectonism along the trend of the fracture zone. This tectonism is still active as several faults disrupt recent ocean floor sediments (Figure 9). At the northern margin, seamounts occur along the fracture zone trend, which appears to terminate at Penguin Island. On Penguin Island there is Quaternary basaltic volcanism. These basalts are unusual in that they are mildly alkaline basalts and have high Ce/Y ratios, suggestive of a deep mantle source (Smellie et al., 1984). We suggest that the occurrence and characteristics of Penguin Island's basalts reflect the activity along the fracture zone which may tap deeper magma sources.

Even though the data do not permit a conclusive analysis, the interpretation indicates that there is a change in the spatial distribution of rift-phase sediments along the basin. To the north of the fracture zone, the interpreted rift sequence lies on the northern edge of the basin, whereas to the south of the fracture zone the rift sequence is at the southern edge of the basin. This is suggestive of a change in the polarity of the rift which would have happened at an

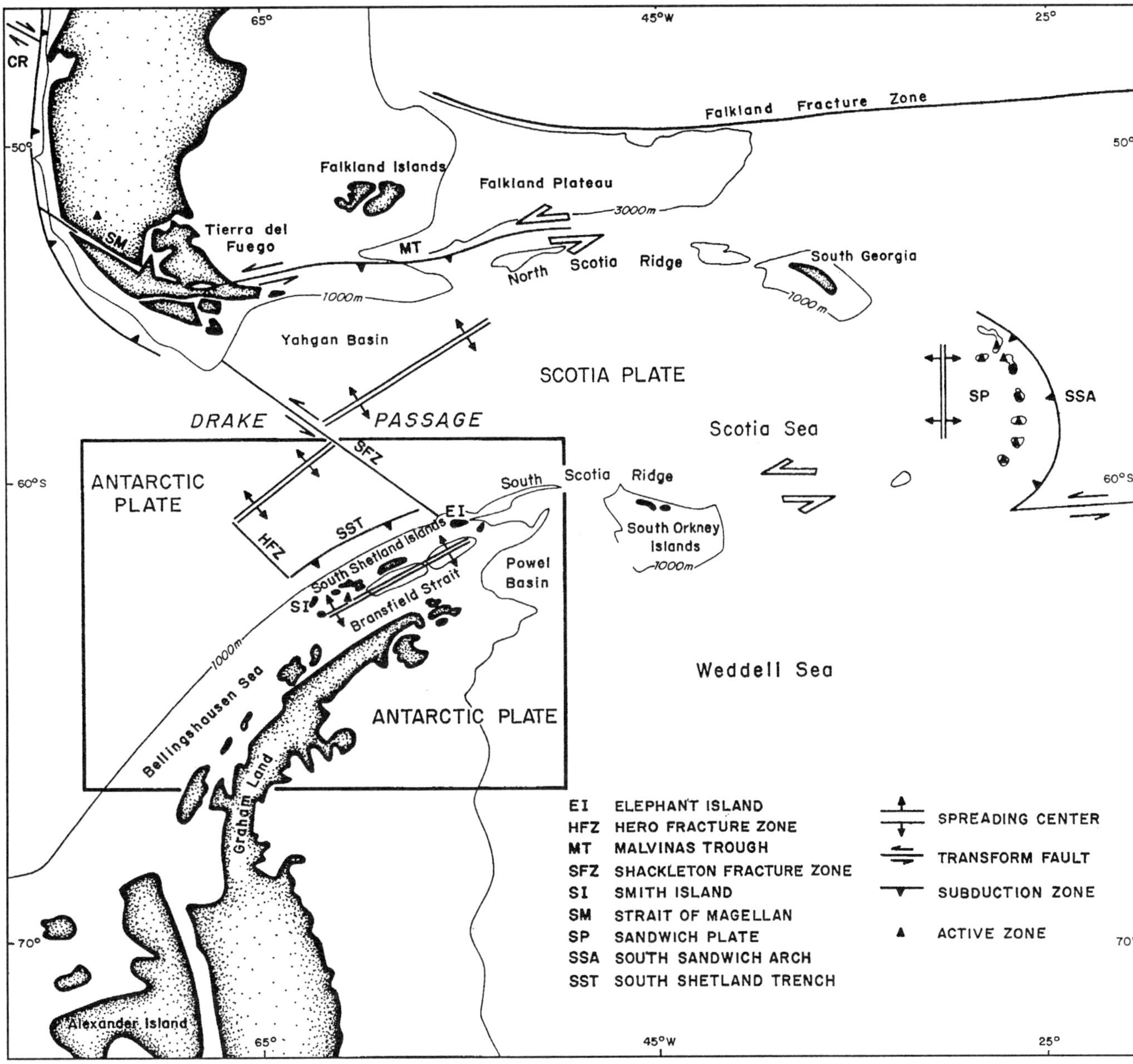

Figure 3—Tectonic setting of the Antarctic Peninsula and Scotia Arc region (modified from Dalziel, 1983). Box shows location of Figure 1.

accommodation zone. The former location of the accommodation zone corresponds approximately to the actual location of the fracture zone.

As a marginal basin, the origin of the Bransfield Basin is linked to the dynamics of the South Shetland Trench, which is bounded to the north and south by the Shackleton and Hero Fracture Zones, respectively. Thus the regional southeasterly trends of these two fracture zones define the limits of the basin. In more detail, as the data permit, the northern boundary is defined by a steep escarpment which marks the southwestern edge of a highly tectonized shallow platform, the Elephant Island Platform (Figures 3, 6). The Shackleton Fracture Zone trend, as mapped in the oceanic realm (Tectonic Map of the Scotia Arc, 1985), is situated at the northern end of this shallow platform (Figure 6). Faulting, folding, and uplift of deeper crustal blocks and topographic features suggest that active faulting occurs in the Elephant Island Platform (Figure 10). Thick pockets of sediments are also observed in the region. The tectonism is particularly intense where the present trend of the fracture zone intersects the platform (near Elephant Island).

The Hero Fracture Zone is a well-defined feature in the southernmost Drake Passage. The fracture zone passes through Smith Island, which, like Elephant Island, apparently is the result of uplift in response to tectonism along a fracture zone. We suggest that this uplift is a consequence of transpression or related vertical movements across these fracture zones, which elevated and eroded the lower levels

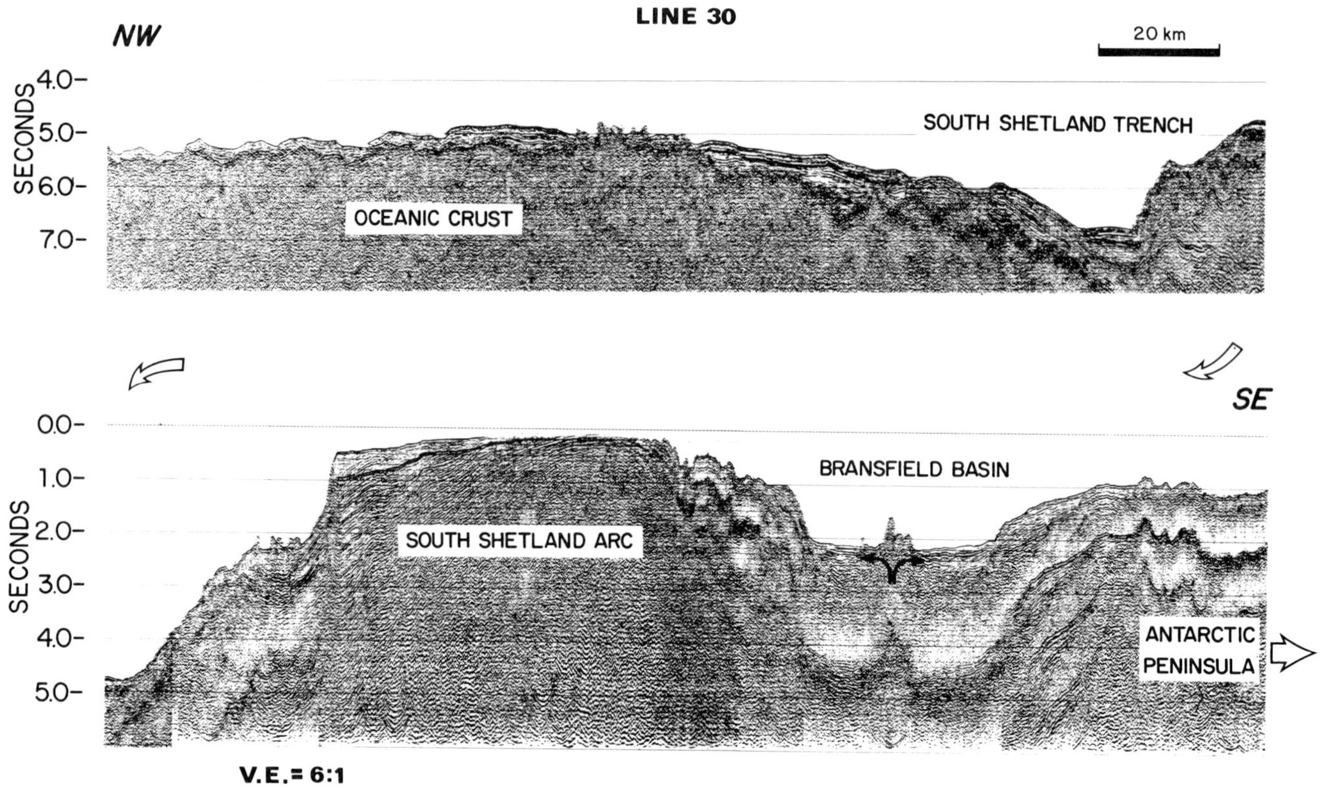

Figure 4—Multichannel seismic profile 30 shot across the South Shetland Trench-Bransfield Basin. Location in Figure 1.

of the accretionary prism, exposing them at Smith and Elephant Islands.

The present crustal dynamics across the South Shetland Trench are not clear. Although a trench morphology is present seaward of the South Shetland Islands (Figure 4), active subduction apparently ceased around 4 Ma (Barker and Burrell, 1977). However, seismological recordings indicate recent occurrence of deep and shallow earthquakes in the region (Forsyth, 1975). The three seismic lines shot across the trench suggest faulting in the forearc region (Figures 1, 11).

Bellingshausen Continental Margin

The Bellingshausen continental margin and its adjacent ocean floor contain an important record of the tectonic evolution of the Antarctic continent. Existing interpretations of the region are based mainly on studies of magnetic anomalies because very few seismic lines have been obtained. A series of magnetic anomalies identified by Herron and Tucholke (1976) indicated that the oceanic crust adjacent to the margin is Eocene to Miocene in age. Because the youngest crust is adjacent to the peninsula, Herron and Tucholke (1976) concluded that the spreading center responsible for the formation of this oceanic crust was consumed by a subduction zone active along the western margin of the peninsula. Later, Herron et al. (1981) and Barker (1982) showed that the subducted spreading center had been divided and offset by fracture zones into limited segments, and that the southern segment was subducted prior to the northern one.

Our seismic profiles across the Bellingshausen margin show a well-developed continental rise, a steep continental slope, and a broad continental shelf (Figure 12), which together form a basin named the Camara Basin by Gambôa et al. (1988). The sediments of the Camara Basin were deposited above an erosional unconformity below which tilted and faulted layers are inferred (Figure 13, forearc material). These sediments extend from outer continental shelf to the adjacent abyssal floor. At the outer shelf the sediments display clinoform reflection patterns typical of prograding continental shelves (Figure 13). A deep-sea fan occurs to the north of Adelaide Island (Figure 13).

An eroded basement high, parallel to the trend of the margin, defines the eastern boundary of the Camara Basin (Figures 6, 14). A closed, elongated basin lies to the east of the high. The sediments of this basin are folded and faulted. Unfortunately our coverage of this basin is limited and only a few seismic profile lines present an entire cross section of the basin, which is interpreted as a paleo backarc basin.

The features interpreted on the seismic data complement well the geologic evolution deduced from the oceanic magnetic anomalies. The seismic data indicate the presence of a volcanic arc and a backarc basin, which existed along this margin prior to the ridge-trench collision indicated by the magnetic data. The reflections below the unconformity shown on Figures 13 and 14 probably correspond to material deposited and/or emplaced during the active subduction phase of the margin, whereas the well-organized reflections above correspond to the sediments deposited during the "passive" phase, when the subduction no longer occurred along the margin.

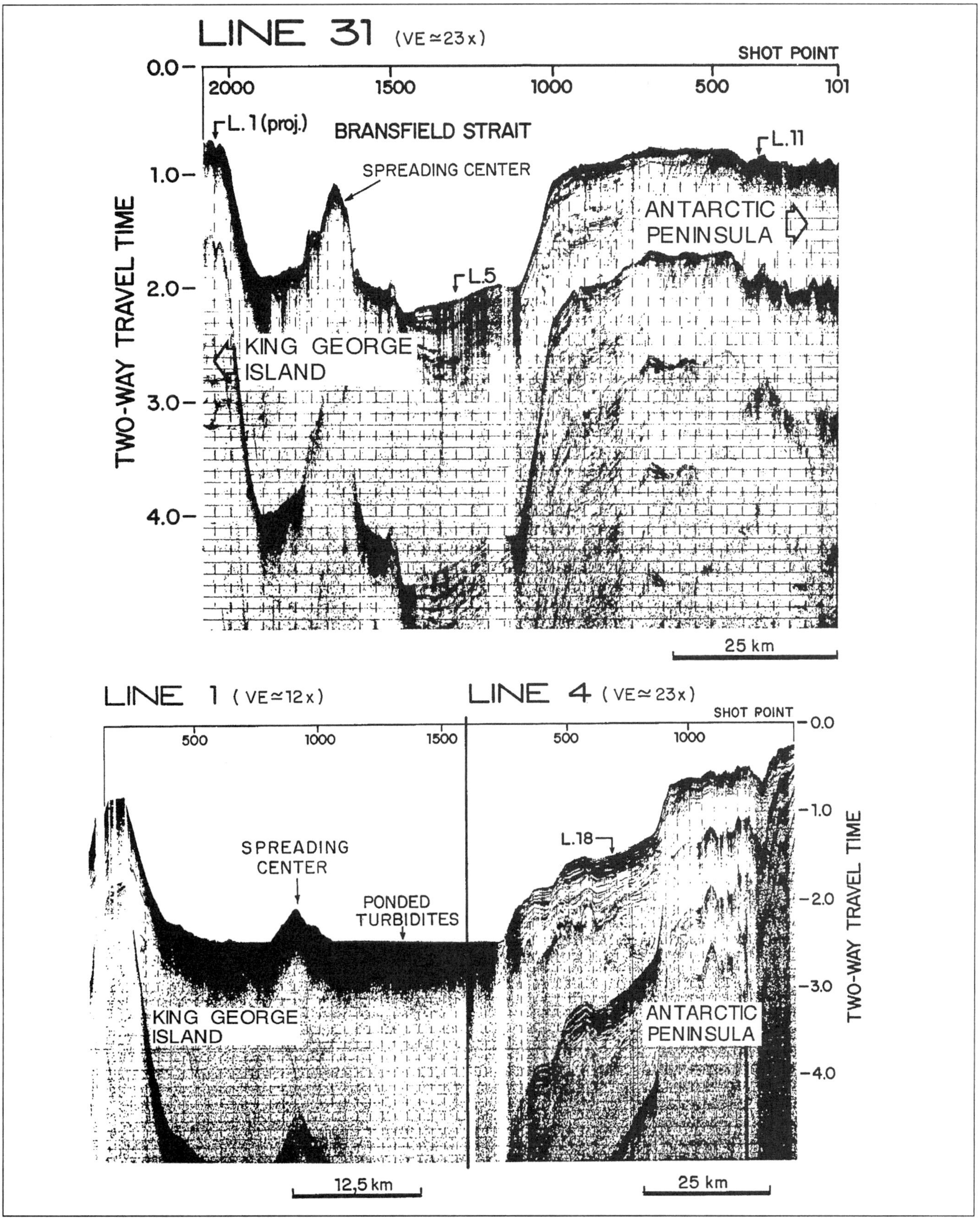

Figure 5—Single-channel seismic profiles across the Bransfield Basin showing the asymmetry, the ponded turbidites and the conspicuous spreading center of the basin. Note the change of horizontal scale at the intersection of Line 1 and 4. Location in Figure 1.

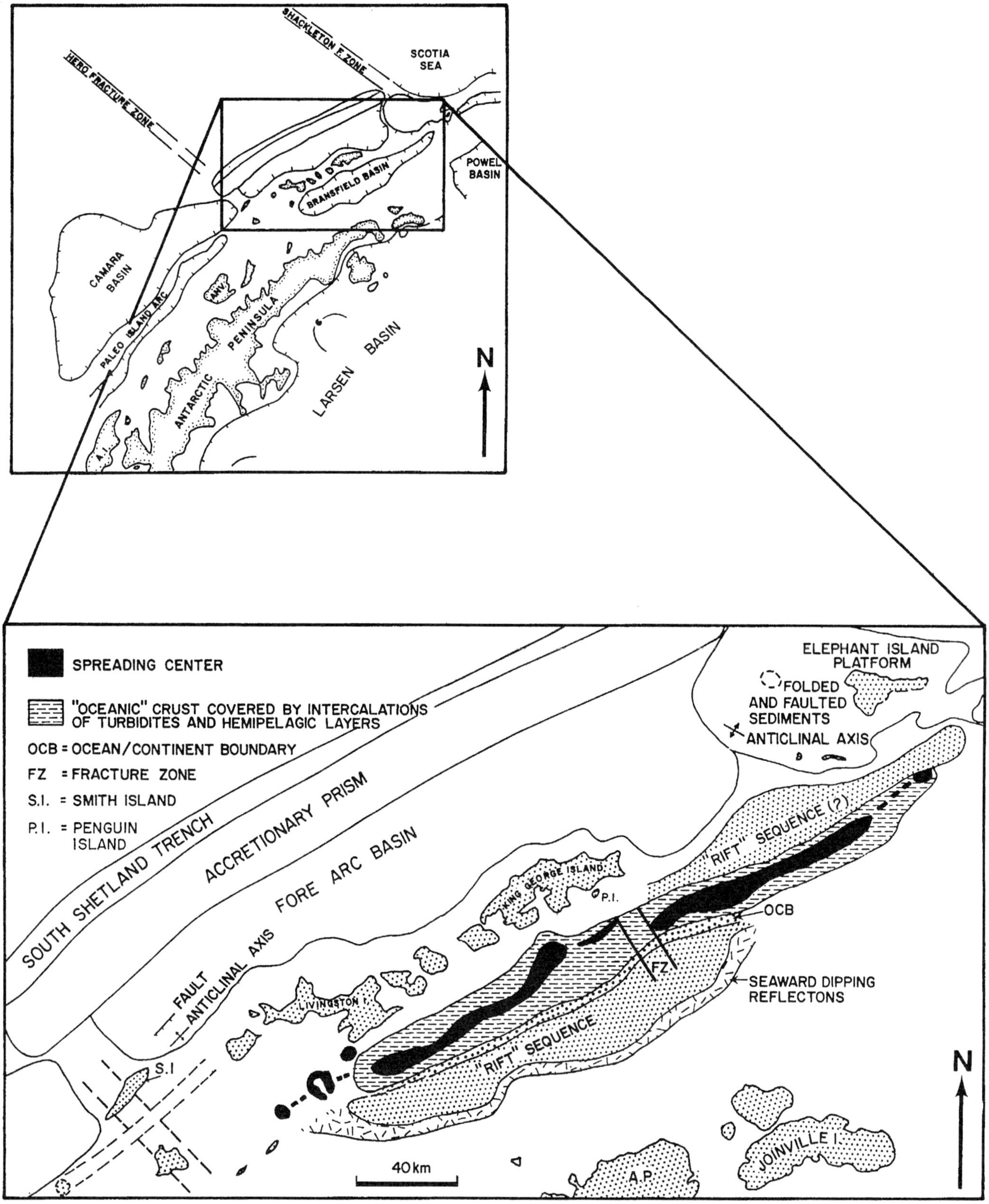

Figure 6—Geologic map of major features in the study region. Note the possible occurrence of the "rift" sequence to the north of King George Island, at the northern margin of the Bransfield Basin. Note also the continuation of the trend of the paleovolcanic arc along the Bellingshausen margin towards the South Shetland Islands.

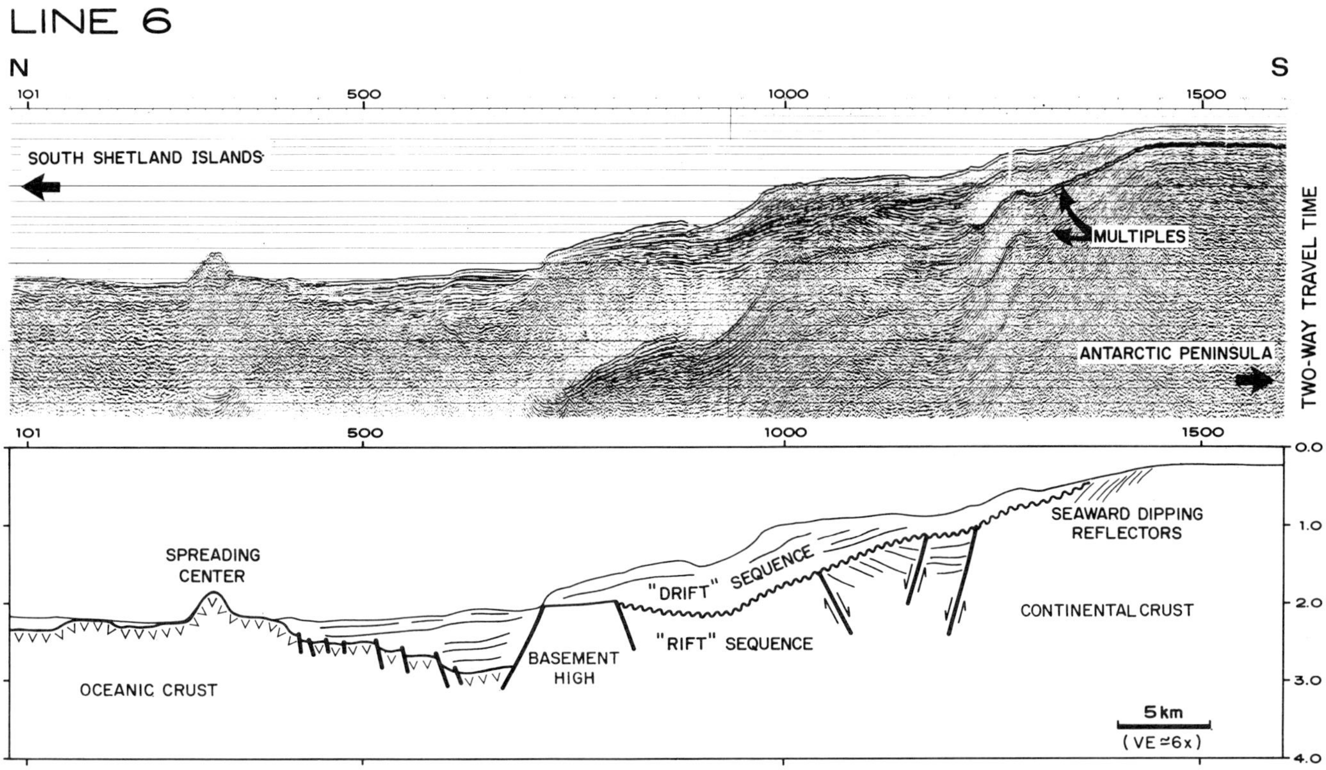

Figure 7—Segment of seismic profile 6 showing the two main sedimentary sequences found along the Antarctic Peninsula margin–Bransfield Basin depocenter. Note the unconformity separating the "rift" from the "drift" sequences, and the basement high at the boundary between "oceanic" and stretched continental crust. Location in Figure 1.

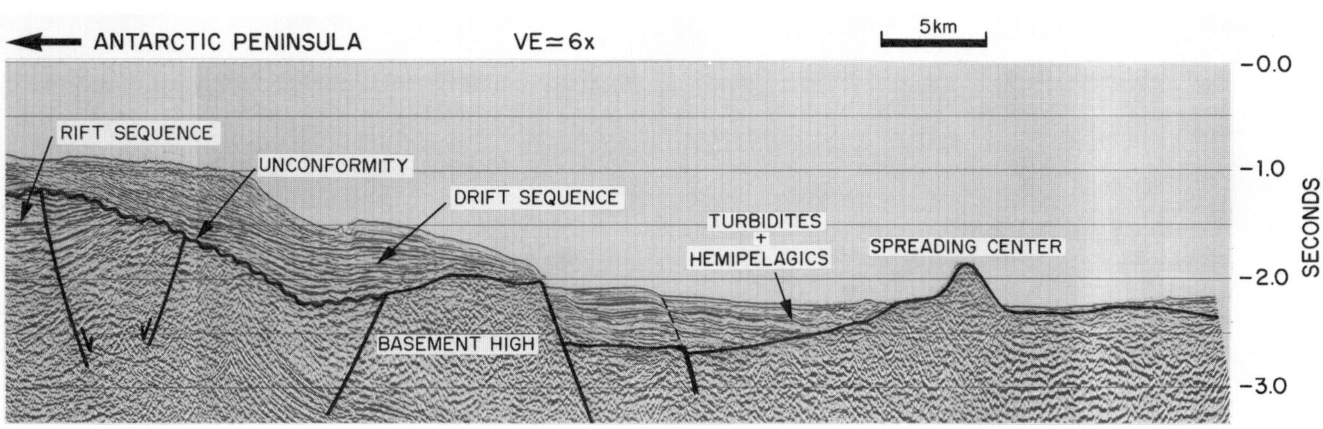

Figure 8—Detail of the profile shown on Figure 7 showing the seismic character of the "rift" and "drift" sequences.

The basement high and the adjacent closed marginal basin can be followed northward and they seem to merge with the South Shetland Island and the Bransfield Basin, forming a similar system, with a possibly related tectonic evolution. Thus, it may also be possible that the Bransfield Basin is as old as Miocene, when the ridge-trench collision, south of the Hero Fracture Zone, closed the southern arc/basin system.

SUMMARY

Reconnaissance seismic profiling off the northwestern coast of the Antarctic Peninsula carried out by the Brazilian Antarctic Program has permitted refinement of the previous tectonic model of the region. The results distinguish two provinces, separated by the northwest–trending Hero Fracture Zone. The province to the northeast of the Hero Fracture

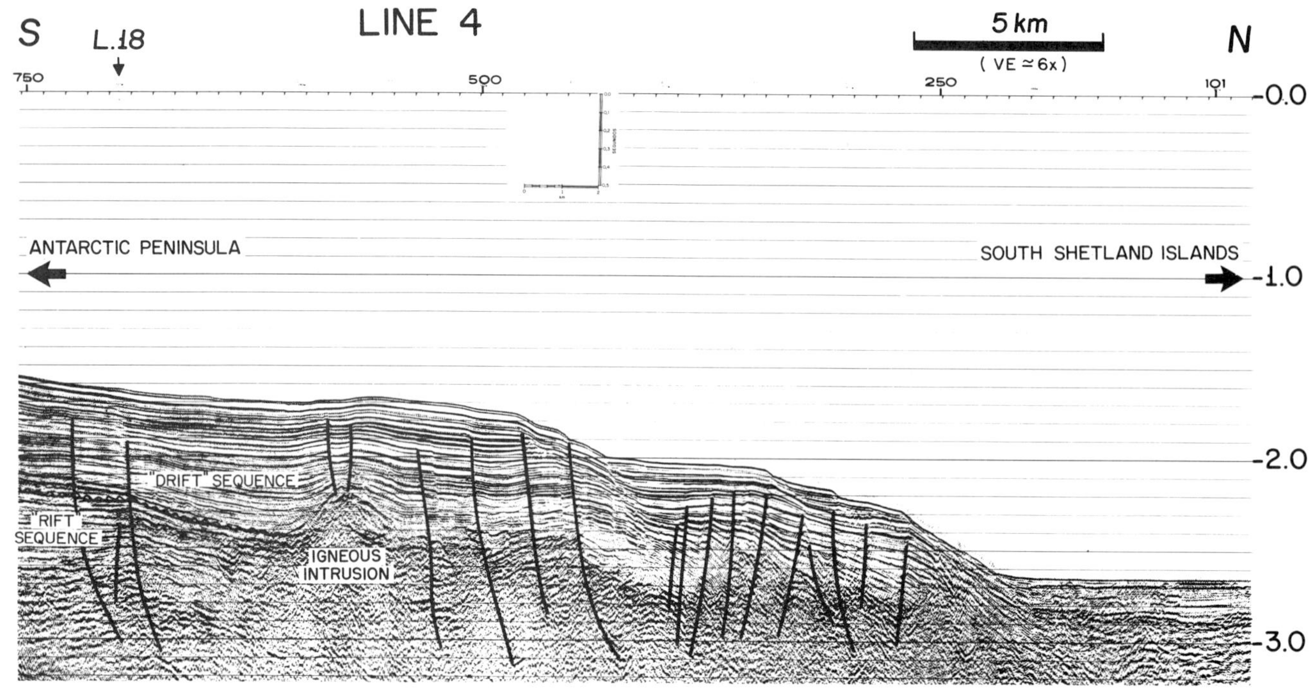

Figure 9—Segment of seismic profile 4, across the northern margin of the Antarctic Peninsula. Note the intense tectonism affecting the drift sequence, and the igneous body, probably also relatively recent, as indicated by the two normal faults at the apex of the intrusion. This line cuts at a very low angle an unnamed fracture zone within the Bransfield Basin.

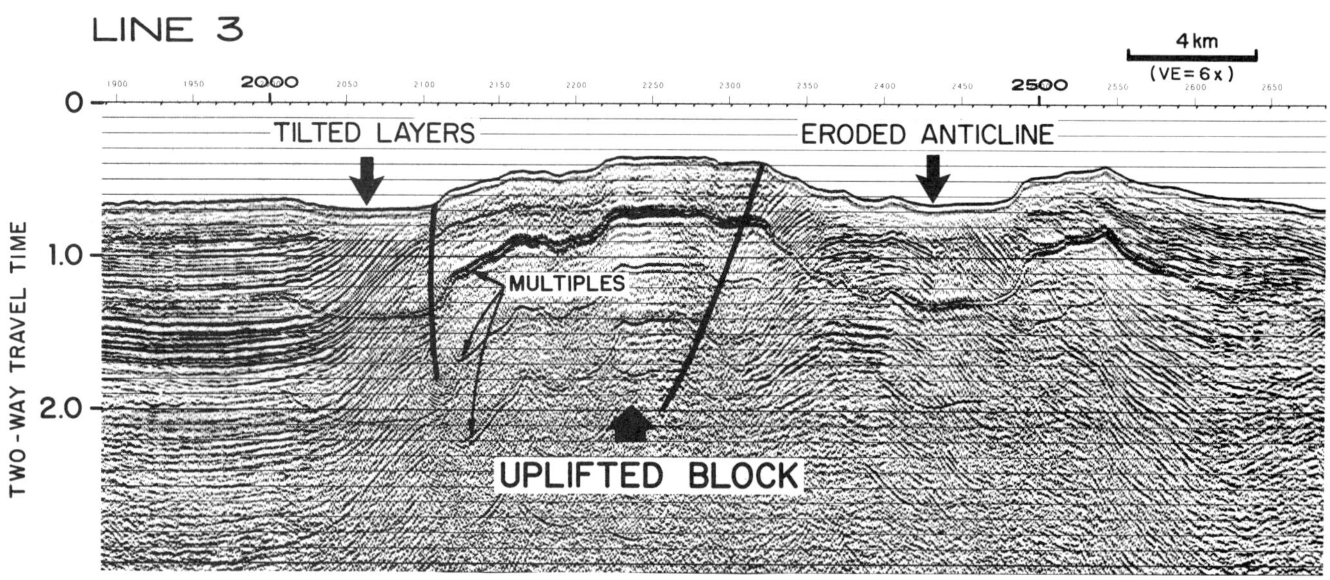

Figure 10—Segment of Line 3, displaying an example of the intense tectonism affecting the Elephant Island Platform.

Zone shows evidence of recent (perhaps contemporary) subduction of the oceanic crust. A partially filled trench, an accretionary prism, a disrupted forearc basin, a volcanic arc (the South Shetland Island arc, which terminates to the northeast at the Shackleton Fracture Zone) and a backarc basin are observed. The forearc basin contains strata derived from the arc; these strata are cut by an array of normal faults, which may have developed as the prism extended to maintain a critical wedge taper.

The backarc basin, which underlies the Bransfield Strait, contains a well-defined spreading ridge. The axis of the spreading ridge is offset by a short transform fault, which appears to have been derived from an accommodation zone linking two rift compartments of opposite polarity. The

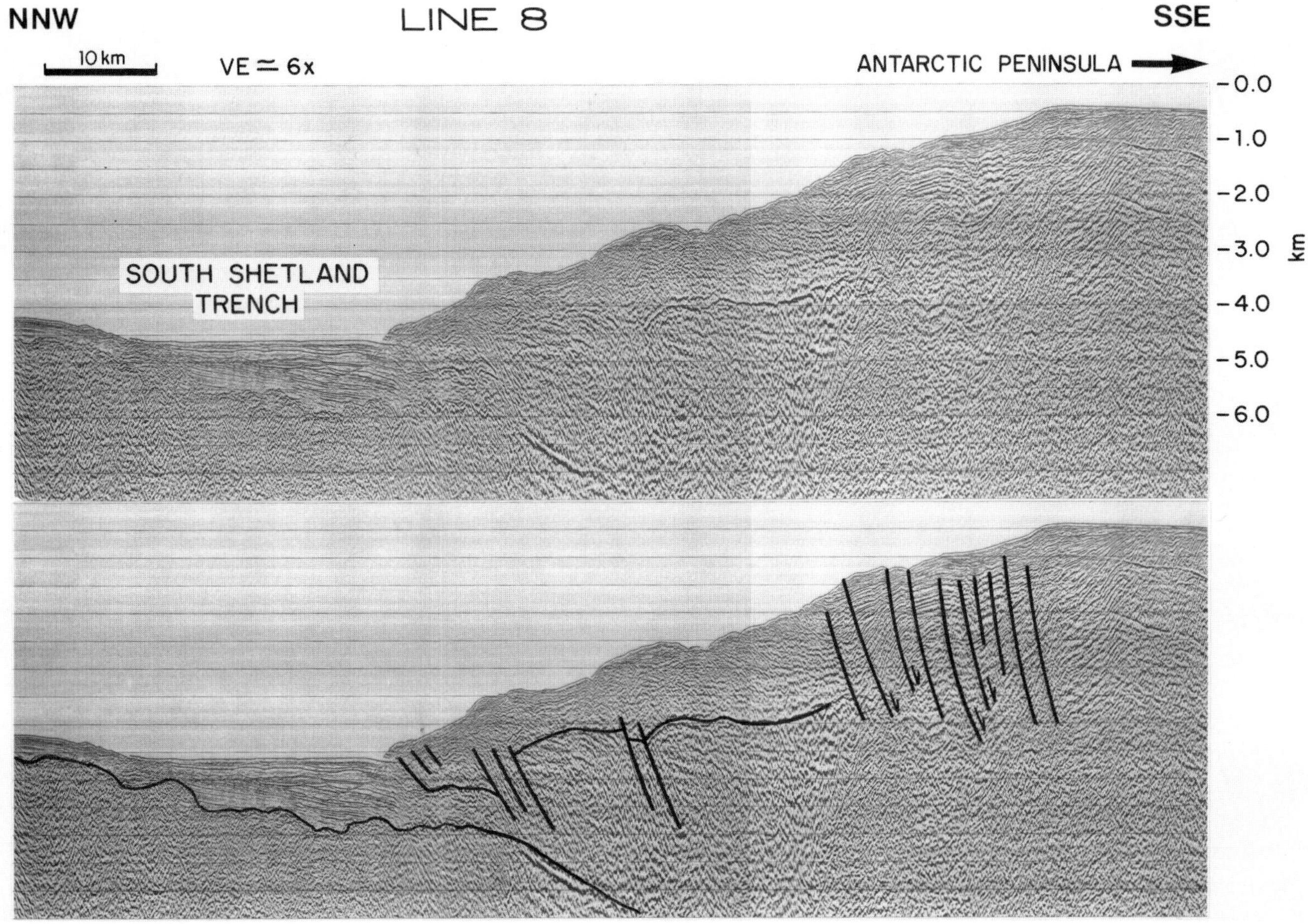

Figure 11—Interpreted (bottom) and uninterpreted (top) segment of seismic profile 8 showing the morphology and characteristics of the South Shetland Trench and associated accretionary prism. Location of profile in Figure 1.

southeastern margin of the backarc basin presents a typical passive margin structure. At depth, we observe a sequence with extensive normal faulting (the "rift" sequence) that is truncated by an unconformity. Above this unconformity the "drift" sequence was deposited and remains unfaulted.

In the province to the southwest of the Hero Fracture Zone, the Aluk spreading ridge collided with the volcanic arc. Our seismic profiles indicate that, at present, the accretionary prism and trench have been buried, so topographically the northeast edge of the Antarctic Peninsula in this region is similar in appearance to an Atlantic-type margin. Beneath these surficial sediments, however, we observe from northwest to southeast: (1) the remnants of the spreading ridge, (2) a buried accretionary prism, (3) the buried core of a volcanic arc, and (4) a closed backarc basin. This closed backarc basin, which may in the past have been continuous with the presently open Bransfield Basin, contains strata folded by the shortening of the basin.

ACKNOWLEDGMENTS

This work was supported by PETROBRÁS and the Brazilian Navy. We gratefully acknowledge the critical reviews of A. Northfleet, S. Marshak, R. Brown, M. Augusta da Silva, P. Zalan and A. C. M. Castro Jr. We thank the captains and crew of the R/S *Almirante Câmara* whose efforts enabled gathering the data presented in this chapter. We thank PETROBRÁS for encouragement and permission to publish these results. The suggestions of J. Damuth and J. Gallagher greatly improved the chapter.

REFERENCES CITED

Ashcroft, W. A., 1972, Crustal structure of the South Shetland Islands and Bransfield Strait: British Antarctic Survey Scientific Reports, n. 66, 43 p.

Barker, P. F., 1970, Plate tectonics of the Scotia Sea region: Nature, v. 228, p. 1293-1296.

Barker, P. F., 1982, The Cenozoic subduction history of the Pacific margin of the Antarctic Peninsula: ridge crest-trench interactions: Journal of Geological Society London, v. 139, p. 787-801.

Barker, P. F. et al., 1976, The evolution of the Southwestern Atlantic Ocean basin: Leg 36 data, *in* P. F. Barker, I. W. D. Dalziel et al., eds, Initial Reports of the Deep Sea Drilling Project, v. 36, Washington (U.S. Government Printing Office), p. 993-1014.

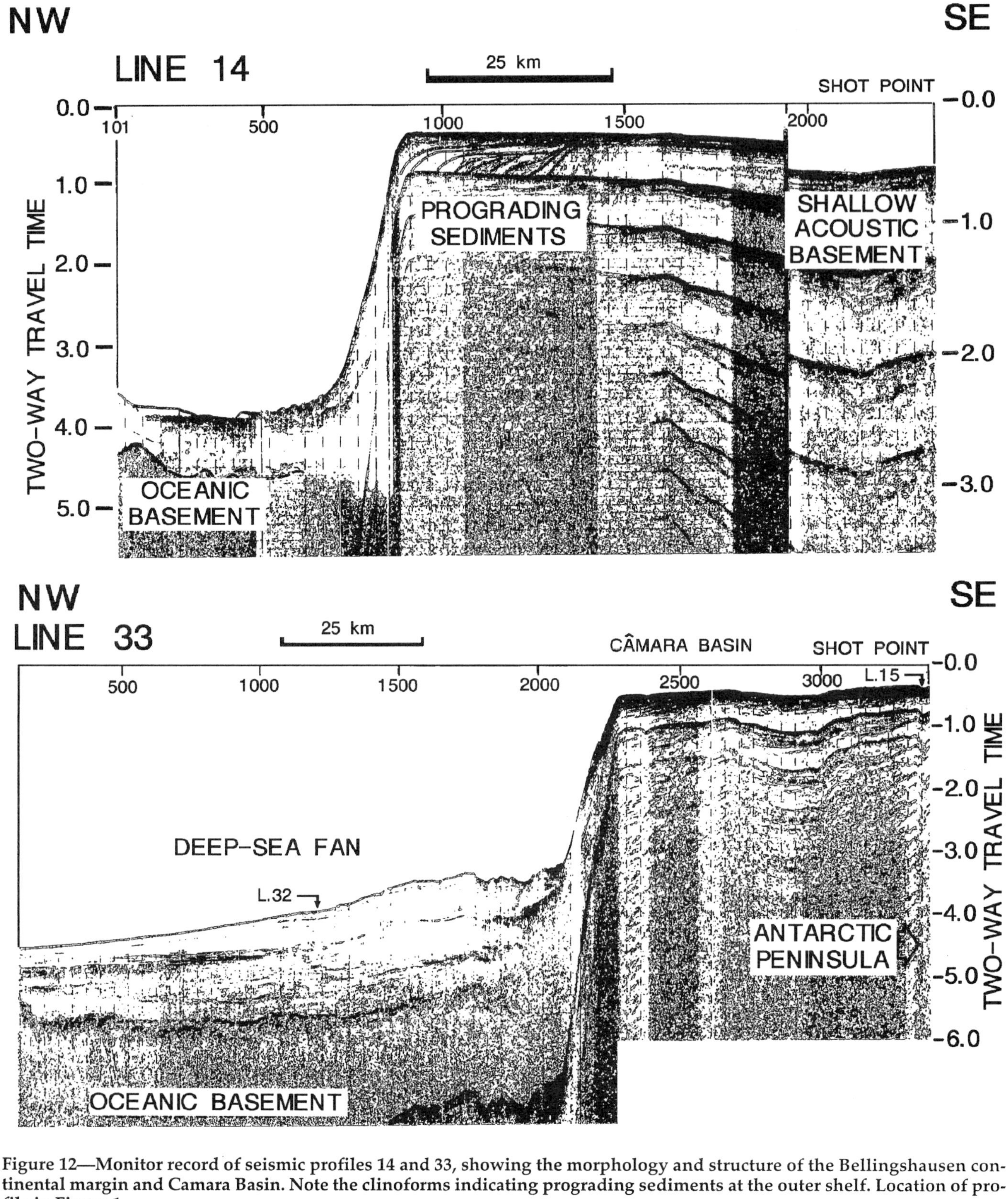

Figure 12—Monitor record of seismic profiles 14 and 33, showing the morphology and structure of the Bellingshausen continental margin and Camara Basin. Note the clinoforms indicating prograding sediments at the outer shelf. Location of profile in Figure 1.

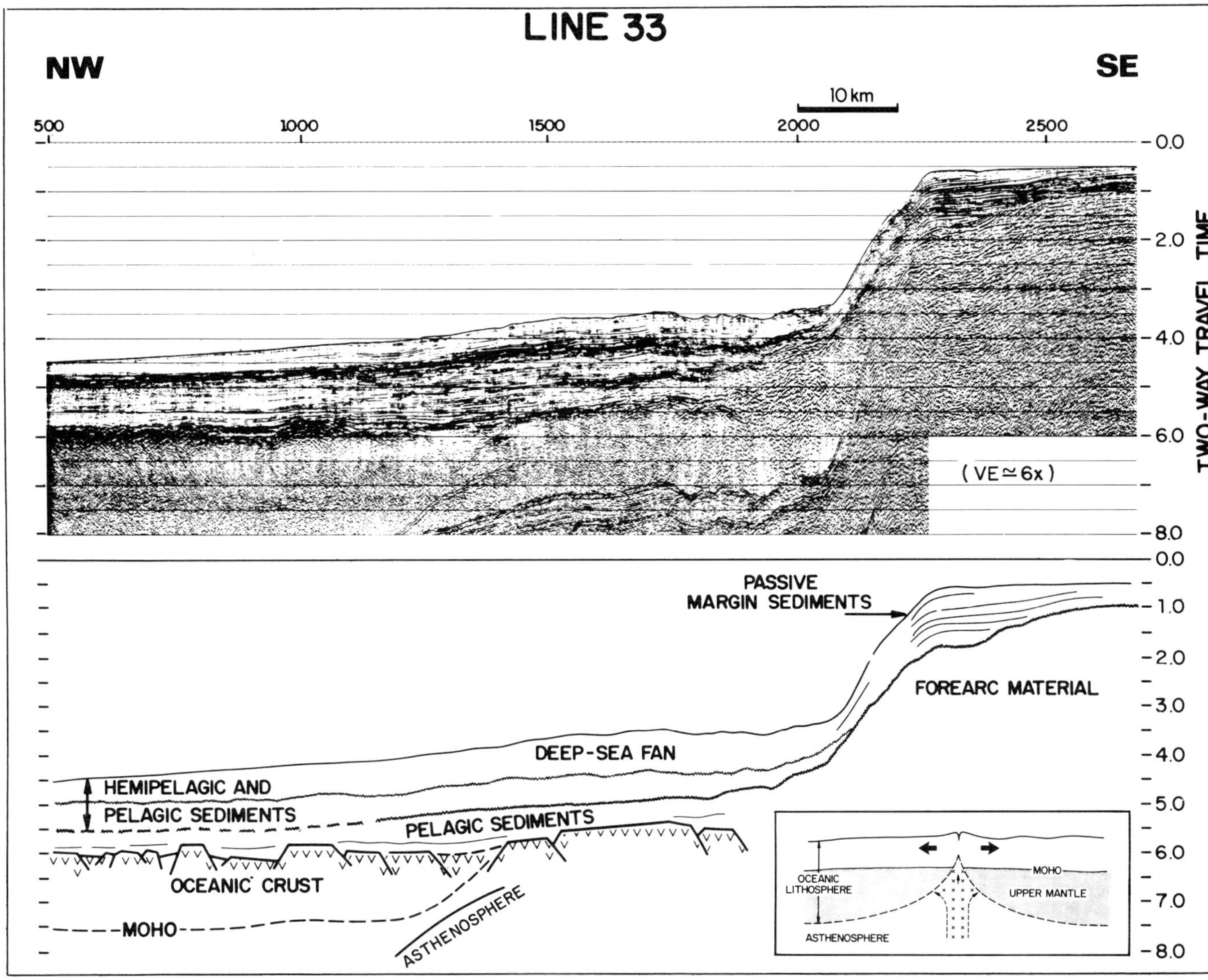

Figure 13—Detail of Line 33, migrated. Note the unconformity separating the forearc material from the passive margin–deposited sediments and the strong reflector interpreted as the asthenosphere/lithosphere contact. Inset showing the relation of Moho/Asthenosphere at spreading centers redrawn from Bott, 1982. Location in Figure 1.

Barker, P. F. and J. Burrell, 1977, The opening of Drake Passage: Marine Geology, v. 25, p. 15-34.

Barker, P. F. and D. H. Griffiths, 1972, The evolution of the Scotia Ridge and Scotia Sea: Philosophical Transactions Royal Society of London, A. 271, p. 151-183.

Bott, M. H. P., 1982, The interior of the Earth: its structure, constitution and evolution: London, Edward Arnold, 403 p.

Cochran, J. R., 1983, A model for development of Red Sea: AAPG Bulletin, v. 67, p. 41-69.

Coleman, R. G. and A. V. McGuire, 1988, Magma systems related to the Red Sea opening: Tectonophysics, v. 150, p. 77-100.

Dalziel, I. W. D., 1983, The evolution of the Scotia Arc: a review, *in* R. L. Oliver, P. R. James and J. B. Jago, eds., Antarctic Earth Science: London, Cambridge University Press, p. 283-288.

Dalziel, I. W. D., 1984, The Scotia Arc: an international geological laboratory: Episodes, v. 7, n. 3, p. 8-13.

Davey, F. J., 1972, Marine gravity measurements in Bransfield Strait and adjacent areas, *in* R.J. Adie, ed., Antarctic Geology and Geophysics: Universitetsforlaget, Oslo, p. 39-45.

De Wit, M. J., 1977, The evolution of the Scotia Arc as a key to the reconstruction of Southwestern Gondwanaland: Tectonophysics, v. 37, p. 53-81.

Dott, R. H., Jr., R. D. Winn Jr. and C. H. L. Smith, 1982, Relationship of late Mesozoic and early Cenozoic sedimentation to the tectonic evolution of Southernmost Andes and Scotia Arc, *in* C. Craddock, ed., Antarctic Geoscience, International Union of Geological Sciences (IUGS), Series B, n. 4, Madison, The University of Wisconsin Press, p. 193-202.

Elliot, D. H., 1983, The Mid-Mesozoic to Mid-Cenozoic active plate margin of the Antarctic Peninsula, *in* R. L. Oliver, P. R. James and J. B. Jago, eds., Antarctic Earth Science: London, Cambridge University Press, p. 347-351.

Farquharson, G. W., 1982, Late Mesozoic sedimentation in the Northern Antarctic Peninsula and its relationship to the Southern Andes: Journal Geological Society of London, v. 139, p. 721-727.

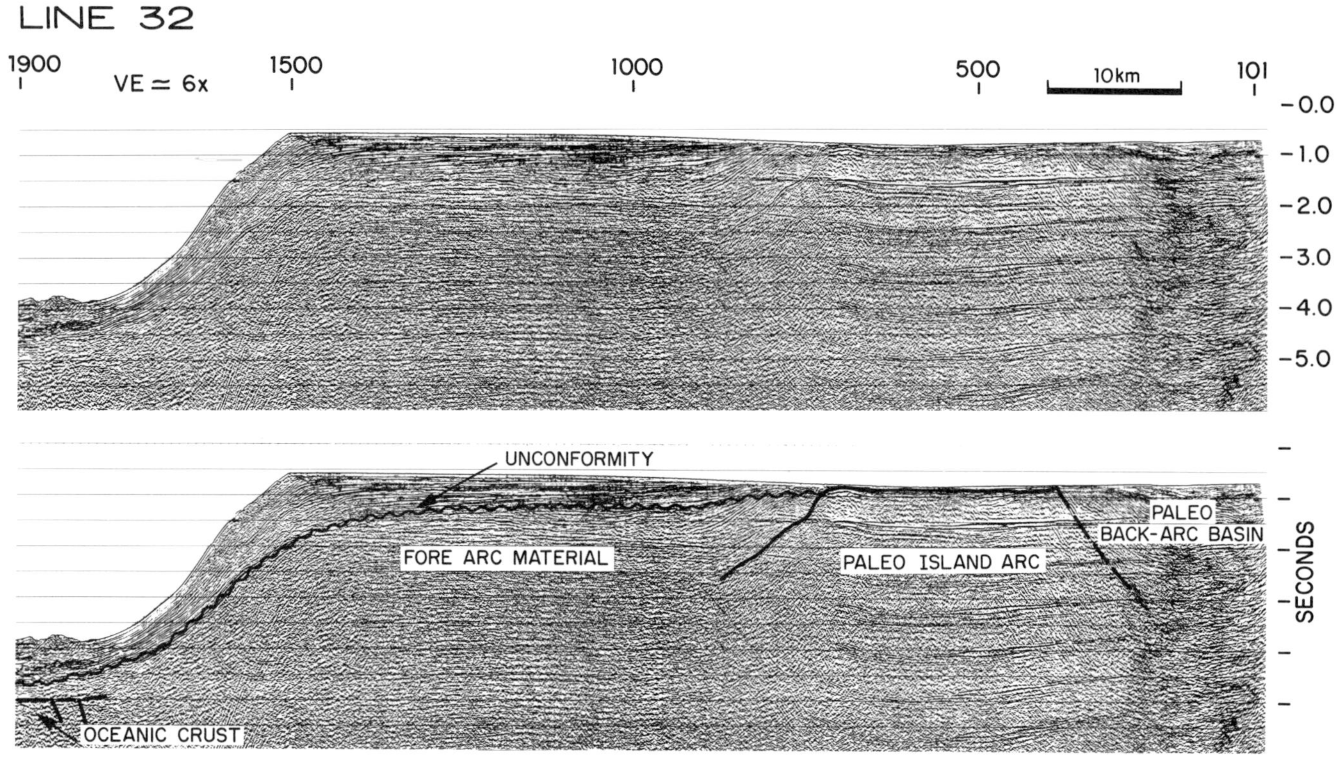

Figure 14—Segment of seismic profile 32 showing the eroded basement high and the closed basin interpreted as a paleo island arc and paleo backarc basin. Location in Figure 1.

Forsyth, D. W., 1975, Fault plane solutions and tectonics of the South Atlantic and Scotia Sea: Journal Geophysical Research, v. 80, p. 1429-1443.

Forsythe, R., 1982, The late Paleozoic to early Mesozoic evolution of Southern South America: a plate tectonic interpretation: Journal Geological Society of London, v. 139, p. 671-682.

Gambôa, L. A. P., S. L. Barrocas and P. R. Maldonado, 1988, Investigações Geofísicas no Estreito de Bransfield e Mar de Bellingshausen-Antártica: Anais do XXXV Congresso Brasileiro de Geologia, Belém, Pará, v. 5, p. 2343-2357.

González-Ferrán, O., 1985, Volcanic and tectonic evolution of the Northern Antarctic Peninsula—Late Cenozoic to Recent: Tectonophysics, v. 114, p. 389-409.

Harrison, C. G. A., E. J. Barron and W. W. Hay, 1979, Mesozoic evolution of the Antarctic Peninsula and the Southern Andes: Geology, v. 7, p. 374-378.

Herron, E. M., S. C. Cande and B. R. Hall, 1981, An active spreading center collides with a subduction zone: A geophysical survey of the Chile Margin triple junction: GSA Memoir 154, p. 683-701.

Herron, E. M. and B. E. Tucholke, 1976, Sea-floor magnetic patterns and basement structure in the Southeastern Pacific, *in* C. D. Hollister, C. Craddock et al., Initial Reports of the Deep Sea Drilling Project, v. 24, Washington (U.S. Government Printing Office), p. 263-278.

Hilde, T. W. C. and C. S. Lee, 1984, Origin and evolution of the West Philippine Basin: a new interpretation: Tectonophysics, p. 102, p. 85-104.

Hinz, K., 1981, A hypothesis on terrestrial catastrophes. Wedges of very thick oceanward dipping layers beneath passive continental margins—their origin and paleoenvironmental significance: Geologisches Jahrbuch, Reihe E., Geophysik, Heft 22, p. 3-28.

Karig, D. E., R. N. Anderson and L. D. Bibee, 1978, Characteristics of back arc spreading in the Mariana Trough: Journal Geophysical Research, v. 83, p. 1213-1226.

La Brecque, J. L. and N. Zitellini, 1985, Continuous sea-floor spreading in the Red Sea: an alternative interpretation of magnetic anomaly pattern: AAPG Bulletin, v. 69, n. 4, p. 513-524.

Lawver, L. A. and C. R. Scotese, 1987, A revised reconstruction of Gondwanaland, *in* G. D. Mackenzie, ed., Gondwana Six: Structure, Tectonics, and Geophysics: American Geophysical Union Geophysical Monograph 40, p. 17-24.

Mrozowski, C. L. and D. E. Hayes, 1979, The evolution of the Parece Vela Basin, East Phillipine sea: Earth Planetary Science Letters, v. 46, p. 49-67.

Ojeda, H. A. O., 1982, Structural framework, stratigraphy and evolution of Brazilian marginal basins: AAPG Bulletin, v. 66, n. 6, p. 732-749.

Rabinowitz, P. R. and J. La Brecque, 1977, The isostatic gravity anomaly: key to the evolution of the ocean-continent boundary at passive continental margins: Earth Planetary Science Letters, v. 35, p. 145-150.

Saunders, A. D. and J. Tarney, 1982, Igneous activity in the Southern Andes and Northern Antarctic Peninsula: a review: Journal Geological Society of London, v. 139, p. 691-700.

Smellie, J. L., R. J. Pankhurst, M. R. A. Thomson and R. E. S. Davies, 1984, The geology of the South Shetland Islands: VI. Stratigraphy, geochemistry and evolution: British Antarctic Survey Scientific Reports, n. 87, 85 p.

Smellie, J. L. and P. D. Clarkson, 1975, Evidence for pre-Jurassic subduction in Western Antarctica: Nature, v. 258, p. 701-702.

Storey, B. C. and S. W. Garrett, 1985, Crustal growth of the Antarctic Peninsula by accretion, magmatism and extension: Geological Magazine, v. 122, p. 5-14.

Tankard, A. J. and H. J. Welsink, 1987, Extensional tectonics and stratigraphy of Hibernia oil field, Grand Banks, Newfoundland: AAPG Bulletin, v. 71, n. 10, p. 1210-1232.

Tectonic Map of the Scotia Arc, 1988, British Antarctic Survey, High Cross, Madingley Road, Cambridge, CB3 OET, England, Sheet BAS (Misc) 3, Edition 1, Scale 1:3,000,000.

Thomson, M. R. A., R. J. Pankhurst and P. D. Clarkson, 1983, The Antarctic Peninsula—a late Mesozoic-Cenozoic arc (review), *in* R. L. Oliver, P. R. James and J. B. Jago, eds., Antarctic Earth Science, London, Cambridge University Press, p. 289-294.

Voggenreiter, W., H. Hotzl and J. Mechie, 1988, Low-angle detachment origin for the Red Sea rift system? Tectonophysics, v. 150, p. 51-75.

Watts, A. B., J. K. Weissel and R. L. Larson, 1977, Sea floor spreading in marginal basins of the Western Pacific: Tectonophysics, v. 37, p. 167-181.

Chapter 10

Petroleum Geology from the CIROS-1 Drill Hole, McMurdo Sound: Implications for the Potential of the Victoria Land Basin, Antarctica

J. D. Collen
P. J. Barrett
Research School of Earth Sciences
Victoria University of Wellington
Wellington, New Zealand

ABSTRACT

CIROS-1, the most recent scientific drill hole in Victoria Land Basin, Antarctica, cored to 702 m below sea floor (mbsf) near the basin's western margin. The cored sequence is largely marine sandy mudstone and diamictite, with lesser sandstone and conglomerate, and shows a glacial influence throughout. The strata range in age from early Miocene near the sea floor to the Eocene/Oligocene boundary at 702 mbsf, some 800 m above acoustic basement.

The strata cored by CIROS-1 have a uniformly low organic carbon content (average total organic carbon 0.34%), little hydrocarbon generating potential (maximum 0.34 mg hydrocarbons per gram of rock), and a low level of thermal alteration (vitrinite reflectance of 0.36% at 670 m). Kerogen is mainly highly oxidized terrestrial organic matter reworked from older strata.

Residual asphaltic oil was found in minute quantities in sandstone at 632 mbsf. Geochemical studies indicate that it formed at greater depth, probably in nearshore clastic sediments with both marine and terrestrially derived organic matter. Similar coarse- to fine-grained feldspathic sandstones are common in the mudstone that forms the lower half of the core, and offer reservoir potential. Calcite and zeolite cements are common but in-situ porosities range up to 22%, with good dissolution porosity in some intervals.

In balance, the possible prospects of Oligocene strata on the margin of the Victoria Land Basin must be considered low. Source characteristics are unfavorable and only minute amounts of hydrocarbons have been encountered. In addition, the up-dip location of CIROS-1 with respect to the basin center suggests that the possible prospects of the western Victoria Land Basin should also be regarded as low.

INTRODUCTION

The Victoria Land Basin covers an area of approximately 45,000 km^2 in the western Ross Sea, Antarctica (Figure 1). It is the westernmost of three basins within the Ross Embayment rift zone (Cooper et al., 1990), also termed the Ross Sea Rift System (Tessensohn, in press) and Transantarctic Rift (Schmidt and Rowley, 1986). The basin trends north-south from Cape Washington in northern Victoria Land to the Ross Ice Shelf, and is bounded to the west by the Transantarctic Mountains. These are block-tilted mountains up to 4500 m high and formed by flexural uplift at the rift margin (Stern and ten Brink, 1989). To the east the basin is bounded by uplifted and eroded basement along the submarine Coulman High. The basin extends under the Ross Ice Shelf and its southern margin is unknown. This chapter discusses the results from recent scientific drilling on the western margin of the basin, and concludes with some inferences on petroleum potential of the basin itself.

Marine geological and geophysical investigations of the basin, which is entirely an offshore feature, have been hampered by three main environmental constraints—sea ice, icebergs, and weather (reviewed by Keys, 1984). Nevertheless a number of geophysical surveys have given information on subsea–floor stratal geometry, beginning with the report by Northey et al. (1975) of a single-channel seismic survey of McMurdo Sound. Prior to 1984, surveys were limited to single-channel marine and sea ice-based seismic reflection, refraction, and gravity studies (brief review in Cooper et al., 1987). These showed that sedimentary strata cover high velocity basement in western McMurdo Sound to a depth of 1.5 to 2 km, thickening to 4 or 5 km just east of Ross Island in the central part of the basin. In 1984, however, an extensive multichannel seismic, gravity, and magnetic survey by the R/V *S.P. Lee* (Davey, 1987) covered for the first time the entire length and breadth of the basin and showed that the sedimentary strata reach a thickness of as much as 14 km.

The early geophysical surveys were in part justified by their value in helping to locate suitable sites for drilling the Victoria Land Basin margin, where strata were expected to provide a record of the uplift of the Transantarctic Mountains and the early history of Antarctic glaciation. Indeed, reflector D of Northey et al. (1975), which coincided with a significant increase in seismic velocity, was interpreted as the boundary between glacial and preglacial strata (Barrett, 1979), providing impetus not only for the first offshore drill hole in the basin (DVDP-15, Dry Valley Drilling Project) but also others that followed, for it was a feasible drilling target, lying only 300 m beneath the sea floor of western McMurdo Sound.

The first drill hole 12 km offshore at DVDP-15 in 1974 was not successful, terminating after only 10 days and 65 m of penetration when methane was encountered and the sea-ice platform cracked (Barrett and Treves, 1981). However, important operational experience was gained, and the scientific questions concerning glacial and uplift history remained. The next attempt to core the basin margin was at MSSTS-1 (McMurdo Sound Sediment and Tectonic Studies) in 1979 and was more successful, reaching a depth of 227 mbsf with a core recovery of 56% (Barrett, 1986; Barrett et al., 1987). A third drilling attempt in 1984 failed because of thin sea ice. The most recent attempt at CIROS-1 (Cenozoic Investigations in the western Ross Sea) was made in 1986, 12 km offshore and 4 km seaward of the western margin of the basin in 200 m of water (Barrett, 1987) (Figure 2). Just over 702 m were drilled of which 685 m were cored with recovery of over 98%. The results, which include downhole logging and organic geochemistry, are reported in detail in Barrett (1989). Most of the data reviewed here come from this report.

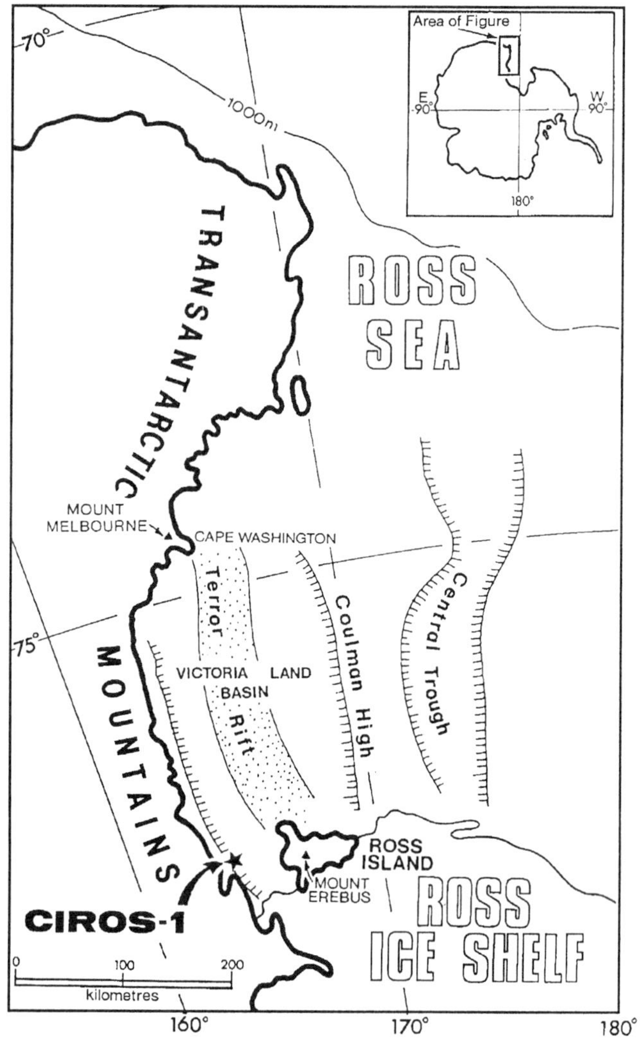

Figure 1—Map of the western Ross Sea, showing the locations of the Victoria Land basin and the CIROS-1 drill hole.

GEOLOGICAL SETTING

The western margin of Victoria Land Basin is marked by a zone of normal faulting close to the coast, with large offset basement faults, horsts and grabens, and probable intrusives (Cooper and Davey, 1985; Cooper et al., 1987). Basement for the region is exposed only in the Transantarctic Mountains, where it comprises late Precambrian to early Paleozoic metasedimentary and granitic rocks overlain by flat-lying nonmarine sedimentary strata of Devonian to Early Jurassic age intruded by Middle Jurassic sills of dolerite. However, these rocks are thought to underlie the basin also (Barrett, 1981).

Internally the basin is structurally complex, with seismic data showing evidence of faulting, uplift, and intrusion of the sedimentary section. Cooper et al. (1987, 1990) describe the basin as a broad half-graben containing up to 14 km of sedimentary rocks and with a north-south trending, active rift zone (the Terror Rift) comprising the sediment-filled Discovery Graben and the adjacent magmatically intruded Lee Arch. The rift is superimposed upon older basin structures and deeply buried strata. These "early-rift" grabens are

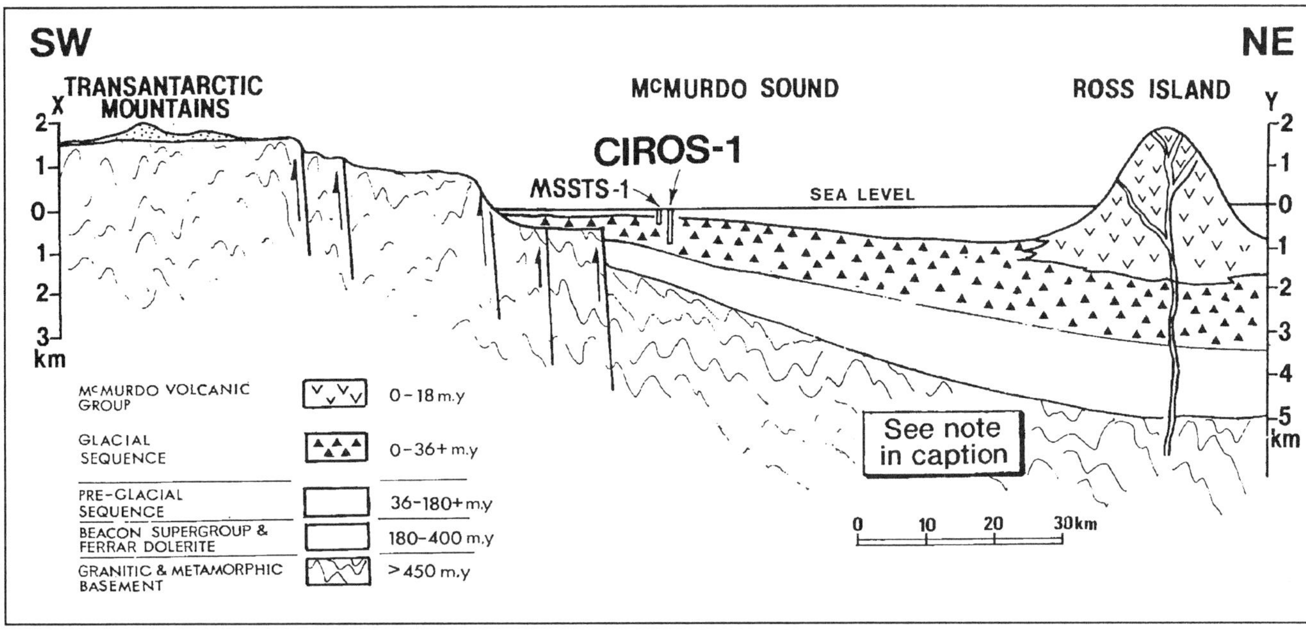

Figure 2—Geological cross section of McMurdo Sound from drill hole data (Barrett, 1987) and seismic data (McGinnis et al., 1985). Note: These rocks have velocities in excess of 5 km/sec and were originally identified as granitic basement. However, the recognition of "intrabasement reflectors" down to 14 km (McGinnis et al., 1985) and reflectors to similar depths in multichannel seismic profiles 100 km to the north (Cooper et al., 1987) suggest that these rocks are sedimentary.

thought to have originated during Middle Jurassic to Early Cretaceous time as part of the Gondwana breakup. A younger phase of rifting and sedimentation began probably during Cretaceous to early Tertiary times, with later deformation coeval with uplift of the Transantarctic Mountains (Cooper et al., 1987; 1990).

The Oligocene to early Miocene strata cored at CIROS-1 represent the younger part of the post early-rift Cenozoic basin fill. They form the upper part of a deltaic sequence of sediments supplied by rivers and glaciers cutting through the Transantarctic Mountains to the west. The mountains were by this time already a major topographic feature, for the dominance of basement debris in the CIROS-1 core shows that they had risen at least 2 km by the earliest Oligocene (Barrett et al., 1989).

The post-early Miocene history of the basin margin is hard to gauge, as there is little sediment from this period beyond scattered glacial deposits in the Transantarctic Mountains (Denton et al., 1984; Webb et al., 1984). However, the high seismic velocities of the CIROS-1 and MSSTS-1 cores (Froggatt, 1986; Davy and Alder, 1989) and relatively high in–situ vitrinite reflectance for a CIROS-1 coal (Lowery, 1989) make it clear that a significant thickness of strata, perhaps a kilometer or more, had been deposited and subsequently eroded (Barrett and McKelvey, 1981; Collen and Froggatt, 1986). The late Cenozoic uplift implied by this erosion may be related to renewed uplift of the Transantarctic Mountains, or to the raised isotherms from late Cenozoic volcanic activity which are a feature of the McMurdo region (Kyle, 1981).

STRATIGRAPHY OF CIROS-1

Coring was continuous in CIROS-1 from 26 mbsf to 702 mbsf, with better than 98% recovery. Core descriptions are given in detail by Robinson et al. (1987), and summarized in Figure 3. The sequence consists of an upper part dominated by diamictite and a lower part dominated by mudstone and with relatively little diamictite.

The upper part (26-366 mbsf) includes a significant proportion (40%) of diamictite, which is a mixture of mud, sand, and stones, with the latter forming more than 1% of the rock and uniformly spread through it. The diamictite is interpreted as lodgement and water-lain till. Four periods of glaciation associated with low relative sea level are recorded, and considered to represent episodes of major ice buildup on Antarctica. Deposits from local ice advances are also recorded (Barrett et al., 1989). The upper part of the CIROS-1 core was deposited from ~22 to 30.5 Ma (early Miocene and late Oligocene), on the evidence of diatom and foraminifer assemblages and magnetostratigraphy (Harwood et al., 1989); strata younger than early Miocene are virtually absent.

The lower part (366-702 mbsf) is largely deep water mudstone with sandstone beds and occasional conglomerate deposited from gravity flows. Strata also contain scattered subangular to subrounded stones, many of which are faceted and striated, indicating some glaciation on land and ice calving at sea level in this region back to earliest Oligocene time. A review of data from diatoms, foraminifers, coccoliths and palynomorphs, and strontium isotopes (Harwood et al., 1989) indicates a major time break from 30.5 to 34.5 Ma at or just above 366 mbsf, and has the lower part of the core deposited entirely from 34.5 to ~ 36 Ma (early Oligocene).

The most prominent seismic reflectors are both in the upper part of the hole, corresponding approximately in the cored sequence to a lithologic change from mudstone to sandstone at 140 mbsf (Reflector Q) and to a mudstone between two thick diamictite beds at 220 mbsf (Reflector R)

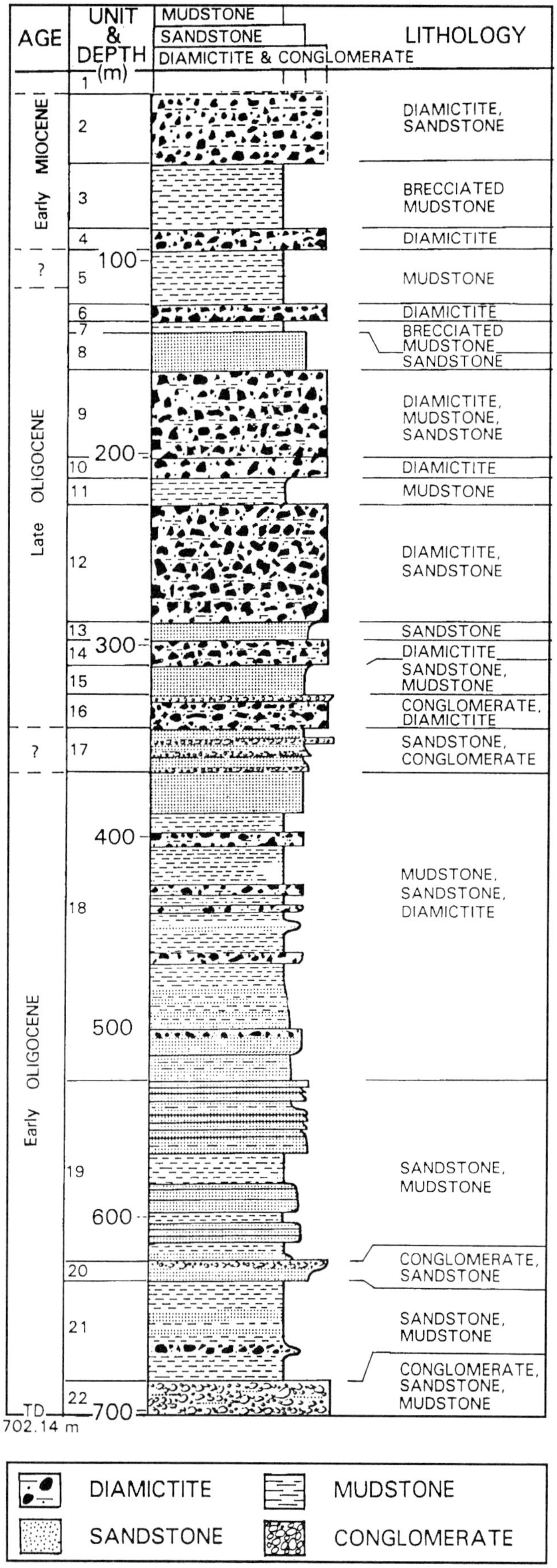

Figure 3—Stratigraphic column for CIROS-1.

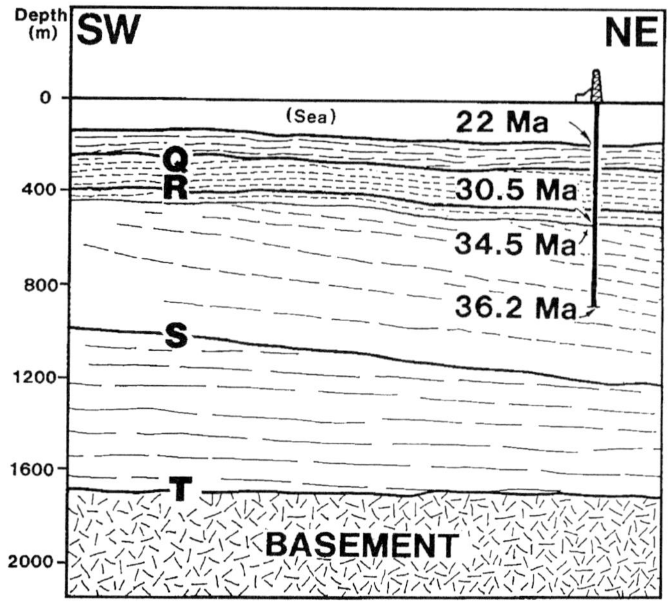

Figure 4—Section through the CIROS-1 drill site based on the seismic reflection data of Davy and Alder (1989). Letters are reflectors labeled in their paper.

(Figure 4). Overall, the sequence is horizontal, with up to 1000 m of sediment missing from the top.

The sequence in nearby MSSTS-1 is essentially similar to that of the upper 200 m of CIROS-1. Entirely glacigenic sediments were cored, mainly sandy mudstone and muddy sandstone with lonestones, but also a number of diamictites considered to have been deposited as basal tills (Barrett, 1986). An unconformity representing possibly a kilometer or more of sediments is present at the top of the sequence here also.

PETROLEUM GEOLOGY OF CIROS-1

Source Rock Characteristics

Source rock characteristics of fine-grained sediments from CIROS-1 were reported by Collen et al. (1989), and are shown here in Figure 5. Total organic carbon (TOC) values are all low, ranging from 0.21 to 0.66% with an average of 0.34%. They increase slightly down the hole but except for the sample at 672 mbsf (0.66% TOC), all have less than the 0.5% TOC generally regarded as the least amount of organic matter necessary for rock to have source potential (Dow, 1977) (Figure 5A). Values for MSSTS-1 were similarly low, ranging from 0.05 to 0.18% (Collen and Froggatt, 1986).

Pyrolysis studies using a Rock-Eval system (Espitalié et al., 1977) have been made of CIROS-1 samples (Collen et al., 1989). The technique involves progressive heating of samples. Hydrocarbons already formed in the rock are mobilized first (S_1 peak). Breakdown of kerogen next generates hydrocarbons (S_2 peak), then CO_2 and water (S_3 peak). Values for S_1, S_2 and S_3, together with indices derived from them, and the temperature (T_{max}) at which maximum generation of hydrocarbons occurs can be used to determine the type of kerogen present, the hydrocarbon generating potential of the rock, and the degree of organic maturation reached.

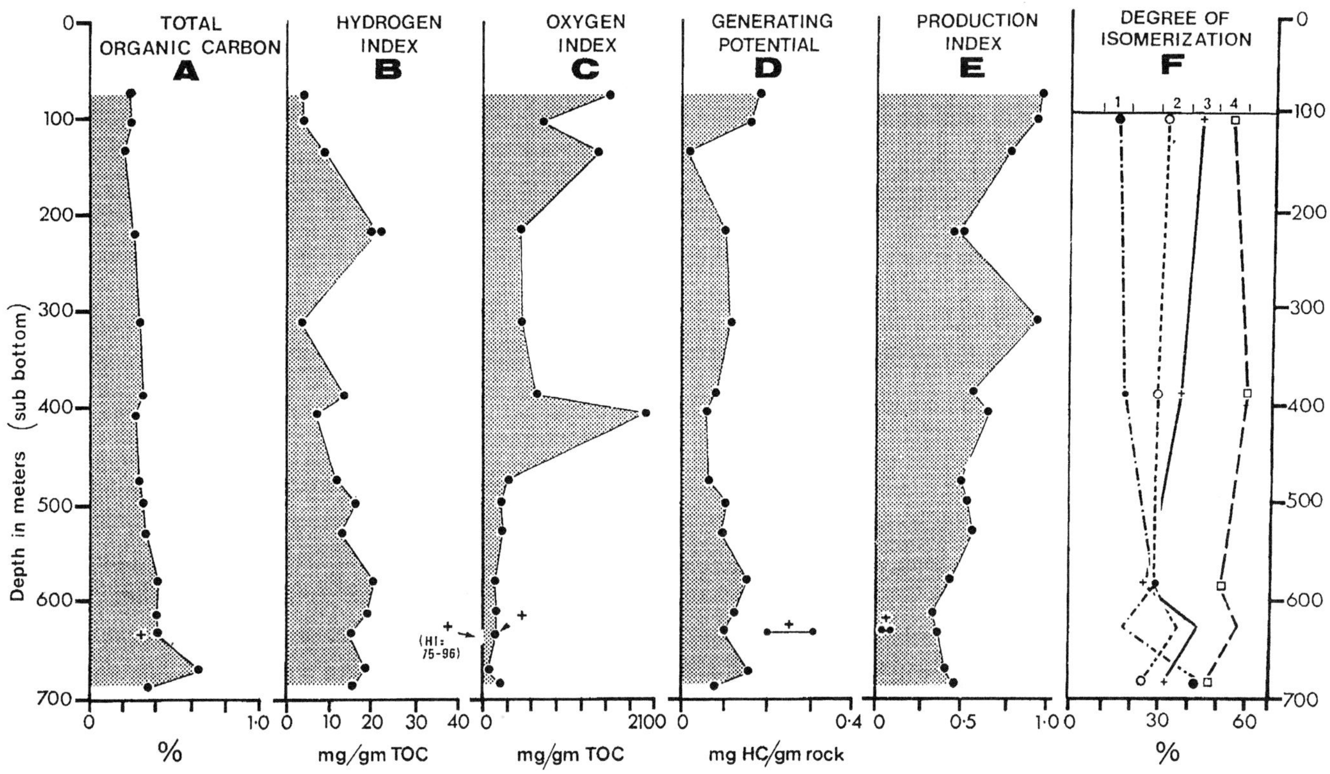

Figure 5—Plot of total organic carbon content (A), indices derived from pyrolysis data (B-E), and biomarker ratios (F) for CIROS-1. + = analyses for oil-impregnated sandstone. Isomerization ratios (F) are: 1—C30 moretane/C30 hopane; 2—C29 steranes S/S + R; 3—% bb steranes; and 4—C32 hopane S/S + R. Partly after Collen et al. (1989).

CIROS-1 samples have very high oxygen indices (45-1942 mg/g TOC) and very low hydrogen indices (3-24 mg/g TOC) (Figure 5B,C). These indices correlate to the O/C ratio and H/C ratio, respectively (Tissot and Welte, 1978) and show the kerogen to be equivalent to the Type III kerogen of Tissot et al. (1974) or an even more oxidized material. Similarly, Lowery (1989) found that much of the particulate organic matter is present as reworked coal (see below). The hydrocarbon generating potential of the samples (calculated by adding the amount of hydrocarbons already present in the rock [S_1] to that formed by kerogen breakdown during pyrolysis [S_2]) is low (0.06-0.34 mg hydrocarbon per gram of rock; Figure 5D), suggesting no oil generating potential. These results suggest that most of the organic matter is derived from terrestrial higher plants, and that the sediments have no oil- and little gas-generating potential. Pyrolysis data for MSSTS-1 similarly indicated no hydrocarbon source potential for that core (Collen and Froggatt, 1986).

Gas chromatograph/mass spectrometer (GCMS) studies of the steranes and triterpanes (Collen et al., 1989) indicate a mixed marine-terrestrial source. The steranes $C_{27}\alpha\alpha\alpha$ 20R and $C_{29}\alpha\alpha\alpha$ 20R are present in similar abundances, and C_{27}/C_{29} = 60-95%. Predominance of C_{27} steranes normally indicates largely marine organic input, whereas predominance of C_{29} steranes indicates a terrigenous source (Huang and Meinschein, 1979). Although Grantham (1986) showed that both C_{27} and C_{29} steranes could occur in rock extracts and oils of marine origin, the presence of both compounds in the CIROS samples probably indicates mixed marine and terrestrial input of organic matter. A wide range of marine microorganisms was reported from CIROS-1 sediments (calcareous nannofossils: Edwards and Waghorn, 1989; dinoflagellates: Wilson, 1989; foraminifera: Webb, 1989; siliceous microfossils: Harwood, 1989). With respect to the terrestrial organic input, Mildenhall (1989) suggested from study of palynofloras that coastal forests of southern beech (*Nothofagus* spp.) with podocarps, proteas, and other shrubby angiosperms existed near the CIROS-1 site in the Oligocene. This is supported by the presence in the core of part of a beech leaf (Hill, 1989).

Organic Maturation

Values for the degree of organic maturation have been obtained from pyrolysis (T_{max} and production index; Figure 5E) and biomarker analysis (Figure 5F) (Collen et al., 1989), and vitrinite reflectance measurements (Lowery, 1989). T_{max} ranges from 393-422°C, which Barker (1974) considers to indicate immaturity for hydrocarbon generation.

Lowery (1989) found that most coal in CIROS-1 samples was detrital, and ranged from lignite to meta-anthracite and natural coke. The medium-volatile bituminous and higher rank material is thermally metamorphosed with high inertinite content, and appears to be detritus from Permian and Triassic coals in the Transantarctic Mountains. The lignite grains, which are low in inertinite, are presumed to have come from post-Ferrar (Middle Jurassic) preglacial (pre-Oligocene) coal measures which could lie beneath the sequence cored in CIROS-1. The existence of such a sequence has been postulated by Cooper et al. (1990) from regional geology considerations.

The only coal considered to be in situ (from thin, carbonaceous laminae at 699.59 mbsf) yielded a mean maximum vitrinite reflectance of 0.36 ± 0.03% (Lowery, 1989), indicating not only that erosion of 500-1000 m or more of sediment (depending on paleogeothermal gradient) has occurred at this site but also again that the remaining sediments are still not thermally mature for hydrocarbon generation. Vitrinite reflectances measured for the MSSTS-1 core were all 0.54% or less (Collen and Froggatt, 1986).

Biomarker studies, particularly the stereochemical reactions of hopanes and steranes, also allow assessment of organic maturity (Seifert and Moldowan, 1978). Particularly important are the isomerization of $C_{30}\beta\alpha$ hopane (moretane) to $C_{30}\alpha\beta$ hopane, and the epimerization reactions $C_{32}\alpha\beta$ hopane 22R to 22S, $C_{29}\alpha\alpha\alpha$ sterane 20R to 20S, and $C_{29}\alpha\alpha\alpha$ sterane (20R and 20S, normal sterane) to $C_{29}\alpha\beta\beta$ sterane (isosterane). Details of the particular reactions involved are given by MacKenzie (1982) and MacKenzie et al. (1982), and their progress is measured by the ratios C_{30} moretane/C_{30} hopane; C_{32} hopanes, S/S+R; C_{29} steranes, S/S+R; and % $\beta\beta\beta$ steranes. With increasing maturity, the moretane/hopane ratio decreases and the other ratios increase.

Biomarkers from a sample at 630 mbsf may not represent the sedimentary organic matter (Collen et al., 1989), as this sample closely overlies the sandstone containing residual oil (see below) and thus may be impregnated with some oil. If biomarker data from this sample are disregarded, then the CIROS-1 samples show a clear trend of decreasing maturity downwards (Collen et al., 1989), with moretane/hopane ratios increasing and epimerization ratios decreasing with depth (Figure 5F). Values for C_{32} hopane S/S+R for samples from the upper part of the core reach the equilibrium value (about 60%; MacKenzie, 1982; MacKenzie et al., 1982), equivalent to about mid-catagenesis of organic matter (MacKenzie and Maxwell, 1981). Samples lower in the section have significantly lower values (47-51%), indicating lesser maturity. Similar results are given by the C_{29} S/S+R sterane ratios (34 and 30% in the upper part of the hole versus 24 and 29% in the lower) and by the C_{30} moretane/hopane ratios (17 and 19% versus 29 and 46%, respectively). The latter ratios show significantly lower maturity values for the deepest sample (673m, 46%) than for a sample at 579 mbsf (29%). The biomarker data for the upper part of the sequence, at least, must therefore come largely from redeposited organic matter.

The $S_1/(S_1 + S_2)$ ratio (the production index or transformation ratio) compares the amount of hydrocarbons already formed to the total amount able to be generated (Tissot and Welte, 1978), and should increase downwards with increasing thermal maturity. Again, in CIROS-1 this ratio decreases downwards (Figure 5E).

The anomalous decrease of organic maturity with depth in CIROS-1 that is indicated by both the hopane and sterane biomarker data and by the production indices may result from the patterns of organic sedimentation and of hydrocarbon migration in the area (Collen et al., 1989). The lesser biomarker isomerization in the lower part of the hole may be due to reworked organic material derived from early erosion of the rising Transantarctic Mountains being deposited first, with higher rank organic matter from deeper erosion being deposited higher in the sequence. Reversed maturity indicated by the production indices and due to larger amounts of generated hydrocarbons in the sediments is more difficult to explain but could result either from hydrocarbons already present in organic matter being redeposited or from migration of small amounts of hydrocarbons into the younger sediments.

Porosity/Permeability

Numerous sandstone beds up to 22 m thick occur in the sequence. These range from well sorted to very poorly sorted, and are coarse- to fine-grained (Bridle and Robinson, 1989). Most are arkoses to subarkoses. Authigenic mineral development is variable, from a variety of zeolites that form relatively minor pore linings to complete pore occlusion by calcite cement. Little authigenic clay is present. Visual porosity ranges from 0 to 15% (Bridle and Robinson, 1989) and corrected log porosities up to 21.6% (White, 1989). A number of sandstones have good secondary porosity resulting from dissolution of carbonate cement, bioclastic debris, and detrital feldspar.

Source of CIROS-1 Hydrocarbons

A sandstone from 632 to 634 mbsf contained traces (less than 1%) of an asphaltic residual oil without associated gas. This sandstone is brownish black, moderately sorted, muddy, and very coarse- to fine-grained, with good visual porosity of about 20%. Hydrocarbons are present as dark flecks with minor fluorescence and only small amounts could be extracted at a slow rate with trichloroethane (Cook and Woolhouse, 1989). Only traces of methane were detected (Wada and Sano, 1989) and, if all the measured TOC is from the oil, residual oil saturation of the porosity is 1.4% or less. Chemical analysis of material extracted from the sandstone (Cook and Woolhouse, 1989) shows it to have the characteristics of a hydrocarbon derived from moderately mature source sediment. Isomerization ratios for steranes and hopanoid terpanes from the oil indicate that the source sediment had at least reached the thermal requirements for the onset of oil generation (R_o = 0.6%) but had yet to reach the peak of hydrocarbon generation (R_o = 1.0%) where equilibrium ratios would be expected. The pristane/phytane ratio of 0.5 is consistent with a highly anoxic environment whereas the pristane/nC_{17} ratio of 0.5 indicates abundant bacterial activity consistent with open water sedimentation. Sterane and triterpane biomarkers indicate the source was probably a nearshore marine clastic sediment which had received both marine and terrestrially derived organic matter. Cook and Woolhouse (1989) thus suggest that the oil was generated elsewhere in a source sediment that was not at the level for peak oil generation, migrated into the 632 mbsf sand in CIROS-1, then migrated farther leaving a residue behind. Analysis of a pyrolysate of the sandstone led these authors to infer the additional presence in that rock of autochthonous organic matter with very low maturity, probably derived from peat swamps. If correct, this means that part of the total organic carbon measured is from kerogen and the residual oil saturation is lower than the value quoted above.

POSSIBLE PETROLEUM PROSPECT OF THE VICTORIA LAND BASIN

Sediments drilled in the Victoria Land Basin to date have low organic content and kerogens are mainly derived from reworked, oxidized organic material. Their generating potential for hydrocarbons is very low to negligible, and all sediments drilled show low values for organic maturity, being well above the top of the oil window despite the loss of perhaps more than a kilometer of sediment from the top of the sequence. Residual oil found only in the 2 m thick sandstone bed in CIROS-1 shows that hydrocarbons have been generated within the basin. However, residual oil saturation in that sandstone is very low (1.4% or less), compared to the values of 20% or more residual saturation often shown in carrier beds after migration of oil (MacKenzie and Quigley, 1988). This observation suggests that only small amounts of hydrocarbon have passed through CIROS-1 sandstones.

Cook and Woolhouse (1989) suggest a dominantly terrestrial source with some marine input for the oil. Around 800 m of sediments have been estimated to lie beneath the bottom of CIROS-1 (Davy and Alder, 1989), but these are unlikely to have reached sufficient maturity to have sourced the oil. Lowery (1989) estimated from reflectance and geothermal gradient data that reflectances of 0.5 to 0.6% (equivalent to the onset of hydrocarbon generation) could only have been attained in sediments now at depths of at least 1500 to 2000 m. Kerogens deposited at this site close to the basin margin in pre-Oligocene times might also contain dominantly terrestrial organic matter, even in a marine setting. The source for the residual oil is therefore most likely to be sediments in more distal parts of the basin that contain richer, less oxidized and possibly marine organic matter and have been within the oil window. The distribution of such sediments cannot be determined at present.

Gravity cores from beneath the Ross Ice Shelf contained terrestrial plant remains and reworked Paleogene palynomorphs in sediments of Miocene or younger age (Wrenn and Beckman, 1982). Hydrocarbon geochemistry of near-surface sediments of the Ross Sea also suggests an important contribution from continentally derived higher plants, most probably reworked from older sediments (Rapp et al., 1987). Sackett et al. (1974) suggest from analyses of stable carbon isotopes that up to 90% of the organic matter in Ross Sea sediments is reworked. These data suggest that similar patterns of organic sedimentation to those inferred for the CIROS-1 area may have been widespread throughout the Victoria Land Basin since early Oligocene time, and that oil-generating capacity in source rocks of this age is limited over the whole basin.

Waples-Lopatin geochemical models produced for the Victoria Land Basin by Cook and Davey (1984) and Cooper et al. (1990) suggest that the lowermost sediments in the basin are within the oil generation zone and that the basin has good modeled potential. However, the values for organic carbon content inferred by Cook and Davey (1984) were considerably higher than those actually measured from the drill holes, and both they and Cooper et al. (1990) used a time-temperature index (TTI) of 15 to predict the onset of hydrocarbon generation and the top of the "oil window." This does not take into account the higher temperatures likely to be required for generation of hydrocarbons from terrestrial kerogen. Cook (1988) and Johnston et al. (1988) have shown that oils generated from terrestrial organic matter in the western basins of New Zealand may not be released from the source rock until temperatures equivalent to vitrinite reflectances of 1.0% or higher are reached. If concentrations of terrestrial organic matter act as a source in the Victoria Land Basin, then similar high temperatures may be necessary for hydrocarbon generation from them.

In summary, the CIROS-1 data show unfavorable source characteristics in the Miocene to early Oligocene sediments. The drill hole was situated updip from the basin center but contained only minute amounts of residual oil in sandstone beds with good porosity and seals. The possible hydrocarbon prospect of the glacial sequence of the basin in the McMurdo Sound area must be assessed as low.

ACKNOWLEDGMENTS

The CIROS project was part of the New Zealand Antarctic Research Programme. It was logistically supported by Antarctic Division, DSIR, and funded by the New Zealand Department of Scientific and Industrial Research and the New Zealand University Grants Committee. We thank A.K. Cooper, F.J. Davey, and K.A. Kvenvolden for their helpful and constructive reviews of the manuscript.

REFERENCES CITED

Barker, C., 1974, Pyrolysis techniques for source-rock evaluation: AAPG Bulletin v. 58, p. 2349-2361.

Barrett, P.J., 1979, Proposed drilling in McMurdo Sound—1979, *in* T. Nagata, ed., Proceedings of the Seminar III on Dry Valley Drilling Project, 1978: Memoirs of the National Institute of Polar Research, special issue 13, p. 231-239.

Barrett, P.J.,1981, History of the Ross Sea region during the deposition of the Beacon Supergroup 400-180 million years ago: Journal Royal Society of New Zealand, v. 11, p. 447-458.

Barrett, P.J.,1986, Introduction, *in* P.J. Barrett, ed., Antarctic Cenozoic history from the MSSTS-1 drillhole, McMurdo Sound: New Zealand Department of Scientific and Industrial Research Bulletin 237, Wellington, p. 5-7.

Barrett, P.J.,1987, Oligocene sequence cored at CIROS-1, western McMurdo Sound: New Zealand Antarctic Record, v. 7, p. 1-7.

Barrett, P.J.,1989, ed., Antarctic Cenozoic history from the CIROS-1 drillhole, McMurdo Sound: New Zealand Department of Scientific and Industrial Research Bulletin 245, Wellington, 254 p.

Barrett, P.J. and B.C. McKelvey, 1981, Cenozoic glacial and tectonic history of the Transantarctic Mountains in the McMurdo Sound region: progress from recent drilling: Polar Record, v. 20, p. 543-548.

Barrett, P.J. and S.B. Treves, 1981, Sedimentology and petrology of core from DVDP-15, western McMurdo Sound, *in* L.D. McGinnis, ed., Dry Valley Drilling Project: American Geophysical Union Antarctic Research Series, v. 33, p. 281-314.

Barrett, P.J., D.P. Elston, D.M. Harwood, B.C. McElevey and P.-N. Webb, 1987, Cenozoic record of glaciation, tectonism and sea level change from MSSTS-1 on the margin of the Victoria Land Basin, Antarctica: Geology, v. 15, p. 634-637.

Barrett, P.J., M.J. Hambrey, D.M. Harwood, A.R. Pyne and P.-N. Webb, 1989, Synthesis, *in* P.J. Barrett, ed., Antarctic Cenozoic history from the CIROS-1 drillhole, McMurdo Sound: New Zealand Department of Scientific and Industrial Research Bulletin 245, Wellington, p. 241-252.

Bridle, I.M. and P.H. Robinson, 1989, Diagenesis, *in* P.J. Barrett, ed., Antarctic Cenozoic history from the CIROS-1 drillhole, McMurdo Sound: New Zealand Department of Scientific and Industrial Research Bulletin 245, Wellington, p.201-208.

Collen, J.D. and P.C. Froggatt, 1986, Depth of burial and hydrocarbon source potential, *in* P.J. Barrett, ed., Antarctic Cenozoic history from the MSSTS-1 drillhole, McMurdo Sound: New Zealand Department of Scientific and Industrial Research Bulletin 237, Wellington, p. 161-164.

Collen, J.D., Yang Xinghua, R.J. Collier and J.H. Johnston, 1989, Hydrocarbon source rock potential and organic maturation, *in* P.J. Barrett, ed., Antarctic Cenozoic history from the CIROS-1 drillhole, McMurdo Sound: New Zealand Department of Scientific and Industrial Research Bulletin 245, Wellington, p. 223-230.

Cook, R.A., 1988, Interpretation of the geochemistry of oils of Taranaki and West Coast Region, western New Zealand: Energy Exploration and Exploitation Journal, v. 6, p. 201-202.

Cook, R.A. and F.J. Davey, 1984, The hydrocarbon exploration of the basins of the Ross Sea, Antarctica, from modelling of the geophysical data: Journal of Petroleum Geology, v. 7, 213-226.

Cook, R.A., and A.D. Woolhouse, 1989, Residual hydrocarbons, *in* P.J. Barrett, ed., Antarctic Cenozoic history from the CIROS-1 drillhole, McMurdo Sound: New Zealand Department of Scientific and Industrial Research Bulletin 245, Wellington, p. 211-218.

Cooper, A.K. and F.J. Davey, 1985, Episodic rifting of Phanerozoic rocks in the Victoria Land Basin, western Ross Sea, Antarctica: Science, v. 229, p. 1085-1087.

Cooper, A.K., F.J. Davey and J.C. Behrendt, 1987, Seismic stratigraphy and structure of the Victoria Land Basin, Western Ross Sea, Antarctica, *in* A.K. Cooper and F.J. Davey, eds., The Antarctic continental margin: geology and geophysics of the western Ross Sea: Circum-Pacific Council Energy and Mineral Resources, Earth Science Series, v. 5B, Houston, p. 27-76.

Cooper, A.K., F.J. Davey and K. Hinz, 1990, Geology and hydrocarbon potential of the Ross Sea, Antarctica: this volume.

Davey, F.J., 1987, Geology and structure of the Ross Sea region, *in* A.K. Cooper and F.J. Davey, eds., The Antarctic continental margin: geology and geophysics of the western Ross Sea: Circum-Pacific Council Energy and Mineral Resources, Earth Science Series, v. 5B, Houston, p. 1-16.

Davy, B. and G. Alder, 1989, Seismic reflection surveys, *in* P.J. Barrett, ed., Antarctic Cenozoic history from the CIROS-1 drillhole, McMurdo Sound: New Zealand Department of Scientific and Industrial Research Bulletin 245, Wellington, p. 15-22.

Denton, G.H., M.L. Prentice, D.E. Kellog and T.B. Kellog, 1984, Late Tertiary history of the Antarctic ice sheet: evidence from the Dry Valleys: Geology, v. 12, p. 263-267.

Dow, W.G., 1977, Kerogen studies and geological interpretations: Journal of Geochemical Exploration, v. 7, p. 79-99.

Edwards, A.R. and D. B. Waghorn, 1989, Calcareous nannofossils, *in* P.J. Barrett, ed., Antarctic Cenozoic history from the CIROS-1 drillhole, McMurdo Sound: New Zealand Department of Scientific and Industrial Research Bulletin 245, Wellington, p. 145-149.

Espitalié, J., J.I. Laporte, M. Madec, F. Marquis, P. Leplat, J. Paulet and A. Boutefeu, 1977, Méthode rapide de caractérisation des roches mères, de leur potentiel pétrolier et de leur degré d'evolution: Revue de l'Institute Français Pétrolier, v. 32, p. 23-42.

Froggatt, P.C., 1986, Seismic velocities, *in* P.J. Barrett, ed., Antarctic Cenozoic history from the MSSTS-1 drillhole, McMurdo Sound: New Zealand Department of Scientific and Industrial Research Bulletin 237, Wellington, p. 165-166.

Grantham, P.J., 1986, The occurrence of unusual C27 and C29 sterane predominances in two types of Oman crude oil: Organic Geochemistry, v. 9, p. 1-10.

Harwood, D.M., 1989, Siliceous microfossils, *in* P.J. Barrett, ed., Antarctic Cenozoic history from the CIROS-1 drillhole, McMurdo Sound: New Zealand Department of Scientific and Industrial Research Bulletin 245, Wellington, p. 67-98.

Harwood, D.M., P.J. Barrett, A.R. Edwards, H.J. Rieck and P.-N. Webb, 1989, Biostratigraphy and chronology, *in* P.J. Barrett, ed., Antarctic Cenozoic history from the CIROS-1 drillhole, McMurdo Sound: New Zealand Department of Scientific and Industrial Research Bulletin 245, Wellington, p. 231-239.

Hill, R.S., 1989, Fossil leaf, *in* P.J. Barrett, ed., Antarctic Cenozoic history from the CIROS-1 drillhole, McMurdo Sound: New Zealand Department of Scientific and Industrial Research Bulletin 245, Wellington, p. 143-144.

Huang, W.-Y. and W.G. Meinschein, 1979, Sterols as ecological indicators: Geochimica et Cosmochimica Acta, v. 43, p. 739-745.

Johnston, J.H., R.J. Collier and J.T. Craig, 1988, Oil-source rock correlations in the Maui-4 oil exploration well, south Taranaki Basin: Energy Exploration and Exploitation Journal, v. 6, p. 233-247.

Keys, J.R., 1984, Antarctic marine environments and offshore oil: New Zealand Commission for the Environment, Wellington, 168 p.

Kyle, P.R., 1981, Glacial history of the McMurdo Sound area as indicated by the distribution and nature of McMurdo Volcanic Group rocks, *in* L.D. McGinnis, ed., Dry Valley Drilling Project: American Geophysical Union Antarctic Research Series, v. 33, p. 403-412.

Lowery, J.H., 1989, Coal particles and laminae, *in* P.J. Barrett, ed., Antarctic Cenozoic history from the CIROS-1 drillhole, McMurdo Sound: New Zealand Department of Scientific and Industrial Research Bulletin 245, Wellington, p. 219-222.

MacKenzie, A.S., 1982, Steroid hydrocarbons and the thermal history of sediments: Nature, v. 295, p. 223-226.

MacKenzie, A.S. and J.R. Maxwell, 1981, Assessment of thermal maturation in sedimentary rocks by molecular measurements, *in* J. Brooks, ed., Organic maturation studies and fossil fuel exploration: London and New York, Academic Press, p. 239-253.

MacKenzie, A.S. and T.M. Quigley, 1988, Principles of geochemical prospect appraisal: AAPG Bulletin, v. 72, p.399-415.

MacKenzie, A.S., S.C. Brassell, G. Eglinton and J.R. Maxwell, 1982, Chemical fossils: the geological fate of steroids: Science, v. 217, p. 491-504.

McGinnis, L.D., R.H. Bowen, J.M. Erikson, B.J. Allred and J.L. Kroner, 1985, East-West Antarctic boundary in McMurdo Sound: Tectonophysics, v. 114, p. 341-356.

Mildenhall, D.C., 1989, Terrestrial palynology, *in* P.J. Barrett, ed., Antarctic Cenozoic history from the CIROS-1 drillhole, McMurdo Sound: New Zealand Department of Scientific and Industrial Research Bulletin 245, Wellington, p. 119-128.

Northey, D.J., C. Brown, D.A. Christoffel, H.K. Wong and P.J. Barrett, 1975, A continuous seismic profiling survey in McMurdo Sound, Antarctica: Dry Valley Drilling Project, Bulletin 5, p. 167-179.

Rapp, J.B., K.A. Kvenvolden and M. Golan-Bac, 1987, Hydrocarbon geochemistry of sediments offshore from Antarctica, *in* A.K. Cooper and F.J. Davey, eds., The Antarctic continental margin: geology and geophysics of the western Ross Sea: Circum-Pacific Council for Energy and Mineral Resources, Earth Science Series, v. 58, Houston, p. 217-224.

Robinson, P.H., A.R. Pyne, M.J. Hambrey, K.J. Hall and P.J. Barrett, 1987, Core log, photographs and grain size analyses from the CIROS-1 drillhole, western McMurdo Sound, Antarctica: Antarctic Research Centre Publication, Victoria University of Wellington, 241 p.

Sackett, W.M., C.W. Poag and B.J. Eadie, 1974, Kerogen recycling in the Ross Sea, Antarctica: Science, v. 185, p. 1045-1047.

Schmidt, D.L. and P.D. Rowley, 1986, Continental rifting and transform faulting along the Jurassic Transantarctic rift, Antarctica: Tectonics, v. 5, p. 279-291.

Seifert, W.K. and J.M. Moldowan, 1978, The applications of steranes, terpanes and monoaromatics to the maturation, migration and source of crude oils: Geochimica et Cosmochimica Acta, v. 42, p. 77-95.

Stern, T. and U. ten Brink, 1989, Flexural uplift of the Transantarctic Mountains: Journal of Geophysical Research, v. 92, p. 10375-10390.

Tessensohn, F., in press, The Cenozoic Ross Sea rift system—definition and comparison with the East African Rift, *in* M.R.A. Thompson, J.A. Crawe and J.W. Thompson, eds., Proceedings of 5th International Antarctic Symposium, Cambridge, August, 1987.

Tissot, B., B. Durand, J. Espitalié and A. Combaz, 1974, Influence of nature and diagenesis of organic matter in formation of petroleum: AAPG Bulletin, v. 58, p. 499-506.

Tissot, B.P. and D.H. Welte, 1978, Petroleum formation and occurrence: New York, Springer-Verlag.

Wada, H. and Y. Sano, 1989, Gas analyses, *in* P.J. Barrett, ed., Antarctic Cenozoic history from the CIROS-1 drillhole, McMurdo Sound: New Zealand Department of Scientific and Industrial Research Bulletin 245, Wellington, p. 209-210.

Webb, P.-N., 1989, Benthic foraminifera, *in* P.J. Barrett, ed., Antarctic Cenozoic history from the CIROS-1 drillhole, McMurdo Sound: New Zealand Department of Scientific and Industrial Research Bulletin 245, Wellington, p. 99-118.

Webb, P.-N., D.N. Harwood, B.C. McKelvey, J.H. Mercer and L.D. Scott, 1984, Cenozoic marine sedimentation and ice-volume variation on the east Antarctic craton: Geology, v. 12, p. 287-291.

White, P., 1989, Down-hole logging, *in* P.J. Barrett, ed., Antarctic Cenozoic history from the CIROS-1 drillhole, McMurdo Sound: New Zealand Department of Scientific and Industrial Research Bulletin 245, Wellington, p. 7-14.

Wilson, G.J., 1989, Marine palynology, *in* P.J. Barrett, ed., Antarctic Cenozoic history from the CIROS-1 drillhole, McMurdo Sound: New Zealand Department of Scientific and Industrial Research Bulletin 245, Wellington, p. 129-134.

Wrenn, J.H. and S.W. Beckman, 1982, Maceral, total organic carbon, and palynological analyses of Ross Ice Shelf Project Site J9 cores: Science, v. 216, p. 187-189.

Index